Mathematics for Technicians New Level I

A. Greer CEng, MRAeS
FORMERLY SENIOR LECTURER

G. W. Taylor BSc (Eng), CEng, MIMechE
PRINCIPAL LECTURER

Gloucestershire College
of Art and Technology

STANLEY THORNES (PUBLISHERS) LTD

First published in 1982 by:
Stanley Thornes (Publishers) Ltd
EDUCA House
Old Station Drive
off Leckhampton Road
CHELTENHAM GL53 0DN
England

British Library Cataloguing in Publication Data

Greer, A.
Mathematics for technicians level 1.
1. Shop mathematics
I. Title II. Taylor, G.W.
510'.246 TJ1165

ISBN 0-85950-352-6

Typeset by
Tech-Set, Felling, Gateshead
Printed and bound in Great Britain at
The Pitman Press, Bath

CONTENTS

AUTHORS' NOTE vii

Chapter 1 **OPERATIONS IN ARITHMETIC** 1

Introduction — some definitions — addition — subtraction — combined addition and subtraction — arithmetical signs, terms and symbols — multiplication — division — tests for divisibility — factors and multiples — lowest common multiple — highest common factor — sequence of arithmetic operations

Chapter 2 **FRACTIONS** 23

Introduction — vulgar fractions — reducing a fraction to its lowest terms — types of fractions — lowest common denominator — addition of fractions — subtraction of fractions — combined addition and subtraction — multiplication — cancelling — the operation 'of' — division of fractions — operations with fractions

Chapter 3 **THE DECIMAL SYSTEM** 39

Introduction — the decimal system — addition and subtraction of decimals — multiplying decimals by 10, 100, etc — dividing decimals by 10, 100, etc. — long multiplication — long division — significant figures — rounding and degrees of accuracy — rough checks for calculations — fraction to decimal conversion — conversion of decimals to fractions

Chapter 4 **RATIO, PROPORTION AND PERCENTAGES** 60

Ratio — proportion — proportional parts — percentages — percentage of a quantity

Chapter 5 **DIRECTED NUMBERS** 70

Introduction — positive and negative numbers — the addition of directed numbers — the addition of numbers having different signs — subtraction of directed numbers — multiplication of directed numbers — division of directed numbers

Chapter 6 **INDICES AND LOGARITHMS** 78

Base, index and power — reciprocal — laws of indices — logarithms — logarithm of a negative number — numbers in standard form — SI units: preferred standard form

Chapter 7 **THE SCIENTIFIC ELECTRONIC CALCULATOR** 94

Keyboard layout — rough checks — worked examples — 'whole number' powers — square root and 'pi' keys — the power key

Chapter 8 **TABLES AND CHARTS** 109

Tables requiring use of positive and negative mean differences — conversion tables: illustrating use of interpolation — parallel scale conversion charts — nomographs — network diagrams

Chapter 9 **INTRODUCTION TO ALGEBRA** 122

Use of symbols — addition and subtraction of algebraic terms — multiplication and division signs — multiplication and division of algebraic quantities — sequence of mixed operations on algebraic quantities — brackets — network diagrams — the product of two binomial expressions — the square of a binomial expression — the product of the sum and difference of two terms — highest common factor — factorising — factorising by grouping — lowest common multiple — handling algebraic fractions — adding and subtracting algebraic fractions — expressing a single fraction as partial fractions — mixed operations with fractions

Chapter 10 **LINEAR EQUATIONS** 146

Identities — equations — solving linear equations — making expressions — constructing simple equations

Chapter 11 **SIMULTANEOUS LINEAR EQUATIONS** 162

Solution of simultaneous linear equations — problems involving simultaneous equations

Chapter 12 **FORMULAE** 169

Evaluating formulae — formulae giving rise to simple equations — transposition of formulae

Chapter 13 **GRAPHS** 178

Areas of reference — coordinates — axes and scales — graphs of simple equations — the law of a straight line — obtaining the straight line law of a graph — graphs of experimental data — direct variation — inverse variation

Chapter 14 **ANGLES AND STRAIGHT LINES** 200

Angles — angular measurement — types of angles — properties of angles and straight lines

Chapter 15 **TRIANGLES** 213

Types of triangles — angle property of triangles — standard notation for a triangle — Pythagoras' theorem — properties of the isosceles triangle — constructing a right angle — similar triangles — congruent triangles — construction of triangles

Chapter 16 **THE CIRCLE** 237

Relation between diameter, radius and circumference — the numerical values of π — angles in circles

Chapter 17 **AREA AND VOLUME** 251

Quadrilaterals — parallelogram — units of length — units of area — the rectangle — the square — the parallelogram — area of a trapezium — area of a triangle — area of a circle — areas of composite figures — the unit of volume — volume of a cylinder — similar shapes — similar solids

Chapter 18 **TRIGONOMETRY** 281

The trigonometrical ratios — the sine of an angle — reading the table of sines of angles — inverse notation — sines from a scientific calculator — the cosine of an angle — the tangent of an angle — trigonometrical ratios for 30°, 60° and 45° — given one ratio to find the others — complementary angles — practical trigonometry problems

Chapter 19 **STATISTICS** 306

Recording information — the population — sampling — frequency distributions — the class width — the histograms — grouped data — discrete and continuous variables — discrete distributions — frequency polygons — cumulative frequency distributions — relative frequency — relative percentage frequency — representing frequencies by pictograms — representing frequencies by means of bar charts

ANSWERS 330

INDEX 341

AUTHORS' NOTE

This volume covers all the objectives included in the standard TEC Mathematics unit U80/683. It also covers the objectives included in the appendix to the unit which are intended to provide revision of preparatory topics.

This book is for students of all disciplines using the standard Mathematics unit. Suitable examples have been included to cover applications for Engineering, Building Construction, and Science Technologies.

Volumes covering the subsequent Levels II and III units have already been prepared and will provide a full course of study for all the Mathematics required for the TEC Certificate.

A Greer
G W Taylor

1. OPERATIONS IN ARITHMETIC

After reaching the end of this chapter you should be able to:

1. *Perform calculations including addition, subtraction, multiplication and division involving positive and negative integers, lowest common multiple and highest common factor.*
2. *Verify numerically the laws:*

$$a + (b + c) = (a + b) + c$$
$$a(bc) = (ab)c$$
$$a + b = b + a$$
$$ab = ba$$
$$a(b + c) = ab + ac$$

INTRODUCTION

In this chapter the basic operations of arithmetic are revised. We will add, subtract, multiply and divide numbers. Since arithmetic is all about numbers let us first consider what numbers are and how they are represented.

SOME DEFINITIONS

Numbers are represented by symbols which are called *digits*. There are nine digits which are 1, 2, 3, 4, 5, 6, 7, 8 and 9. We also use the symbol 0 (i.e. zero) where no digit exists. Digits and zero may be combined together to represent any number.

Numeration expresses numbers in words — zero, one, two, three, four, five, six, seven, eight and nine.

Notation expresses numbers in figures or symbols (0, 1, 2, 3, 4, 5, 6, 7, 8, 9). These are all unit figures. The next number, ten, is 10 which is a combination of one and zero. 11 (eleven) is the combination of one and one and it equals ten plus one. 20 (twenty) is the combination of two and zero and it equals two tens. Ten tens are one hundred and ten hundreds are one thousand and so on.

100 indicates one hundred

900 indicates nine hundreds

954 indicates nine hundreds, five tens and four units

1000 indicates one thousand

8000 indicates eight thousands

9999 indicates nine thousands, nine hundreds, nine tens and nine units

Note that in the case of the number 9999:

9 9 9 9

$d\ c\ b\ a$

a	is a units figure and equals	9
b	is a tens figure and equals	90
c	is a hundreds figure and equals	900
d	is a thousands figure and equals	9000

In arithmetic the sign + means plus or add and the sign = means equals.

Thus $7 + 2 = 9$

The number $9999 = 9000 + 900 + 90 + 9$

If we want to write six hundreds, five tens and seven units then we write 657. If we want to write four hundred and seven units we write 407; the zero keeps the place for the missing tens.

eight thousands and thirty five is written 8035

eight thousand and nine is written 8009

ten thousand is written 10 000

one hundred thousand is written 100 000

one thousand thousand is called one million which is written 1 000 000

8 000 000 indicates eight million

37 895 762 indicates thirty seven million, eight hundred and ninety five thousand, seven hundred and sixty two.

In the number 37 895 762 we have grouped the digits in threes with a space between them. This space takes the place of the comma which was traditionally used to group the figures of a number into threes. The change has taken place because many foreign countries use a comma instead of a decimal point.

Exercise 1.1

Write the following in figures:

1) Four hundred and fifty seven.

2) Nine thousand, five hundred and thirty six.

3) Seven thousand, seven hundred and seventy seven.

4) Three thousand and eight.

5) Seven hundred and five.

6) Thirty thousand and twenty eight.

7) Five thousand and ninety.

8) Four thousand, nine hundred and four.

9) One hundred and twenty five thousand, nine hundred and six.

10) Three million, eight hundred thousand and seven.

11) Ninety five million, eight hundred and twenty seven thousand.

12) Three hundred million and nine.

Write the following numbers in words:

13) 225	14) 8321	15) 3017
16) 3960	17) 1807	18) 20 004
19) 17 000	20) 198 376	21) 200 005
22) 7 365 235	23) 27 000 309	

ADDITION

When adding numbers together place the figures in columns making sure that all the units figures are placed under one another, that all the tens figures are placed beneath each other and so on. Thus all the figures having the sample place value fall in the same column.

EXAMPLE 1.1

Add together 4219, 583, 98 and 1287.

```
4219
 583
  98
1287
----
6187
----
```

Start off by adding the units column. Thus 7 and 8 make 15, and 3 makes 18, and 9 makes 27. Place the 7 in the units column of the answer and carry the 2 forward to the tens colum. Adding this we have 2 and 8 is 10 and 9 is 19 and 8 is 27 and 1 is 28. Place the 8 in the tens column of the answer and carry the 2 forward to the hundreds column which we now add. 2 and 2 is 4 and 5 is 9 and 2 is 11. We write a 1 in the hundreds column of the answer and carry 1 forward to the thousands column which we now add. 1 and 1 is 2 and 4 is 6. Writing the 6 in the thousands column of the answer we see that the answer to the addition is 6187.

EXAMPLE 1.2

Find the value of $17\,638 + 108\,749 + 1011 + 2\,345\,008$

The + sign simply means add the numbers together and our problem is to add the four numbers. As before we write them in a column so that digits having the same place values are written beneath each other.

$$\begin{array}{r} 17\,638 \\ 108\,749 \\ 1\,011 \\ 2\,345\,008 \\ \hline 2\,472\,406 \\ \hline \end{array}$$

Add up the units column from bottom to top, saying audibly, 9, 18, 26. Write 6 in the units column of the answer and carry the 2 forwards to the tens column. Add the tens column as 2, 3, 7 and 10. Write 0 in the tens column of the answer and carry the 1 forward to the hundreds column. Carry on in this way with the remaining columns until the answer is obtained.

EXAMPLE 1.3

Fig. 1.1 shows a shaft. Calculate the dimension *L*.

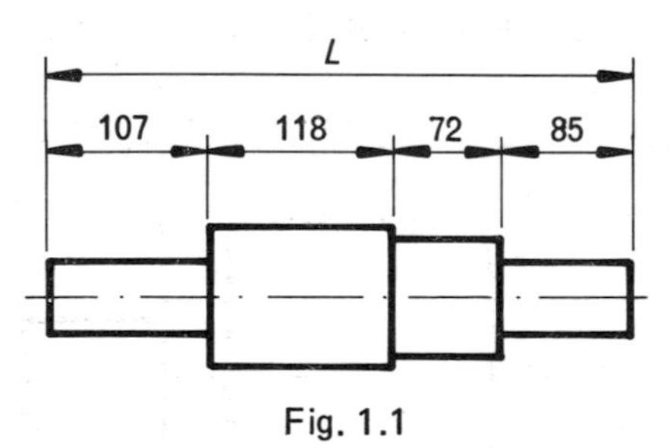

Fig. 1.1

$$\text{Dimension } L = 107 + 118 + 72 + 85$$

$$= 382\,\text{mm}$$

$$\begin{array}{r} 107 \\ 118 \\ 72 \\ 85 \\ \hline 382 \\ \hline \end{array}$$

Exercise 1.2

Find the values of each of the following:

1) $96 + 247 + 8$

2) $109 + 57 + 3478 + 926$

3) $35\,068 + 21\,007 + 905 + 1178 + 32$

4) $23\,589 + 7\,987\,432 + 234\,068 + 9871 + 324\,689$

5) $15\,437 + 1344 + 1626 + 107\,924$

6) Five resistors are placed in series. Their values are 898, 763, 1175, 72 and 196 ohms. Their total resistance is found by adding these five values. Calculate the total resistance.

7) Fig. 1.2 shows a shaft used in an electrical machine. Calculate the length of the shaft.

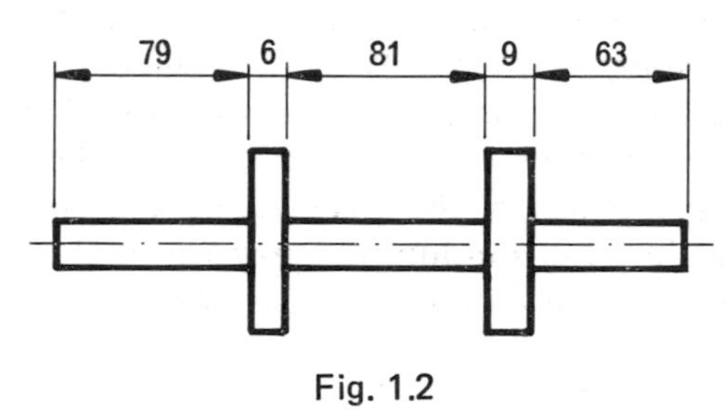

Fig. 1.2

8) Calculate the overall length of the shaft shown in Fig. 1.3.

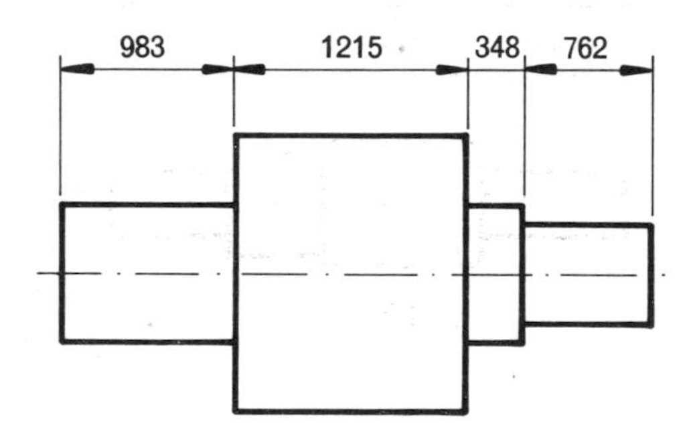

Fig. 1.3

SUBTRACTION

Subtraction means taking away. Let us take 5 from 6. We know that 1 is left. We write $6 - 5 = 1$ which we read as six minus five equals 1.

EXAMPLE 1.4

Subtract 17 from 59

59 17 — 42 —	Place 17 under 59. 7 from 9 leaves 2. Write 2 in the units column of the answer and then 1 from 5 leaves 4. Writing 4 in the tens column of the answer we see that $59-17 = 42$.

There are two methods by which subtraction can be performed. Consider

$$15-8 = 7$$

1st method: Take 8 from 15. We have 7 left.

2nd method: If to 7 we add 8 then we obtain 15. 7 is therefore the difference between 15 and 8.

EXAMPLE 1.5

Find the difference between 32 and 17

Which is the greater of 32 and 17? Clearly 32 is the greater. Therefore we subtract 17 from 32.

32 17 — 15 —	In the units column we cannot take 7 from 2. However if we borrow 1 from the tens place and put it before the 2 we get 12, the 3 in the tens column becoming 2. Now 7 from 12 leaves 5. We write 5 in the tens column of the answer and take 1 from 2 in the tens column leaving 1.

Many people find it easier to work the borrowing method the other way round and to write the subtraction out in this way:

32 17 1 — 15 —	We say that 7 from 2 will not go, so we take 7 from 12 giving 5 which we write in the units column of the answer. We now increase the 1 in the tens column by 1 making it 2 (the smaller figure 1 is a useful aid to the memory until practice makes it unnecessary). Finally we take 2 from 3 leaving 1 which is written in the tens column of the answer.

EXAMPLE 1.6

Subtract 1835 from 5423

5423 1835 1 1 1 —— 3588 ——	In the units column 5 from 3 will not go, so take 5 from 13 leaving 8. Increase the 3 on the bottom of the tens column by 1 making it 4. 4 from 2 will not go, so take 4 from 12 leaving 8 and increase the 8 on the bottom of the hundreds column by making it 9. 9 from 4 will not go, so

take 9 from 14 leaving 5. Finally increase the 1 on the bottom of the thousands column and take 2 from 5 leaving 3.

EXAMPLE 1.7

A bar of mild steel whose diameter is 120 mm is to be turned down to 87 mm diameter. How much metal is to be turned off the diameter?

$$\text{Amount turned off diameter} = 120-87 = 33\text{ mm}$$

Exercise 1.3

1) Find the difference between 27 and 59

2) Subtract 258 from 593

3) Find the value of $53-39$

4) Subtract 7693 from 9251

5) What is the difference between 336 and 9562?

6) Two holes are drilled in a plate as shown in Fig. 1.4. Calculate dimension A.

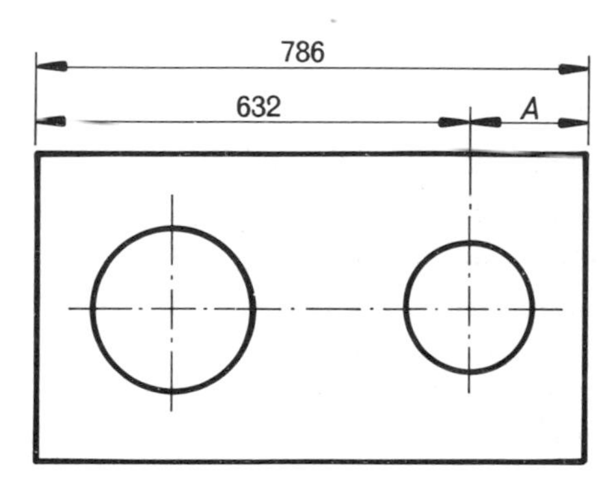

Fig. 1.4

7) A casting has a mass of 138 kilograms. A piece having a mass of 9 kilograms is removed. What is the mass of the casting remaining? remaining?

8) A small tank holds 798 litres of oil when full. If 39 litres are run off, how much oil remains in the tank?

COMBINED ADDITION AND SUBTRACTION

Suppose we want to find the value of

$$18+7-5+3-16+8$$

we pick out all the numbers preceded by a plus sign and add them together. Thus

$$18+7+3+8 = 36$$

(Note that the first number, it is 18, has no sign in front of it. When this happens a plus sign is always assumed.)

Next we pick out all the numbers preceded by a minus sign and add these together. Thus

$$-5-16 = -21$$

Finally we subtract 21 from 36 to give 15. Hence

$$18+7-5+3-16+8 = 36-21 = 15$$

EXAMPLE 1.8

Find the value of $2+6-3+9-5+11$

$$2+6+9+11 = +28 \qquad 28$$

$$-3-5 = -8 \qquad 8$$

Subtracting: $28 - 8 = 20$

EXAMPLE 1.9

What is the dimension X shown in Fig. 1.5?

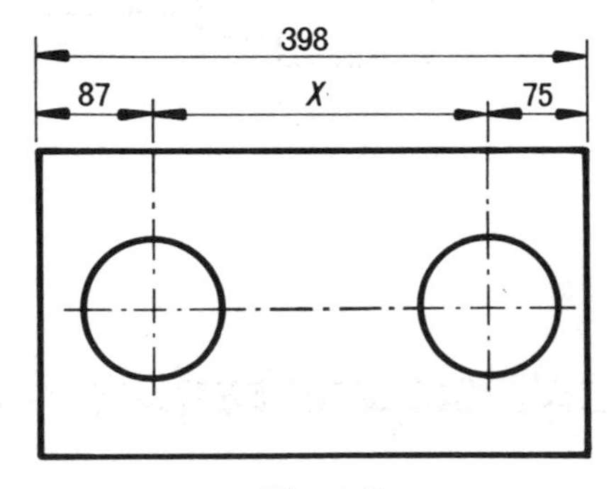

Fig. 1.5

$$\text{Dimension } X = 398-87-75 = 398-162 = 236\,\text{mm}$$

Exercise 1.4

Find the value of each of the following:

1) $8-6+7-5+9-2$

2) $21+32-63-58+79+32-11$

3) $152-78+43-81$

4) $27+45+9+7-15-23-41-8+17$

5) Calculate dimension A in Fig. 1.6.

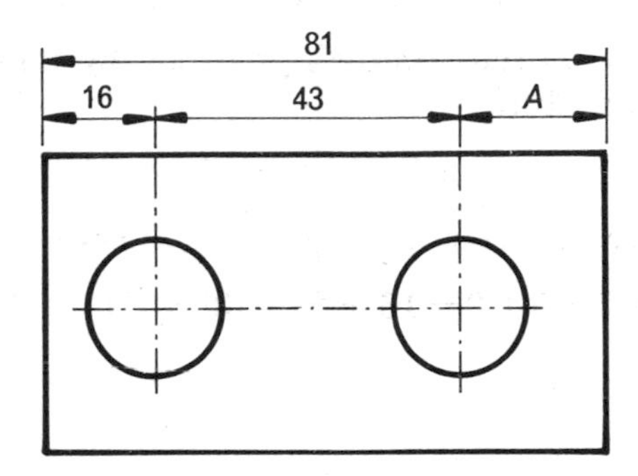

Fig. 1.6

6) Fig. 1.7 shows a shaft which is used in a diesel engine. Calculate the dimension marked X.

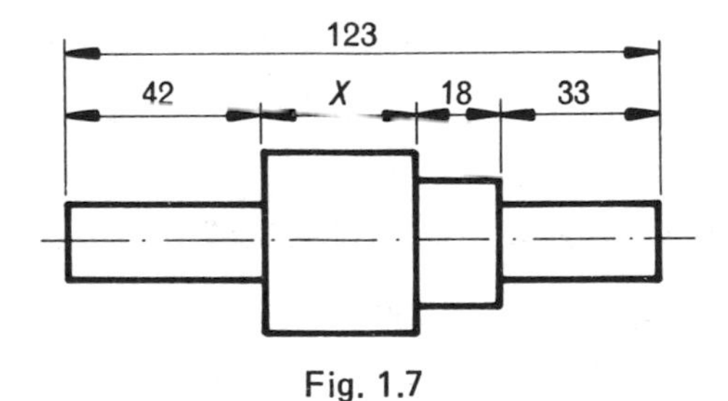

Fig. 1.7

ARITHMETICAL SIGNS, TERMS AND SYMBOLS

The result obtained by adding numbers is called the *sum*. Thus the sum of 9 and 6 is 15.

The result obtained by subtracting one number from another is called the *difference*. The difference between 19 and 8 is $19-8=11$

The sign = is the sign of *equality* and means equal to. Thus 4 hours = 240 minutes.

+ is the *addition* sign meaning plus. Thus $4+5=9$

− is the *subtraction* sign meaning minus. Thus $9-5=4$

× is the *multiplication* sign meaning multiplied by, or times. Thus $6\times 8=48$

÷ is the *division* sign meaning divided by. There are several ways of indicating division which are as follows:

(1) $6\div 3=2$ This reads six divided by three equals two.

(2) $\frac{6}{3}=2$ This reads six over three (or six divided by three) equals two.

(3) 3)6(2 This reads three into six goes two (or six divided by three equals two).

MULTIPLICATION

We can find the value of $6+6+6+6$ by adding the four sixes together. The answer is 24. We could, however, do this more rapidly by using the multiplication tables because we know that $4\times 6=24$

When two numbers are multiplied together the result is called the *product*. Thus the product of 5 and 9 is $5\times 9=45$

EXAMPLE 1.10

Multiply 236 by 7

$$\begin{array}{r} 236 \\ 7 \\ \hline 1652 \\ \hline \end{array}$$

7 times 6 is 42. Place the 2 in the answer and carry the 4. 7 times 3 is 21, plus the 4 carried, is 25. Place 5 in the answer and carry the 2. 7 times 2 is 14, plus the 2 carried is 16

EXAMPLE 1.11

Multiply 369 by 527

369 527 ——— 2583 738 1845 ——— 194463 ———	Write the two numbers with their respective units figures directly underneath each other. Start by multiplying 369 by 7 giving 2583. Write the 3 directly beneath the units figures of the two numbers to be multiplied together. Now multiply 369 by 2 (which is really 20) giving 738. Make sure that the figures obtained by multiplying are this time moved one place to the left. Finally, when multiplying by 5 (which is really 500) it is again necessary to move one further place to the left. We now add the three sets of figures obtained by multiplication, the result being 194 463

Alternatively if you wish, you can start with the left-hand figure in the multiplifer as shown below.

369 527 ——— 184500 7380 2583 ——— 194463 ———	First multiply 369 by 500 giving 184 500. Then multiply 369 by 20 giving 7380 and finally multiply 369 by 7 giving 2583. To obtain the product, add these three sets of figures obtained by multiplication.

EXAMPLE 1.12

A casting has a mass of 24 kg. What is the mass of 73 such castings?

Mass of 73 castings = 24×73

```
  24
  73
 ----
  72
 168
 ----
1752
 ----
```

Hence the mass of 73 castings is 1752 kg.

Exercise 1.5

Obtain the following products:

1) 29×32

2) 359×26

3) 3149×321

4) 5683×789

5) $17\,632 \times 58$

6) A screw has a mass of 12 grams. What is the mass of 1250 such screws?

7) 9 hole centres are to be marked off 34 mm apart. What is the distance between the first and last hole centres?

8) A forging required for an electrical machine has a mass of 198 kg. 24 such forgings are required by the firm manufacturing the machines. What is the mass of these 24 forgings?

DIVISION

Division consists of finding how many times one number is contained in another number.

The *dividend* is the number to be divided.

The *divisor* is the number by which the dividend is divided.

The *quotient* is the result of the division.

Thus $$\frac{\text{dividend}}{\text{divisor}} = \text{quotient}$$

Short Division

If the divisor is less than 10 it is usual to work by short division.

EXAMPLE 1.13

Divide 2625 by 7

```
7)2625(375
  21
  ---
   52
   49
   ---
    35
    35
    --
    ..
```

7 will not divide into 2. Next try 7 into 26. It goes 3 and a remainder of 5. Carry the remainder so that the next number to be divided is 52. 7 goes into 52 7 times and remainder 3. Carry the 3 so that the next number to be divided is 35. 7 into 35 goes 5 exactly.

EXAMPLE 1.14

Divide 1979 by 9

```
9)1979(219
  18    remainder 8
  --
   17
    9
   --
    89
    81
    --
     8
```

9 will not divide into 1 so try dividing 9 into 19. It goes 2 remainder 1. Carry the 1 so that the next number to be divided is 17. 9 into 17 goes 1 remainder 8. Carry the 8 so that the next number to be divided is 89. 9 goes into 89 9 remainder 8. There are no more numbers to divide so the answer is 219 remainder 8.

Exercise 1.6

Work out the answers to the following:

1) $1968 \div 8$ 2) $392 \div 7$ 3) $2168 \div 5$

4) $7369 \div 4$ 5) $5621 \div 9$

Long Division

The method is shown in the next example.

EXAMPLE 1.15

Divide 3024 by 36

```
36)3024(84
   288
   ———
    144
    144
    ———
    ...
```

36 consists of two digits. Look at the first two digits in the dividend, i.e. 30. 36 will not divide into 30 because 36 is the larger number. Next look at the first three figures of the dividend. They are 302. Will 36 divide into 302? It will because 302 is the larger number. How many times will it go? Let us multiply 36 by 9 the result is 324 which is greater than 302. Now try 36×8. The result is 288 which is less than 302. Place 8 in the answer (i.e. the quotient) and write the 288 under the 302. Subtracting 288 from 302 we get a remainder of 14. Now bring down the next figure in the dividend, which is 4. Now divide 36 into 144. The result is 4 exactly because $4 \times 36 = 144$. Write 4 in the quotient and we see that $3024 \div 36 = 84$ exactly.

EXAMPLE 1.16

Divide 1 000 000 by 250

```
250)1000000(4000
     1000
     ————
     ....
```

250 will not divide into the first three figures of the dividend (100) so we try 250 into 1000. It goes 4 times exactly leaving no remainder. To obtain the quotient the remaining three zeros are written in the quotient giving 4000

TESTS FOR DIVISIBILITY

A number is divisible by:

2 if it is an even number,

3 if the sum of the digits is divisible by 3 (3156 is divisible by 3 because $3 + 1 + 5 + 6 = 15$ which is divisible by 3),

4 if its last two figures are divisible by 4 (3024 is divisible by 4 because 24 divided by 4 is 6 exactly),

5 if the last figure is a zero or a five (3265 and 4280 are both divisible by 5),

10 if the last figure is a zero (198 630 is divisible by 10).

EXAMPLE 1.17

A steel bar is 936 mm long. How many lengths each 32 mm long can be cut from the bar and what length remains?

The problem is $936 \div 32$

```
32)936(29 remainder 8
   64
   ---
   296
   288
   ---
     8
```

Hence 29 complete lengths can be cut from the bar and a length of 8 mm remains.

Exercise 1.7

Work out the answers to the following:

1) $4918 \div 9$ 2) $7584 \div 6$ 3) $1237 \div 4$

4) $10\,001 \div 11$ 5) $15\,352 \div 17$ 6) $45\,927 \div 27$

7) $2\,093\,595 \div 35$ 8) $290\,227 \div 49$

9) A bar 150 mm radius is to be reduced to 108 mm radius. How many cuts each 6 mm deep are required?

10) 98 forgings have a total mass of 8134 kg. How much is the mass of each forging?

11) A steel bar 918 mm long is to be divided into 51 equal parts. What is the length of each part?

12) 27 resistors all having the same value are arranged in series and their total resistance is 459 ohms. What is the resistance in ohms of each?

FACTORS AND MULTIPLES

If one number divides exactly into a second number the first number is said to be a *factor* of the second. Thus

$35 = 5 \times 7$ . . . 5 is a factor of 35 and so is 7

$240 = 3 \times 8 \times 10$ . . . 3, 8 and 10 are all factors of 240

$63 = 3 \times 21 = 7 \times 9$

63 is said to be a *multiple* of any of the numbers 3, 7, 9 and 21 because each of them divides exactly into 63.

Every number has itself and 1 as factors. If a number has no other factors, apart from these, it is said to be a *prime number.* Thus 2, 3, 7, 11, 13, 17 and 19 are all prime numbers.

A factor which is a prime number is called a *prime factor.*

Exercise 1.8

1) What numbers are factors of:

(a) 24 (b) 56 (c) 42?

2) Which of the following numbers are factors of 12:

2, 3, 4, 5, 6, 12, 18 and 24?

Which of them are multiples of 6?

3) Write down all the multiples of 3 between 10 and 40

4) Express as a product of prime factors:

(a) 24 (b) 36 (c) 56 (d) 132

5) Write down the two prime numbers next larger than 19

LOWEST COMMON MULTIPLE (LCM)

The LCM of a set of numbers is the smallest number into which each of the given numbers will divide exactly.

Thus the LCM of 4, 5 and 10 is 20, because 20 is the smallest number into which 4, 5 and 10 will divide exactly.

The LCM may often be found by inspection, or alternatively the method shown in the following example may be used.

EXAMPLE 1.18

Find the LCM of the numbers 36, 54, 60 and 80

We express each number as the product of its prime factors.

Thus $36 = 2 \times 2 \times 3 \times 3$

and $54 = 2 \times 3 \times 3 \times 3$

and $60 = 2 \times 2 \times 3 \times 5$

and $80 = 2 \times 2 \times 2 \times 2 \times 5$

We now note the greatest number of times each prime factor occurs in any one particular line.

Now factor 2 occurs four times in the line for 80,

and factor 3 occurs three times in the line for 54,

and factor 5 occurs once in either of the lines for 60 or 80.

The product of these gives the required LCM.

Thus $\text{LCM} = (2 \times 2 \times 2 \times 2) \times (3 \times 3 \times 3) \times (5)$

$= 720$

Exercise 1.9

Find the LCM of the following sets of numbers:

1) 8 and 12
2) 3, 4 and 5
3) 2, 6 and 12
4) 3, 6 and 8
5) 2, 8 and 10
6) 20 and 25
7) 20 and 32
8) 10, 15 and 40
9) 12, 42, 60 and 70
10) 18, 30, 42 and 48

HIGHEST COMMON FACTOR (HCF)

The HCF of a set of numbers is the greatest number which is a factor of each of the numbers. Thus 12 is the HCF of 24, 36 and 60. Also 20 is the HCF of 40, 60 and 80.

The HCF may often be found by inspection, or alternatively the method shown in the following example may be used.

EXAMPLE 1.19

Find the HCF of the numbers 36, 60, 108 and 240

We express each number as the product of its prime factors.

Thus $36 = 2 \times 2 \times 3 \times 3$

and $60 = 2 \times 2 \times 3 \times 5$

and $108 = 2 \times 2 \times 3 \times 3 \times 3$

and $240 = 2 \times 2 \times 2 \times 2 \times 3 \times 5$

We now note the prime factors which are common to each of the lines. Factor 2 is common twice and factor 3 once. Factor 5 is present only in the second and fourth lines and is not, therefore, a common factor.

The product of these common factors gives the required HCF.

Thus
$$\text{HCF} = (2 \times 2) \times (3) = 12$$

Exercise 1.10

Find the HCF of each of the following sets of numbers:

1) 8 and 12
2) 24 and 36
3) 10, 15 and 30
4) 26, 39 and 52
5) 18, 30, 12 and 42
6) 28, 42, 84, 98 and 112

SEQUENCE OF ARITHMETICAL OPERATIONS

When numbers are added together the order in which they occur is unimportant.

Thus $7+8 = 8+7 = 15$

$$5+6+9 = 6+5+9 = 9+5+6 = 6+9+5 = 20$$

When numbers are multiplied together the order in which they occur is unimportant.

Thus $3 \times 5 = 5 \times 3 = 15$

$$6 \times 8 \times 7 = 8 \times 6 \times 7 = 8 \times 7 \times 6 = 6 \times 7 \times 8 = 336$$

Correct Sequence for Mixed Operations

Numbers are often combined in a series of arithmetical operations. When this happens a definite sequence must be observed.

(1) Brackets are used if there is any danger of ambiguity. The contents of the bracket must be evaluated before performing any other operation.

Thus
$$2 \times (7+4) = 2 \times 11 = 22$$
$$15-(8-3) = 15-5 = 10$$

(2) In calculations with fractions the word 'of' may appear. It should always be taken as meaning 'multiply'. Thus:

$$\frac{1}{2} \text{ of } 20 = \frac{1}{2} \times 20 = 10$$

Thus operation 'of' should be performed immediately after the contents of brackets have been evaluated.

(3) Multiplication and division must be done before addition and subtraction.

Thus $5 \times 8+7 = 40+7 = 47$ (not 5×15)

$8 \div 4+9 = 2+9 = 11$ (not $8 \div 13$)

$$5 \times 4-12 \div 3+7 = 20-4+7 = 27-4 = 23$$

To remember the correct sequence of operations it helps to think of the word BODMAS. This word is made up from the initial letters of the correct sequence of operations, namely: Brackets, Of, Divide, Multiply, Add, Subtract.

Thus
$$\begin{aligned} 8 \div 2+6 \times (4-1)-\frac{1}{2} \text{ of } 6 &= 8 \div 2+6 \times 3-\frac{1}{2} \text{ of } 6 \\ &= 8 \div 2+6 \times 3-3 \\ &= 4+6 \times 3-3 \\ &= 4+18-3 \\ &= 22-3 \\ &= 19 \end{aligned}$$

EXAMPLE 1.20

15 hole centres are to be marked off 39 mm apart. If 18 mm is to be allowed between the centres of the end holes and the edges of the plate, find the total length of plate required.

$$\begin{aligned} \text{Length of plate required} &= 14 \times 39+2 \times 18 \\ &= 546+36 \\ &= 582 \text{ mm} \end{aligned}$$

Exercise 1.11

Find values for the following:

1) $3+5\times 2$

2) $3\times 6-8$

3) $7\times 5-2+4\times 6$

4) $8\div 2+3$

5) $7\times 5-12\div 4+3$

6) $2+8\times(3+6)$

7) $11-9\div 3+7$

8) $17-2\times(5-3)$

9) $3\times(8+7)$

10) $11-12\div 4+3\times(6-2)$

11) 16 holes spaced 48 mm apart are to be marked off on a sheet metal detail. 17 mm is to be allowed between the centres of the end holes and the edges of the plate. Calculate the total length of metal required.

12) In the first 2 hours of a shift an operator makes 32 soldered joints per hour. In the next 3 hours the operator makes 29 joints per hour. In the final two hours 26 joints are made per hour. How many soldered joints are made in the 7 hours?

13) A machinist makes 3 parts in 15 minutes. How many parts can he make in an 8 hour shift allowing 20 minutes for starting and 10 minutes for finishing the shift.

14) The length of a plate detail is 891 mm. Rivets are placed 45 mm apart and the distance between the centres of the end rivets and the edges of the plate is 18 mm. How many rivets are required?

15) 32 pins each 61 mm long are to be turned in a lathe. If 2 mm is allowed on each pin for parting off what total length of material is required to make the pins?

Self-Test 1

In questions 1 to 4 state the letter corresponding to the correct answer.

1) Thirty thousand and four in figures is:

a 3004 b 30 004 c 34 000
b 30 400 e 300 004

2) Five thousand and fifteen in figures is:

a 5015 b 5150 c 5115
d 50 015 e 515

3) One hundred and six thousand and sixteen in figures is:

a 116 000 b 106 000 c 106 160
d 116 160 e 106 016

4) Ten million, seventeen thousand and six in figures is:

a 10 017 006 b 10 170 006 c 10 017 060
d 10 017 600 e 10 170 060

5) Add up the three sets of figures below:

(a)	(b)	(c)
5 018	3 263	528
362	8 783	5 079
12 894	35	9 867
268	357	61
4 134	10 089	356
______	______	______
______	______	______

6) State the letter corresponding to the correct answer to the problems shown below.

(a) $107-104+63-48+137+50-149$

a 65 b 56 c 66

(b) $368-55+378-286+245-254$

a 416 b 395 c 396

(c) $45-764+418-382+1049-689+1000$

a 677 b 605 c 687

7) Subtract the following:

(a) $2092-987$ (b) $2315-999$ (c) $7005-889$
(d) $958-697$ (e) $432-318$ (f) $23\,301-22\,398$

8) Multiply the following:

(a) $16\,398 \times 7$ (b) $635\,489 \times 12$ (c) $93\,081 \times 407$
(d) $51\,365 \times 450$ (e) 9457×6003 (f) $68\,859 \times 836$

9) Divide the following:

(a) $46\,348 \div 4$ (b) $32\,340 \div 60$ (c) $1536 \div 32$
(d) $17\,280 \div 960$

10) In each of the following division problems there is a remainder. Perform the division and state the remainder:

(a) $685\,329 \div 64$ (b) $61\,385 \div 13$ (c) $969\,234 \div 19$

(d) $17\,432 \div 560$

11) Each of the following numbers is divisible by either 3, 4, 5, 9, 10 or 11. State which number will divide exactly into the given number:

(a) 1404 (b) 8615 (c) 4167

(d) 1564 (e) 102 971 (f) 73 216

12) State the letter corresponding to the correct answer to the problems shown below.

(a) $3 + 7 \times 4$

a 40 b 31 c 84

(b) $6 \times 5 - 2 + 4 \times 6$

a 52 b 42 c 18

(c) $7 \times 6 - 12 \div 3 + 1$

a 40 b 39 c 21

(d) $17 - 2 \times (6 - 4)$

a 30 b 1 c 13

(e) $3 \times 5 - 12 \div (3 + 1)$

a 12 b 10 c 8

In questions 13 to 20 decide if the answer is true or false and write the correct answer in your notebook.

13) $7 + 5 \times 3 = 22$

14) $7 - 2 \times 3 = 15$

15) $6 \times 5 - 3 + 2 \times 7 = 26$

16) $10 \div 2 + 3 = 8$

17) $7 \times 5 - 12 \div 4 + 2 = 10$

18) $18 - 10 \div 2 + 3 \times (5 - 2) = 22$

19) $36 - 27 + 54 - 58 = 15$

20) The exact value of $(312 \times 11 \times 19) \div 39$ is 1672

2. FRACTIONS

After reaching the end of this chapter you should be able to:

1. *Define numerator and denominator of a fraction.*
2. *Simplify fractions by cancellation.*
3. *Recognise that cancellation or division by zero is not permissible.*
4. *Add, subtract, multiply and divide fractions and simplify results.*
5. *Solve problems involving more than one of the operations in 4.*
6. *Simplify expressions involving fractions using precedence rules for brackets and basic arithmetic operations.*

INTRODUCTION

In this chapter we deal with the rules for the addition, subtraction, multiplication and division of fractions.

VULGAR FRACTIONS

The circle in Fig. 2.1 has been divided into eight equal parts. Each part is called one-eighth of the circle and is written as $\frac{1}{8}$. The number 8 below the line shows how many equal parts there are and it is called the *denominator*. The number above the line shows how many of the equal parts are taken and it is called the *numerator*. If five of the eight equal parts are taken then we have taken $\frac{5}{8}$ of the circle.

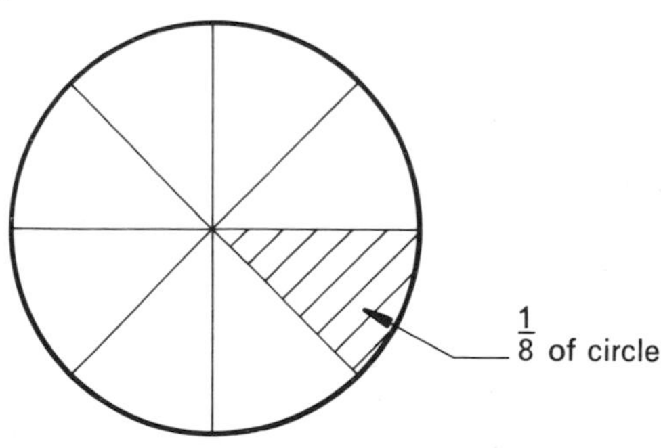

Fig. 2.1

From what has been said above we see that a fraction is always a part of something. The number below the line (the denominator) gives the fraction its name and tells us the number of equal parts into which the whole has been divided. The top number (the numerator) tells us the number of these equal parts that are to be taken. For example, the fraction $\frac{3}{4}$ means that the whole has been divided into four equal parts and that three of these parts are to be taken.

The value of a fraction is unchanged if we multiply or divide both its numerator and denominator by the same amount.

$\frac{3}{5} = \frac{12}{20}$ (by multiplying the numerator and denominator by 4)

$\frac{2}{7} = \frac{10}{35}$ (by multiplying the numerator and denominator by 5)

$\frac{12}{32} = \frac{3}{8}$ (by dividing the numerator and denominator by 4)

$\frac{16}{64} = \frac{1}{4}$ (by dividing the numerator and denominator by 16)

Note that it is not permissible to divide numerator and denominator by 0 because dividing by zero gives an indeterminate number (i.e. a number whose value cannot be stated).

EXAMPLE 2.1

Write down the fraction $\frac{2}{7}$ with a denominator of 28

In order to make the denominator 28, we must multiply the original denominator of 7 by 4 because $7 \times 4 = 28$. Remembering that to leave the value of the fraction unchanged we must multiply both numerator and denominator by the same amount, then

$$\frac{2}{7} = \frac{2 \times 4}{7 \times 4} = \frac{8}{28}$$

Exercise 2.1

Write down the following fractions with the denominator stated:

1) $\frac{3}{4}$ with denominator 28
2) $\frac{3}{5}$ with denominator 20
3) $\frac{5}{6}$ with denominator 30
4) $\frac{1}{9}$ with denominator 63

5) $\frac{2}{3}$ with denominator 12

6) $\frac{1}{6}$ with denominator 24

7) $\frac{3}{8}$ with denominator 64

8) $\frac{5}{7}$ with denominator 35

REDUCING A FRACTION TO ITS LOWEST TERMS

Fractions like $\frac{3}{8}$, $\frac{7}{16}$ and $\frac{5}{32}$ are said to be in their *lowest terms* because it is impossible to find a number which will divide exactly into both the numerator and denominator. However, fractions like $\frac{9}{18}$, $\frac{8}{12}$, and $\frac{21}{24}$ are not in their lowest terms because they can be reduced further by dividing both numerator and denominator by some number which divides exactly into both of them. Thus

$$\frac{9}{18} = \frac{1}{2} \quad \text{(by dividing both numerator and denominator by 9)}$$

$$\frac{8}{12} = \frac{2}{3} \quad \text{(by dividing both numerator and denominator by 4)}$$

$$\frac{21}{24} = \frac{7}{8} \quad \text{(by dividing both numerator and denominator by 3)}$$

Sometimes we can divide the numerator and denominator by the same number several times.

EXAMPLE 2.2

Reduce $\frac{210}{336}$ to its lowest terms.

$$\frac{210}{336} = \frac{105}{168} \quad \text{(by dividing top and bottom by 2)}$$

$$= \frac{35}{56} \quad \text{(by dividing top and bottom by 3)}$$

$$= \frac{5}{8} \quad \text{(by dividing top and bottom by 7)}$$

Hence $\frac{210}{336}$ reduced to its lowest terms is $\frac{5}{8}$.

Exercise 2.2

Reduce the following fractions to their lowest terms:

1) $\frac{8}{16}$ 2) $\frac{9}{15}$ 3) $\frac{8}{64}$ 4) $\frac{15}{25}$ 5) $\frac{42}{48}$

6) $\frac{180}{240}$ 7) $\frac{210}{294}$ 8) $\frac{126}{245}$ 9) $\frac{132}{198}$ 10) $\frac{210}{315}$

TYPES OF FRACTIONS

If the numerator of a fraction is less than its denominator the fraction is called a *proper fraction*. Thus $\frac{2}{3}$, $\frac{5}{8}$ and $\frac{3}{4}$ are all proper fractions. Note that a proper fraction has a value which is less than 1.

If the numerator of a fraction is greater than its denominator then the fraction is called an *improper fraction* or a *top-heavy fraction*. Thus $\frac{5}{4}$, $\frac{3}{2}$ and $\frac{9}{7}$ are all top heavy, or improper, fractions. Note that all top heavy fractions have a value which is greater than 1.

Every top-heavy fraction can be expressed as a whole number and a proper fraction. These are sometimes called *mixed numbers*. Thus $1\frac{1}{2}$, $5\frac{1}{3}$, and $9\frac{3}{4}$ are all mixed numbers.

With the introduction of decimals to most technology measurements, the use of mixed numbers has virtually disappeared. If, however, they are encountered they may be converted to improper fractions as shown in the example which follows.

Alternatively a mixed number may be expressed as a decimal number (see p. 54).

EXAMPLE 2.3

Express the mixed number $3\frac{5}{8}$ as an improper fraction.

Now $3\frac{5}{8}$ represents (3 whole units) plus $\left(\frac{5}{8}\text{ of a unit}\right)$.

Thus $3\frac{5}{8} = 3+\frac{5}{8} = \frac{3\times 8}{8}+\frac{5}{8} = \frac{24}{8}+\frac{5}{8} = \frac{24+5}{8} = \frac{29}{8}$

The above sequence is often shortened to

$$3\frac{5}{8} = \frac{(3 \times 8) + 5}{8} = \frac{24 + 5}{8} = \frac{29}{8}$$

Exercise 2.3

Express each of the following as top-heavy fractions:

1) $2\frac{3}{8}$ 2) $5\frac{1}{10}$ 3) $8\frac{2}{3}$ 4) $6\frac{7}{20}$ 5) $4\frac{3}{7}$

LOWEST COMMON DENOMINATOR

When we wish to compare the values of two or more fractions the easiest way is to express the fractions with the same denominator. This common denominator should be the LCM of the denominators of the fractions to be compared and it is called the *lowest common denominator.*

EXAMPLE 2.4

Arrange the fractions $\frac{3}{4}, \frac{5}{8}, \frac{7}{10}$ and $\frac{11}{20}$ in order of size starting with the smallest.

The lowest common denominator of 4, 8, 10 and 20 is 40. Expressing each of the given fractions with a denominator of 40 gives:

$$\frac{3}{4} = \frac{3 \times 10}{4 \times 10} = \frac{30}{40} \qquad \frac{5}{8} = \frac{5 \times 5}{8 \times 5} = \frac{25}{40}$$

$$\frac{7}{10} = \frac{7 \times 4}{10 \times 4} = \frac{28}{40} \qquad \frac{11}{20} = \frac{11 \times 2}{20 \times 2} = \frac{22}{40}$$

Therefore the order is $\frac{22}{40}, \frac{25}{40}, \frac{28}{40}, \frac{30}{40}$ or $\frac{11}{20}, \frac{5}{8}, \frac{7}{10}$ and $\frac{3}{4}$

Exercise 2.4

Arrange the following sets of fractions in order of size, beginning with the smallest:

1) $\frac{1}{2}, \frac{5}{6}, \frac{2}{3}, \frac{7}{12}$ 2) $\frac{9}{10}, \frac{3}{4}, \frac{6}{7}, \frac{7}{8}$

3) $\frac{13}{16}, \frac{11}{20}, \frac{7}{10}, \frac{3}{5}$

4) $\frac{3}{4}, \frac{5}{8}, \frac{3}{5}, \frac{13}{20}$

5) $\frac{11}{16}, \frac{7}{10}, \frac{9}{14}, \frac{3}{4}$

6) $\frac{3}{8}, \frac{4}{7}, \frac{5}{9}, \frac{2}{5}$

ADDITION OF FRACTIONS

The steps when adding fractions are as follows:

(1) Find the lowest common denominator of the fractions to be added.

(2) Express each of the fractions with this common denominator.

(3) Add the numerators of the new fractions to give the numerator of the answer. The denominator of the answer is the lowest common denominator found in (1).

EXAMPLE 2.5

Find the sum of $\frac{2}{7}$ and $\frac{3}{4}$

First find the lowest common denominator (this is the LCM of 7 and 4). It is 28. Now express $\frac{2}{7}$ and $\frac{3}{4}$ with a denominator of 28.

$$\frac{2}{7} = \frac{2 \times 4}{7 \times 4} = \frac{8}{28} \qquad \frac{3}{4} = \frac{3 \times 7}{4 \times 7} = \frac{21}{28}$$

Adding the numerators of the new fractions:

$$\frac{2}{7} + \frac{3}{4} = \frac{8}{28} + \frac{21}{28} = \frac{8+21}{28} = \frac{29}{28}$$

A better way of setting out the work is as follows:

$$\frac{2}{7} + \frac{3}{4} = \frac{2 \times 4 + 3 \times 7}{28} = \frac{8+21}{28} = \frac{29}{28}$$

EXAMPLE 2.6

Simplify $\frac{3}{4} + \frac{2}{3} + \frac{7}{10}$

The LCM of the denominators 4, 3 and 10 is 60

$$\frac{3}{4} + \frac{2}{3} + \frac{7}{10} = \frac{3 \times 15 + 2 \times 20 + 7 \times 6}{60} = \frac{45+40+42}{60} = \frac{127}{60}$$

EXAMPLE 2.7

Add together 5, $\frac{8}{3}$ and $\frac{2}{5}$

The LCM of the denominators 3 and 5 is 15

The whole number 5 may be considered as the fractions

$$\frac{5}{1} \quad \text{or} \quad \frac{5\times15}{15}$$

Thus

$$5+\frac{8}{3}+\frac{2}{5} = \frac{5\times15+8\times5+2\times3}{15} = \frac{75+40+6}{15} = \frac{121}{15}$$

Exercise 2.5

Add together:

1) $\frac{1}{2}+\frac{1}{3}$
2) $\frac{2}{5}+\frac{9}{10}$
3) $\frac{3}{4}+\frac{3}{8}$
4) $\frac{3}{10}+\frac{1}{4}$
5) $\frac{1}{2}+\frac{3}{4}+\frac{7}{8}$
6) $\frac{1}{8}+\frac{2}{3}+\frac{3}{5}$
7) $\frac{11}{8}+\frac{57}{16}$
8) $\frac{23}{3}+\frac{33}{5}$
9) $3+\frac{2}{7}$
10) $\frac{9}{2}+3+\frac{1}{3}$
11) $\frac{59}{8}+\frac{11}{4}+\frac{7}{8}+\frac{5}{16}$
12) $\frac{23}{3}+\frac{2}{5}+\frac{3}{10}+2$

SUBTRACTION OF FRACTIONS

The method is similar to that used in addition. Find the common denominator of the fractions and after expressing each fraction with this common denominator, subtract.

EXAMPLE 2.8

Simplify $\frac{5}{8}-\frac{2}{5}$

The LCM of the denominators is 40.

$$\frac{5}{8}-\frac{2}{5} = \frac{5\times5-2\times8}{40} = \frac{25-16}{40} = \frac{9}{40}$$

EXAMPLE 2.9

Simplify $\frac{2}{3}-\frac{15}{16}$

The LCM of the denominators 3 and 16 is 48

Thus

$$\frac{2}{3}-\frac{15}{16}=\frac{2\times16-15\times3}{48}=\frac{32-45}{48}=\frac{-13}{48}=-\frac{13}{48}$$

Exercise 2.6

Simplify the following:

1) $\frac{1}{2}-\frac{1}{3}$
2) $\frac{1}{3}-\frac{1}{5}$
3) $\frac{2}{3}-\frac{1}{2}$
4) $\frac{7}{8}-\frac{3}{8}$
5) $\frac{7}{8}-\frac{5}{6}$
6) $\frac{1}{4}-\frac{3}{8}$
7) $3-\frac{5}{7}$
8) $\frac{3}{8}-3$
9) $\frac{29}{10}-\frac{43}{8}$

COMBINED ADDITION AND SUBTRACTION

EXAMPLE 2.10

Simplify $\frac{43}{8}-\frac{5}{4}+2-\frac{7}{16}$

The LCM of the denominators 8, 4 and 16 is 16

Thus $\frac{43}{8}-\frac{5}{4}+2-\frac{7}{16}=\frac{43\times2-5\times4+2\times16-7\times1}{16}$

$$=\frac{86-20+32-7}{16}$$

$$=\frac{91}{16}$$

Exercise 2.7

Simplify the following:

1) $\frac{5}{2}+\frac{1}{4}-\frac{3}{8}$
2) $\frac{1}{10}-\frac{7}{2}+\frac{5}{4}$

3) $\frac{3}{8}-\frac{5}{2}-5$

4) $7-\frac{17}{5}-\frac{3}{2}$

5) $\frac{1}{2}+\frac{1}{3}-\frac{1}{4}-\frac{1}{5}$

6) $\frac{9}{7}-\frac{4}{5}+6-\frac{3}{15}$

7) $\frac{21}{7}+\frac{7}{3}-10$

8) $4-\frac{2}{7}+\frac{7}{2}-7$

MULTIPLICATION

When multiplying together two or more fractions we first multiply all the numerators together and then we multiply all the denominators together.

EXAMPLE 2.11

Simplify $\frac{5}{8}\times\frac{3}{7}$

$$\frac{5}{8}\times\frac{3}{7} = \frac{5\times 3}{8\times 7} = \frac{15}{56}$$

EXAMPLE 2.12

Simplify $\frac{2}{5}\times\frac{7}{3}\times 4$

The whole number 4 is treated as $\frac{4}{1}$

Thus $$\frac{2}{5}\times\frac{7}{3}\times 4 = \frac{2\times 7\times 4}{5\times 3\times 1} = \frac{56}{15}$$

Exercise 2.8

Simplify the following:

1) $\frac{2}{3}\times\frac{4}{5}$

2) $\frac{3}{4}\times\frac{5}{7}$

3) $\frac{2}{9}\times\frac{7}{3}$

4) $\frac{5}{9}\times\frac{11}{4}$

5) $\frac{3}{7}\times\frac{2}{5}\times 6$

6) $\frac{1}{11}\times\frac{9}{5}\times\frac{3}{2}$

CANCELLING

EXAMPLE 2.13

Simplify $\frac{2}{3} \times \frac{15}{8}$

$$\frac{2}{3} \times \frac{15}{8} = \frac{2 \times 15}{3 \times 8} = \frac{30}{24} = \frac{5}{4}$$

The step of reducing $\frac{30}{24}$ to its lowest terms has been done by dividing 6 into both the numerator and denominator.

The work can be made easier by *cancelling* before multiplication as shown below.

$$\frac{\overset{1}{\cancel{2}}}{\underset{1}{\cancel{3}}} \times \frac{\overset{5}{\cancel{15}}}{\underset{4}{\cancel{8}}} = \frac{1 \times 5}{1 \times 4} = \frac{5}{4}$$

We have divided a numerator 2 into a denominator 8, and also we have divided a numerator 15 by a denominator 3. You will see that we have divided the numerators and the denominators by the same amount. Notice carefully that we can only cancel between a numerator and a denominator.

EXAMPLE 2.14

Simplify $\frac{16}{25} \times \frac{7}{8} \times \frac{35}{4}$

$$\frac{\overset{\overset{1}{\cancel{2}}}{\cancel{16}}}{\underset{5}{\cancel{25}}} \times \frac{7}{\underset{1}{\cancel{8}}} \times \frac{\overset{7}{\cancel{35}}}{\underset{2}{\cancel{4}}} = \frac{1 \times 7 \times 7}{5 \times 1 \times 2} = \frac{49}{10}$$

THE OPERATION 'OF'

Sometimes in calculations with fractions the word 'of' appears. It should always be taken as meaning multiply. Thus

$$\frac{4}{5} \text{ of } 20 = \frac{4}{\underset{1}{\cancel{5}}} \times \frac{\overset{4}{\cancel{20}}}{1} = \frac{4 \times 4}{1 \times 1} = \frac{16}{1} = 16$$

Exercise 2.9

Simplify the following:

1) $\frac{3}{4}\times\frac{16}{9}$ 2) $\frac{26}{5}\times\frac{10}{13}$ 3) $\frac{13}{8}\times\frac{7}{26}$

4) $\frac{3}{2}\times\frac{2}{5}\times\frac{5}{2}$ 5) $\frac{5}{8}\times\frac{7}{10}\times\frac{2}{21}$ 6) $2\times\frac{3}{2}\times\frac{4}{3}$

7) $\frac{15}{4}\times\frac{8}{5}\times\frac{9}{8}$ 8) $\frac{15}{32}\times\frac{8}{11}\times\frac{121}{5}$ 9) $\frac{3}{4}$ of 16

10) $\frac{5}{7}$ of 140 11) $\frac{2}{3}$ of $\frac{9}{2}$ 12) $\frac{4}{5}$ of $\frac{5}{2}$

DIVISION OF FRACTIONS

To divide by a fraction, all we have to do is to invert it and multiply.

Thus $$\frac{3}{5}\div\frac{2}{7} = \frac{3}{5}\times\frac{7}{2} = \frac{3\times7}{5\times2} = \frac{21}{10}$$

EXAMPLE 2.15

Divide $\frac{9}{5}$ by $\frac{7}{3}$

$$\frac{9}{5}\div\frac{7}{3} = \frac{9}{5}\times\frac{3}{7} = \frac{27}{35}$$

Exercise 2.10

Simplify the following:

1) $\frac{4}{5}\div\frac{4}{3}$ 2) $2\div\frac{1}{4}$ 3) $\frac{5}{8}\div\frac{15}{32}$

4) $\frac{15}{4}\div\frac{5}{2}$ 5) $\frac{5}{2}\div\frac{15}{4}$ 6) $5\div\frac{26}{5}$

7) $\frac{46}{15}\div\frac{23}{9}$ 8) $\frac{23}{10}\div\frac{3}{5}$

OPERATIONS WITH FRACTIONS

The sequence of operations when dealing with fractions is the same as those used with whole numbers. They are, in order:

(1) Work out brackets.

(2) Multiply and divide.

(3) Add and subtract.

EXAMPLE 2.16

Simplify $\frac{1}{5} \div \left(\frac{1}{3} \div \frac{1}{2}\right)$

$$\frac{1}{5} \div \left(\frac{1}{3} \div \frac{1}{2}\right) = \frac{1}{5} \div \left(\frac{1}{3} \times \frac{2}{1}\right) = \frac{1}{5} \div \frac{2}{3} = \frac{1}{5} \times \frac{3}{2} = \frac{3}{10}$$

EXAMPLE 2.17

Simplify $\dfrac{\frac{14}{5} + \frac{5}{4}}{\frac{18}{5}} - \dfrac{5}{16}$

With problems of this kind it is best to work in stages as shown below:

$$\frac{14}{5} + \frac{5}{4} = \frac{14 \times 4 + 5 \times 5}{20} = \frac{56 + 25}{20} = \frac{81}{20}$$

$$\frac{\frac{81}{20}}{\frac{18}{5}} = \frac{81}{20} \div \frac{18}{5} = \frac{81}{20} \times \frac{5}{18} = \frac{9}{8}$$

$$\frac{9}{8} - \frac{5}{16} = \frac{18 - 5}{16} = \frac{13}{16}$$

Exercise 2.11

Simplify the following:

1) $\frac{45}{14} + \left(\frac{50}{49} \times \frac{7}{10}\right)$ 2) $\frac{1}{4} \div \left(\frac{1}{8} \times \frac{2}{5}\right)$ 3) $\frac{5}{3} \div \left(\frac{3}{5} \div \frac{9}{10}\right)$

4) $\left(\frac{15}{8}\times\frac{12}{5}\right)-\frac{11}{3}$

5) $\dfrac{\frac{8}{3}+\frac{6}{5}}{\frac{29}{5}}$

6) $\frac{11}{3}\div\left(\frac{2}{3}+\frac{4}{5}\right)$

7) $\dfrac{\frac{28}{5}-\frac{7}{2}\times\frac{2}{3}}{\frac{7}{3}}$

8) $\frac{2}{5}\times\left(\frac{2}{3}-\frac{1}{4}\right)+\frac{1}{2}$

9) $\dfrac{\frac{57}{16}\times\frac{4}{9}}{2+\frac{25}{4}\times\frac{6}{5}}$

10) $\dfrac{\frac{5}{9}-\frac{7}{15}}{1-\left(\frac{5}{9}\times\frac{7}{15}\right)}$

SUMMARY

a) The denominator (bottom number) gives the fraction its name and gives the number of equal parts into which the whole has been divided. The numerator (top number) gives the number of equal parts that are to be taken.

b) The value of a fraction remains unaltered if both the numerator and the denominator are multiplied or divided by the same number.

c) The LCM of a set of numbers is the smallest number into which each of the numbers of the set will divide exactly.

d) To compare the values of fractions which have different denominators express all the fractions with the lowest common denominator and then compare the numerators of the new fractions.

e) To add fractions express each of them with their lowest common denominators and then add the resulting numerators.

f) To multiply fractions multiply the numerators together and then multiply the denominators together.

g) To divide, invert the divisor and then proceed as in multiplication.

h) The sequence of operations when dealing with fractions is: (i) work out brackets; (ii) 'of', multiply and divide; (iii) add and subtract.

Self-Test 2

In questions 1 to 15, state the letter, or letters, corresponding to the correct answer or answers.

1) When the fraction $\frac{630}{1470}$ is reduced to its lowest terms the answer is:

a $\frac{63}{147}$ b $\frac{3}{7}$ c $\frac{21}{49}$ d $\frac{9}{20}$

2) Which of the following fractions is equal to $\frac{4}{9}$?

a $\frac{12}{27}$ b $\frac{4}{36}$ c $\frac{36}{4}$ d $\frac{20}{90}$ e $\frac{52}{117}$

3) The fraction $\frac{3}{4}$ when written with denominator 56 is the same as:

a $\frac{3}{56}$ b $\frac{56}{12}$ c $\frac{42}{56}$ d $\frac{56}{42}$

4) The LCM of 5, 15, 40 and 64 is:

a 960 b 192 000 c 640 d 64

5) The mixed number $3\frac{5}{6}$ is equal to:

a $\frac{15}{6}$ b $\frac{5}{18}$ c $\frac{15}{18}$ d $\frac{23}{6}$

6) The mixed number $2\frac{3}{7}$ is equal to:

a $\frac{5}{7}$ b $\frac{3}{9}$ c $\frac{17}{7}$ d $\frac{13}{7}$

7) $\frac{1}{4}+\frac{2}{3}+\frac{3}{5}$ is equal to:

a $\frac{1}{2}$ b $\frac{1}{10}$ c $\frac{91}{60}$ d $1\frac{60}{91}$

8) $\frac{11}{8}+\frac{17}{6}+\frac{13}{4}$ is equal to:

a $\frac{58}{9}$ b $\frac{41}{18}$ c $\frac{35}{24}$ d $\frac{21}{40}$ e $\frac{179}{24}$

9) $\frac{9}{8}+\frac{13}{6}+\frac{15}{4}$ is equal to:

a $\frac{37}{18}$ b $\frac{169}{24}$ c $\frac{25}{24}$ d $\frac{174}{24}$

10) $\frac{7}{8}\times\frac{3}{5}$ is equal to:

a $\frac{10}{13}$ b $\frac{4}{3}$ c $\frac{35}{24}$ d $\frac{21}{40}$

11) $\frac{5}{8}\times\frac{4}{15}$ is equal to one of the following, when the answer is expressed in its lowest terms:

a $\frac{20}{120}$ b $\frac{1}{6}$ c $\frac{32}{75}$ d $\frac{9}{23}$

12) $\frac{3}{4}\div\frac{8}{9}$ is equal to:

a $\frac{24}{36}$ b $\frac{2}{3}$ c $\frac{27}{32}$ d $\frac{3}{2}$

13) $\frac{58}{9}\div\frac{11}{3}$ is equal to:

a $\frac{8}{3}$ b $\frac{638}{27}$ c $\frac{58}{33}$ d $\frac{69}{27}$

14) $3\times\left(\frac{1}{2}-\frac{1}{3}\right)$ is equal to:

a $\frac{3}{2}-\frac{1}{3}$ b $\frac{1}{2}$ c $3\times\frac{1}{2}-3\times\frac{1}{3}$

15) $\frac{5}{8}+\frac{1}{2}\times\frac{1}{4}$ is equal to:

a $\frac{9}{32}$ b $\frac{3}{4}$ c $\frac{9}{16}$ d $\frac{3}{32}$

In questions 16 to 25 decide whether the answer given is true or false.

16) In a fraction the number above the line is called the numerator.

17) In a proper fraction the numerator is always greater than the denominator.

18) If the numerator of a fraction is greater than its denominator then the fraction is called an improper fraction.

19) An improper fraction always has a value greater than 1.

20) The fraction $6\frac{2}{5}$ is called a mixed number.

21) When the fractions $\frac{5}{8}$, $\frac{3}{4}$, $\frac{7}{10}$ and $\frac{3}{5}$ are put in order of size the result is $\frac{3}{5}$, $\frac{5}{8}$, $\frac{7}{10}$ and $\frac{3}{4}$.

22) $\frac{27}{9}$ is the same as $27 \div 9$.

23) $\frac{28}{35} \div \frac{7}{16}$ is the same as $\frac{28}{35} \times \frac{16}{7}$.

24) $3 \times (\frac{1}{2} + \frac{1}{4})$ is the same as $3 \times \frac{3}{4}$.

25) $\frac{5}{8} + \frac{1}{4} \times \frac{1}{2}$ is the same as $\frac{7}{8} \times \frac{1}{2}$.

THE DECIMAL SYSTEM

After reaching the end of this chapter you should be able to:

1. *Recognise a decimal fraction.*
2. *Convert a fraction to a decimal and vice versa.*
3. *Recognise a recurring decimal and distinguish between terminating and non-terminating decimals.*
4. *Reduce a decimal correct to a specified number of decimal places.*
5. *State a decimal number correct to a given number of significant figures.*
6. *Add and subtract decimal numbers.*
7. *Multiply and divide two decimal numbers giving answers to an appropriate number of decimal places.*
8. *Estimate the approximate value of arithmetic expressions.*
9. *Limit any answer to a given number of significant figures.*
10. *Reject any answer which is not a feasible solution.*

INTRODUCTION

In this chapter we first deal with the addition, subtraction, multiplication and division of decimal numbers. Then rough checks for calculations are discussed and finally the conversion of fractions to decimals and vice versa are discussed.

THE DECIMAL SYSTEM

The decimal system is an extension of our ordinary number system. When we write the number 666 we mean $600+60+6$. Reading from left to right each figure 6 is ten times the value of the next one.

We now have to decide how to deal with fractional quantities, that is, quantities whose values are less than one. If we regard 666.666 as meaning $600+60+6+\frac{6}{10}+\frac{6}{100}+\frac{6}{1000}$ then the dot, called the decimal point, separates the whole numbers from the fractional parts. Notice that with the fractional, or decimal parts, e.g. .666, each figure 6 is ten times the value of the following one, reading from left to right. Thus $\frac{6}{10}$ is ten times as great as $\frac{6}{100}$, and $\frac{6}{100}$ is ten times as great as $\frac{6}{1000}$ and so on.

Decimals then are fractions which have denominators of 10, 100, 1000 and so on, according to the position of the figure after the decimal point. If we have to write six hundred and five we write 605; the zero keeps the place for the missing tens. In the same way if we want to write $\frac{3}{10}+\frac{5}{1000}$ we write .305; the zero keeps the place for the missing hundredths. Also $\frac{6}{100}+\frac{7}{1000}$ would be written .067; the zero in this case keeps the place for the missing tenths.

When there are no whole numbers it is usual to insert a zero in front of the decimal point so that, for instance, .35 would be written 0.35

Exercise 3.1

Read off as decimals:

1) $\frac{7}{10}$ 2) $\frac{3}{10}+\frac{7}{100}$ 3) $\frac{5}{10}+\frac{8}{100}+\frac{9}{1000}$

4) $\frac{9}{1000}$ 5) $\frac{3}{100}$ 6) $\frac{1}{100}+\frac{7}{1000}$

7) $8+\frac{6}{100}$ 8) $24+\frac{2}{100}+\frac{9}{10\,000}$ 9) $50+\frac{8}{1000}$

ADDITION AND SUBTRACTION OF DECIMALS

Adding or subtracting decimals is done in exactly the same way as for whole numbers. Care must be taken, however, to write the decimal points directly underneath one another. This makes sure that all the figures having the same place value fall in the same column.

EXAMPLE 3.1

Simplify $11.36+2.639+0.047$

$$\begin{array}{r} 11.36 \\ 2.639 \\ 0.047 \\ \hline 14.046 \\ \hline \end{array}$$

EXAMPLE 3.2

Find the length of the shaft shown in Fig. 3.1.

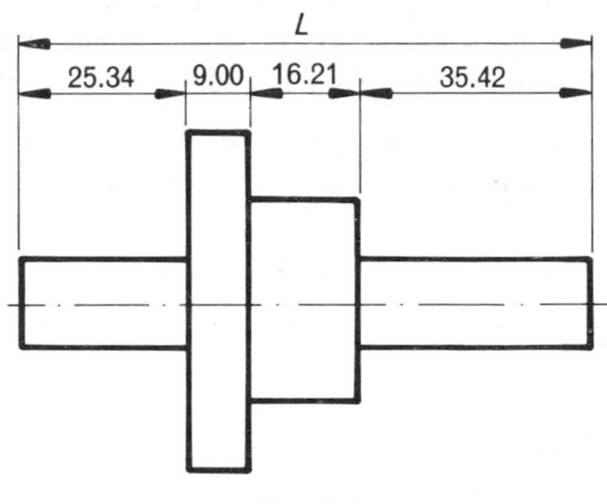

Fig. 3.1

$$L = 25.34 + 9.00 + 16.21 + 35.42 = 85.97\,\text{mm}$$

$$\begin{array}{r} 25.34 \\ 9.00 \\ 16.21 \\ 35.42 \\ \hline 85.97 \\ \hline \end{array}$$

EXAMPLE 3.3

Subtract 8.567 from 19.126

$$\begin{array}{r} 19.126 \\ 8.567 \\ \hline 10.559 \\ \hline \end{array}$$

EXAMPLE 3.4

Fig. 3.2 shows one way of checking a tapered hole using two balls of different sizes. Find the dimension marked A, which is the distance apart of the centres of the two balls.

In this problem we shall use the face XX as the datum face.

Distance from XX to centre of the 30.00 mm diameter ball $= 15.00 - 5.92 = 9.08\,\text{mm}$

$$\begin{array}{r} 15.00 \\ 5.92 \\ \hline 9.08 \\ \hline \end{array}$$

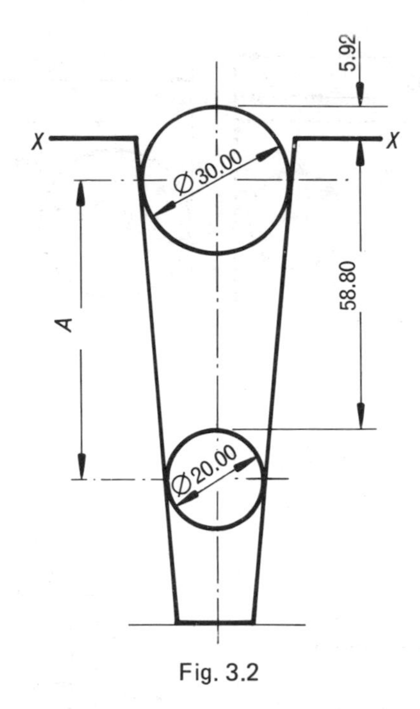

Fig. 3.2

Distance from XX to centre of the 20.00 mm

$$\text{diameter ball} = 58.80 + 10.00 = 68.80\text{ mm}$$

$$\therefore \quad A = 68.80 - 9.08 = 59.72\text{ mm}$$

58.80
10.00
68.80
9.08
59.72

Exercise 3.2

Write down the values of:

1) $2.375 + 0.625$
2) $4.25 + 7.25$
3) $3.196 + 2.475 + 18.369$
4) $38.267 + 0.049 + 20.3$
5) $27.418 + 0.967 + 25 + 1.467$
6) $12.48 - 8.36$
7) $19.215 - 3.599$
8) $2.237 - 1.898$
9) $0.876 - 0.064$
10) $5.48 - 0.0691$
11) Find dimension A shown in Fig. 3.3.
12) Find dimensions A and B shown in Fig. 3.4.

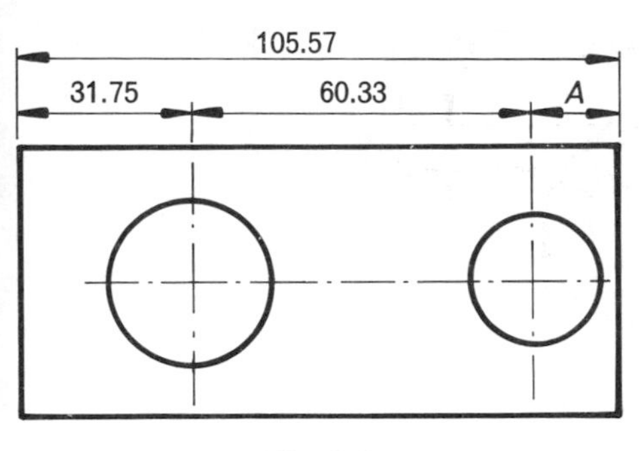

Fig. 3.3

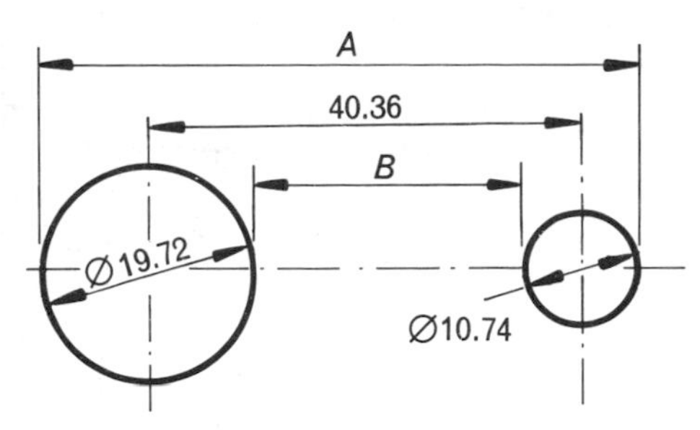

Fig. 3.4

13) Find dimension X in Fig. 3.5.

14) Find dimension A in Fig. 3.6.

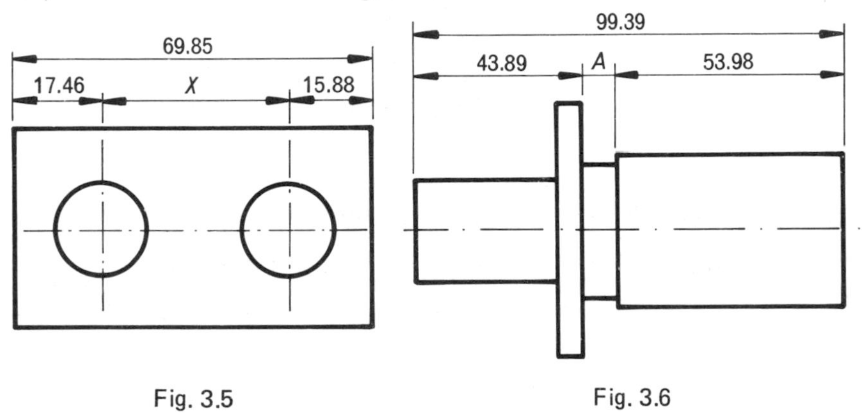

Fig. 3.5

Fig. 3.6

15) Find dimensions A, B and C in Fig. 3.7.

16) Find dimension A in Fig. 3.8.

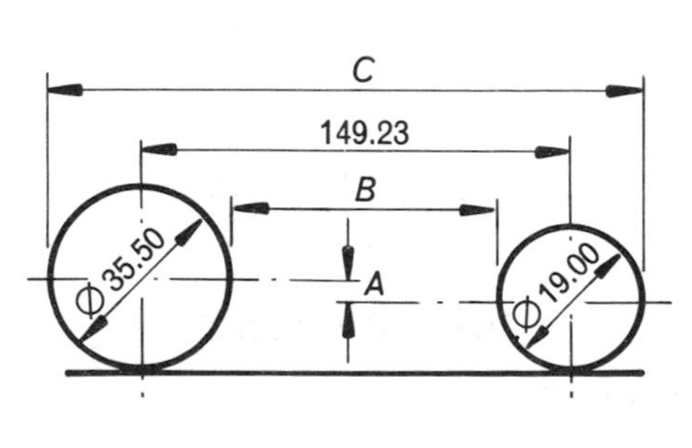

Fig. 3.7

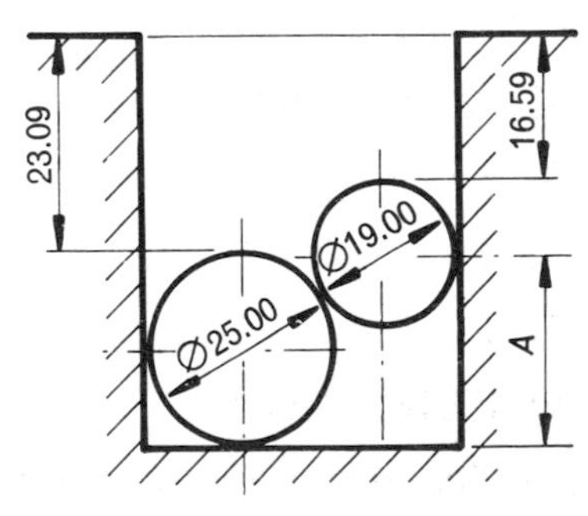

Fig. 3.8

MULTIPLYING DECIMALS BY 10, 100, ETC.

One of the advantages of decimals is the ease with which they may be multiplied or divided by 10, 100, 1000 etc.:

EXAMPLE 3.5

Find the value of 1.4×10

$$1.4 \times 10 = 1 \times 10 + 0.4 \times 10 = 10 + \frac{4}{10} \times 10 = 10 + 4 = 14$$

EXAMPLE 3.6

Find the value of 27.532×10

$$\begin{aligned} 27.532 \times 10 &= 27 \times 10 + 0.5 \times 10 + 0.03 \times 10 + 0.002 \times 10 \\ &= 270 + \frac{5}{10} \times 10 + \frac{3}{100} \times 10 + \frac{2}{1000} \times 10 \\ &= 270 + 5 + \frac{3}{10} + \frac{2}{100} = 275.32 \end{aligned}$$

In both of the above examples you will notice that the figures have not been changed by the multiplication; only the *positions* of the figures have been changed. Thus in Example 3.5, $1.4 \times 10 = 14$, that is the decimal point has been moved one place to the right. In Example 3.6, $27.532 \times 10 = 275.32$; again the decimal point has been moved one place to the right.

To multiply by 10, then, is the same as shifting the decimal point one place to the right. In the same way to multiply by 100 means shifting the decimal point two places to the right and so on.

EXAMPLE 3.7

$17.369 \times 100 = 1736.9$

The decimal point has been moved two places to the right.

EXAMPLE 3.8

$0.078\,95 \times 1000 = 78.95$

The decimal point has been moved three places to the right.

Exercise 3.3

Multiply each of the following numbers by 10, 100 and 1000.

1) 4.1 2) 2.42 3) 0.046

4) 0.35 5) 0.1486 6) 0.001 753

Write down the values of:

7) 0.4853×100 8) 0.009×1000 9) 170.06×10

10) $0.563\,95 \times 10\,000$

DIVIDING DECIMALS BY 10, 100, 1000, ETC.

When dividing by 10 the decimal point is moved one place to the left, by 100, two places to the left and so on. Thus:

$$154.26 \div 10 = 15.426$$

The decimal point has been moved one place to the left.

$$9.432 \div 100 = 0.094\,32$$

The decimal point has been moved two places to the left.

$$35 \div 1000 = 0.035$$

The decimal point has been moved three places to the left.

In the above examples note carcfully that use has been made of zeros following the decimal point to keep the placcs for the missing tenths.

Exercise 3.4

Divide each of the following numbers by 10, 100 and 1000.

1) 3.6 2) 64.198 3) 0.07

4) 510.4 5) 0.352

Give the value of:

6) $5.4 \div 100$ 7) $2.05 \div 1000$ 8) $0.04 \div 10$

9) $0.0086 \div 1000$ 10) $627.428 \div 10\,000$

LONG MULTIPLICATION

EXAMPLE 3.9

Find the value of 36.5×3.504

First disregard the decimal points and multiply 365 by 3504

$$
\begin{array}{r}
365 \\
3\,504 \\
\hline
1\,095\,000 \\
182\,500 \\
1\,460 \\
\hline
1\,278\,960 \\
\hline
\end{array}
$$

Now count up the total number of figures following the decimal points in both numbers (i.e. $1+3=4$). In the answer to the multiplication (the product), count this total number of figures from the right and insert the decimal point. The product is then 127.8960 or 127.896 since the zero is not a significant figure.

EXAMPLE 3.10

Rivets are placed 35.2 mm apart. If 28 rivets are placed in an assembly calculate the length between the first and last rivets.

$$
\begin{array}{r}
352 \\
27 \\
\hline
2464 \\
704 \\
\hline
9504 \\
\hline
\end{array}
$$

Length between first and last rivets $= 35.2 \times 27$

$= 950.4$ mm

(Note that there are 27 spaces when there are 28 rivets.)

Exercise 3.5

Find the values of the following:

1) 25.42×29.23 2) 0.3618×2.63 3) 0.76×0.38

4) 3.025×2.45 5) 0.043×0.032

6) In a production process it takes 0.74 minutes to make a soldered joint. Calculate the total time to make 379 such joints.

7) 1 turn of a screw drives it 0.35 mm. How far is it driven in 37 turns?

8) Holes are drilled 39.7 mm apart in a plate. If a row of 31 holes are drilled, what is the distance between the first and last holes?

LONG DIVISION

EXAMPLE 3.11

Find the value of $19.24 \div 2.6$

First convert the divisor (2.6) into a whole number by multiplying it by 10. To compensate multiply the dividend (19.24) by 10 also so that we now have $192.4 \div 26$. Now proceed as in ordinary division.

```
26)192.4(7.4
   182     — this line 26×7
   ———
    104    — 4 brought down from above. Since 4 lies to the right
    104      of the decimal point in the dividend insert a decimal
   ————      point in the answer (the quotient).
    . . .
   ————
```

Notice carefully how the decimal point in the quotient was obtained. The 4 brought down from the dividend lies to the right of the decimal point. Before bringing this down put a decimal point in the quotient immediately following the 7.

The division in this case is exact (i.e. there is no remainder) and the answer is 7.4. Now let us see what happens when there is a remainder.

EXAMPLE 3.12

Find the value of $15.187 \div 3.57$

As before, make the divisor into a whole number by multiplying it by 100 so that it becomes 357. To compensate multiply the dividend also by 100 so that it becomes 1518.7. Now divide.

```
357)1518.7(4.254 06
    1428        — this line 357×4
    ----
     90 7       — 7 brought down from the dividend. Since it
     71 4         lies to the right of the decimal point insert a
     ----         decimal point in quotient
     19 30      — bring down a zero as all the figures in the divi-
     17 85        dend have been used up.
     -----
      1 450
      1 428
      -----
        2200    — Bring down a zero. The divisor will not go
        2142      into 220 so place 0 in the quotient and bring
        ----      down another zero.
          58
```

The answer to 5 decimal places is 4.254 06. This is not the correct answer because there is a remainder. The division can be continued in the way shown to give as many decimal places as desired, or until there is no remainder.

It is important to realise what is meant by an answer given to so many decimal places. It is the number of figures which follow the decimal point which give the number of decimal places. If the first figure to be discarded is 5 or more the previous figure is increased by 1. Thus:

85.7684 = 85.8 correct to 1 decimal place
= 85.77 correct to 2 decimal places
= 85.768 correct to 3 decimal places

Notice carefully that zeros must be kept:

0.007 362 = 0.007 correct to 3 decimal places
= 0.01 correct to 2 decimal places
7.601 = 7.60 correct to 2 decimal places
= 7.6 correct to 1 decimal place

If an answer is required correct to 3 decimal places the division should be continued to 4 decimal places and the answer corrected to 3 decimal places.

EXAMPLE 3.13

The resistance of a copper wire 0.020 diameter and 27.6 metres long is 1515.24 ohms. What is the resistance per metre?

Resistance per metre = $1515.24 \div 27.6 = 54.9$ ohms

```
276)1512.4(54.9
    1380
    ----
     1352
     1104
     ----
      2484
      2484
      ----
      ....
      ----
```

Exercise 3.6

Find the value of:

1) $18.89 \div 14.2$ correct to 2 decimal places,

2) $0.0396 \div 2.51$ correct to 3 decimal places,

3) $7.21 \div 0.038$ correct to 2 decimal places,

4) $13.059 \div 3.18$ correct to 4 decimal places,

5) $0.1382 \div 0.0032$ correct to 1 decimal place.

6) A wire made from Manganin is 0.080 mm diameter and 15.3 metres long. Its resistance is 1263.78 ohms. What is its resistance per metre?

7) A consignment of screws has a total mass of 1757 grams. If one screw has a mass of 2.51 grams, calculate the number of screws in the consignment.

8) A certain size of steel angle bar have a mass of 800.1 kg for a length of 42 metres. What is the mass of the steel angle bar per metre length?

9) A sack of chemical has a mass of 510 kg. How many packets having a mass of 15 grams can be made from the sack?

10) An oil tank contains 1750 litres. How many tins containing 0.6 litres can be filled from this tank?

SIGNIFICANT FIGURES

Instead of using the number of decimal places to express the accuracy of an answer, significant figures can be used. The number 39.38 is correct to 2 decimal places but it is also correct to 4 significant figures since the number contains four figures. The rules regarding significant figures are as follows:

(1) If the first figure to be discarded is 5 or more the previous figure is increased by 1

$$
\begin{aligned}
8.1925 &= 8.193 \text{ correct to 4 significant figures} \\
&= 8.19 \text{ correct to 3 significant figures} \\
&= 8.2 \text{ correct to 2 significant figures}
\end{aligned}
$$

(2) Zeros must be kept to show the position of the decimal point, or to indicate that the zero is a significant figure

$$
\begin{aligned}
24\,392 &= 24\,390 \text{ correct to 4 significant figures} \\
&= 24\,400 \text{ correct to 3 significant figures} \\
0.0858 &= 0.086 \text{ correct to 2 significant figures} \\
425.804 &= 425.80 \text{ correct to 5 significant figures} \\
&= 426 \text{ correct to 3 significant figures}
\end{aligned}
$$

Exercise 3.7

Write down the following numbers correct to the number of significant figures stated:

1) 24.865 82 (i) to 6 (ii) to 4 (iii) to 2
2) 0.008 357 1 (i) to 4 (ii) to 3 (iii) to 2
3) 4.978 48 (i) to 5 (ii) to 3 (iii) to 1
4) 21.987 to 2
5) 35.603 to 4
6) 28 387 617 (i) to 5 (ii) to 2
7) 4.149 76 (i) to 5 (ii) to 4 (iii) to 3
8) 9.2048 to 3

ROUNDING AND DEGREES OF ACCURACY

When a number is written to a certain number of decimal places or significant figures it is said to be *rounded.*

Suppose we measure the length of a part and state the measurement to be 23.42 mm to the nearest $\frac{1}{100}$ mm. We mean that the dimension is greater than 23.415 mm and less than 23.425 mm. If it were less than 23.415 we should state the measurement as 23.41 mm and if it were greater than 23.425 mm the measurement would be stated as 23.43 mm. This means that the greatest error in the measurement is 0.005 mm too large or too small. We may say, then, that the measurement is 23.42 ± 0.005 mm, meaning that

$$\text{greatest possible dimension} = 23.42 + 0.005 = 23.425\,\text{mm}$$

$$\text{least possible dimension} = 23.42 - 0.005 = 23.415\,\text{mm}$$

This does not mean that the error is bound to be as great as 0.005 mm — it may be considerably less, but it cannot possibly be any more. To illustrate what happens when using rounded numbers in calculations consider the following examples.

EXAMPLE 3.14

The lengths of three bars are measured to an accuracy of $\frac{1}{10}$ mm and the lengths are found to be 24.7 mm, 35.2 mm and 61.8 mm. Find the total length of the three bars.

Greatest possible dimensions		Least possible dimensions	
$24.7 + 0.05 =$	24.75	$24.7 - 0.05 =$	24.65
$35.2 + 0.05 =$	35.25	$35.2 - 0.05 =$	35.15
$61.8 + 0.05 =$	61.85	$61.8 - 0.05 =$	61.75
	121.85		121.55

The sum of the lengths therefore lies somewhere between 121.55 mm and 121.85 mm. If we simply add $24.7 + 35.2 + 61.8$ we get 121.7. We see that the final figure 7 is not used as we can only quote the answer as 122 (to three significant figures), although we started with dimensions correct to one decimal place.

EXAMPLE 3.15

The dimensions of a rectangular plate were measured and found to be 2.16 m and 3.28 m correct to 3 significant figures. What is the area of the plate?

$$\text{Apparent area of the plate} = 2.16 \times 3.28 = 7.0848$$

$$\text{Greatest possible area} = 2.165 \times 3.285 = 7.112\,025$$

$$\text{Least possible area} = 2.155 \times 3.275 = 7.057\,625$$

We see that it is not possible to state the area of the plate more accurately than 7.1 m^2 correct to 2 significant figures.

Generally the answer should not contain more significant figures than the *least* number of significant figures given amongst the numbers used.

EXAMPLE 3.16

Find the value of $7.231 \times 1.24 \times 1.3$. The numbers are correct to the number of significant figures shown.

$$7.231 \times 1.24 \times 1.3 = 12$$

(since only 2 significant figures can be used in the answer, because of the 1.3 given).

Exercise 3.8

1) The side of a square plate is measured as 25 mm correct to the nearest 1 mm. Find the greatest and least possible area of the plate.

2) The sides of a rectangle are measured as 214 mm and 371 mm correct to the nearest mm. Find the greatest and least possible area of the rectangle.

The numbers used in the following calculations are correct to the number of significant figures shown. Calculate the answers to a suitable number of significant figures.

3) 1.7×2.6

4) $3.79 \times 2.411 \times 0.007\,67$

5) $11.36 \div 0.27$

6) $\dfrac{11.3 \times 7.265}{8.2}$

ROUGH CHECKS FOR CALCULATIONS

The worst mistake that can be made in a calculation is that of misplacing the decimal point. To place it wrongly, even by one place, makes the answer ten times too large or ten times too small.

To prevent this occurring it is always worth while doing a rough check by using approximate numbers. When doing these rough checks always try to select numbers which are easy to multiply or which will cancel.

EXAMPLE 3.17

a) 0.23×0.56

For a rough check we will take 0.2×0.6

Product roughly $= 0.2 \times 0.6 = 0.12$

Correct product $= 0.1288$

(The rough check shows that the answer is 0.1288 not 1.288 or 0.012 88.)

b) $173 \div 27.8$

For a rough check we will take $180 \div 30$

Quotient roughly $= 6$

Correct quotient $= 6.23$

(Note the rough check and the correct answer are of the same order.)

c) $\dfrac{8.198 \times 19.56 \times 30.82 \times 0.198}{6.52 \times 3.58 \times 0.823}$

Answer roughly $= \dfrac{8 \times 20 \times 30 \times 0.2}{6 \times 4 \times 1} = 40$

Correct answer $= 50.9$

(Although there is a big difference between the rough answer and the correct answer, the rough check shows that the answer is 50.9 and not 509 or 5.09)

Exercise 3.9

Find rough checks for the following:

1) 223.6×0.0048

2) 32.7×0.259

3) $0.682 \times 0.097 \times 2.38$

4) $78.41 \div 23.78$

5) $0.059 \div 0.002\,68$

6) $33.2 \times 29.6 \times 0.031$

7) $\dfrac{0.728 \times 0.006\,25}{0.0281}$

8) $\dfrac{27.5 \times 30.52}{11.3 \times 2.73}$

FRACTION TO DECIMAL CONVERSION

We found, when doing fractions, that the line separating the numerator and the denominator of a fraction takes the place of a division sign. Thus:

$$\frac{17}{80} \text{ is the same as } 17 \div 80$$

Therefore to convert a fraction into a decimal we divide the denominator into the numerator.

EXAMPLE 3.18

Convert $\frac{27}{32}$ to decimals.

$\frac{27}{32} = 27 \div 32$

```
32)27.0(0.843 75
   25 6
   ----
    1 40
    1 28
    ----
      120
       96
     ----
       240
       224
      ----
        160
        160
        ---
        ...
        ---
```

Therefore $\frac{27}{32} = 0.843\ 75$

EXAMPLE 3.19

Convert $2\frac{9}{16}$ into decimals.

When we have a mixed number to convert into decimals we need only deal with the fractional part. Thus to convert $2\frac{9}{16}$ into decimals we only have to deal with $\frac{9}{16}$.

$\frac{9}{16} = 9 \div 16$

```
16)9.0(0.5625
   80
   ---
   100
    96
    --
     40
     32
     --
      80
      80
      --
      ..
      --
```

The division shows that $\frac{9}{16} = 0.5625$ and hence $2\frac{9}{16} = 2.5625$

Sometimes a fraction will not divide out exactly as shown in Example 3.20.

EXAMPLE 3.20

Convert $\frac{1}{3}$ to decimals.

$\frac{1}{3} = 1 \div 3$

```
3)1.0(0.333
    9
    --
   10
    9
    --
    10
     9
    --
     1
```

It is clear that all we shall get from the division is a succession of threes.

This is an example of a recurring decimal and in order to prevent endless repetition the result is written $0.\dot{3}$, therefore $\frac{1}{3} = 0.\dot{3}$

Some further examples of recurring decimals are:

$\frac{2}{3} = 0.\dot{6}$ (meaning 0.6666 ... etc.)

$\frac{1}{6} = 0.1\dot{6}$ (meaning 0.1666 . . . etc.)

$\frac{5}{11} = 0.\dot{4}\dot{5}$ (meaning 0.454545 . . . etc.)

$\frac{3}{7} = 0.\dot{4}2857\dot{1}$ (meaning 0.428571428571 . . . etc.)

For all practical purposes we never need recurring decimals; what we need is an answer given to so many significant figures or decimal places. Thus

$\frac{2}{3} = 0.67$ (correct to 2 decimal places)

$\frac{5}{11} = 0.455$ (correct to 3 significant figures).

Exercise 3.10

Convert the following to decimals correcting the answers, where necessary to 4 decimal places:

1) $\frac{1}{4}$ 2) $\frac{3}{4}$ 3) $\frac{3}{8}$ 4) $\frac{11}{16}$ 5) $\frac{1}{2}$

6) $\frac{2}{3}$ 7) $\frac{21}{32}$ 8) $\frac{29}{64}$ 9) $1\frac{5}{6}$ 10) $2\frac{7}{16}$

Write down the following recurring decimals correct to 3 decimal places.

11) $0.\dot{3}$ 12) $0.\dot{7}$ 13) $0.1\dot{3}$ 14) $0.1\dot{8}$ 15) $0.3\dot{5}$

16) $0.\dot{2}\dot{3}$ 17) $0.5\dot{2}$ 18) $0.\dot{3}\dot{8}$ 19) $0.\dot{3}2\dot{8}$ 20) $0.\dot{5}67\dot{1}$

CONVERSION OF DECIMALS TO FRACTIONS

We know that decimals are fractions with denominators 10, 100, 1000, etc. Using this fact we can always convert a decimal to a fraction.

EXAMPLE 3.21

Convert 0.32 to a fraction.

$$0.32 = \frac{32}{100} = \frac{8}{25}$$

When comparing decimals and fractions it is best to convert the fraction into a decimal.

EXAMPLE 3.22

Find the difference between $1\frac{3}{16}$ and 1.1632

$$1\frac{3}{16} = 1.1875$$

$$1\frac{3}{16} - 1.1632 = 1.1875 - 1.1632 = 0.0243$$

Exercise 3.11

Convert the following to fractions in their lowest terms:

1) 0.2	2) 0.45	3) 0.3125
4) 2.55	5) 0.0075	6) 2.125

7) What is the difference between 0.281 35 and $\frac{9}{32}$?

8) What is the difference between $\frac{19}{64}$ and 0.295?

SUMMARY

a) Decimals are fractions with denominators of 10, 100, 1000, etc. The decimal point separates the whole numbers from the fractional parts.

b) When adding or subtracting decimal numbers the decimal points are written under one another.

c) To multiply by 10 move the decimal point one place to the right, to multiply by 100 move the decimal point two places to the right, etc.

d) To divide by 10 move the decimal point one place to the left, to divide by 100 move the decimal point two places to the left, etc.

e) When multiplying first disregard the decimal points and multiply the two numbers as though they were whole numbers. To place the decimal point in the product, count up the total number of figures after the decimal point in both numbers and then in the product count this number of figures starting from the extreme right.

f) When dividing first make the divisor into a whole number and compensate the dividend.

g) Significant figures and decimal places are used to denote the accuracy of a number.

h) Before multiplying or dividing always perform a rough check which will ensure that the decimal point is placed correctly.

i) To convert a fraction into a decimal divide the numerator by the denominator.

j) When comparing fractions and decimals, convert the fraction into a decimal.

Self-Test 3

In questions 1 to 10 state the letter, or letters, corresponding to the correct answer or answers.

1) The number 0.028 57 correct to 3 places of decimals is:

a 0.028 b 0.029 c 0.286 d 0.0286

2) The sum of $5 + \frac{1}{100} + \frac{7}{1000}$ is:

a 5.17 b 5.017 c 5.0107 d 5.107

3) 13.0063×1000 is equal to:

a 13.063 b 1300.63 c 130.063 d 13 006.3

4) $1.5003 \div 100$ is equal to:

a 0.015 003 b 0.150 03 c 0.153 d 1.53

5) $18.2 \times 0.013 \times 5.21$ is equal to:

a 12.326 86 b 123.2686 c 1.232 686 d 0.123 268 6

6) The number 158 861 correct to 2 significant figures is:

a 15 b 150 000 c 16 d 160 000

7) The number 0.081 778 correct to 3 significant figures is:

a 0.082 b 0.081 c 0.0818 d 0.0817

8) The number 0.075 538 correct to 2 decimal places is:

a 0.076 b 0.075 c 0.07 d 0.08

9) The number $0.1\dot{6}$ correct to 4 significant figures is:

a 0.1616 b 0.1617 c 0.1667 d 0.1666

10) $0.017 \div 0.027$ is equal to (correct to 2 significant figures):

a 0.63 b 6.3 c 0.063 d 63

In questions 11 to 20 the answer is either true or false. State which.

11) $\frac{5}{100} \times \frac{5}{10\,000} = 0.0505$

12) $5 + \frac{1}{10} + \frac{3}{1000} = 5.13$

13) $8.26 - 1.38 - 2.44 = 4.44$

14) $11.011 \times 100 = 1111$

15) $0.101\,01 \div 100 = 0.010\,101$

16) $0.0302 = \frac{3}{100} + \frac{2}{1000}$

17) 20 963 = 21 000, correct to 2 significant figures

18) 0.099 83 = 0.10, correct to 2 significant figures

19) 0.007 891 = 0.008, correct to 3 decimal places

20) $0.5 \div 0.2 = 2.5$

RATIO, PROPORTION AND PERCENTAGES

After reaching the end of this chapter you should be able to:

1. *Relate ratio and proportion to fractions.*
2. *Convert fractions and decimals to percentages.*

RATIO

A ratio is a comparison between two similar quantities. If the length of a ship is 200 m and a model of it is 1 m long then the length of the model is $\frac{1}{200}$th of the length of the ship. In making the model all the dimensions of the ship are reduced in the ratio of 1 to 200.

The ratio 1 to 200 is usually written 1 : 200.

A ratio can also be written as a fraction, as indicated above, and a ratio of 1 : 200 means the same as the fraction $\frac{1}{200}$.

Before we can state a ratio the units must be the same. We can state a ratio between 3 mm and 2 m provided we bring both lengths to the same units. Thus if we convert 2 m to 2000 mm the ratio between the two lengths is 3 : 2000.

EXAMPLE 4.1

Express the following ratios as fractions reduced to their lowest terms: a) 40 mm to 2.2 m, b) 800 g to 1.66 kg.

a) $$2.2\text{ m} = 2200\text{ mm}$$

Thus $$40:2200 = \frac{40}{2200} = \frac{1}{55}$$

b) $$1.6\text{ kg} = 1600\text{ g}$$

Thus $$800:1600 = \frac{800}{1600} = \frac{1}{2}$$

PROPORTION

Direct Proportion

If 5 litres of oil have a mass of 4 kg, then 10 litres of the same oil will have a mass of 8 kg. That is, if we double the quantity of oil its mass is also doubled. Now $2\frac{1}{2}$ litres of oil will have a mass of 2 kg. That is, if we halve the quantity of oil we halve its mass. This is an example of direct proportion. As the quantity of oil increases the mass increases in the *same proportion*. As the quantity of oil decreases the mass decreases in the *same proportion*.

EXAMPLE 4.2

The electrical resistance of a wire 150 mm long is 2 ohms. Find the resistance of a similar wire which is 1 m long.

The lengths of the two wires are increased in the ratio of 1000 : 150. The resistance will also increase in the ratio 1000 : 150.

Thus Resistance of wire 1 m long $= 2 \times \frac{1000}{150} = 13.3$ ohms.

Inverse Proportion

A motor car will travel 30 km in 1 hour if its speed is 30 km per hour. If its speed is increased to 60 km per hour the time taken to travel 30 km will be $\frac{1}{2}$ hour. That is when the speed is doubled the time taken is halved. This is an example of *inverse proportion*. When we multiplied the speed by 2 we divided the time taken by 2.

EXAMPLE 4.3

Two pulleys of 150 mm and 50 mm diameter respectively are connected by a belt. If the larger pulley revolves at 80 rev/min find the speed of the smaller pulley.

The smaller pulley must revolve faster than the larger pulley and hence the quantities, speed and diameter, are in inverse proportion. The pulley diameters are *decreased* in the ratio 50 : 150, or 1 : 3. The speed will be *increased* in the ratio 3 : 1.

Therefore:

$$\text{Speed of smaller pulley} = 80 \times \frac{3}{1} = 240 \text{ rev/min}$$

PROPORTIONAL PARTS

The diagram (Fig. 4.1) shows the line AB whose length represents 10 m divided into two parts in the ratio 2:3. As can be seen from the diagram the line has been divided into a total of 5 parts. The length AC contains 2 parts and the length BC contains 3 parts. Each part is 2 m long; hence AC is 4 m long and BC is 6 m long.

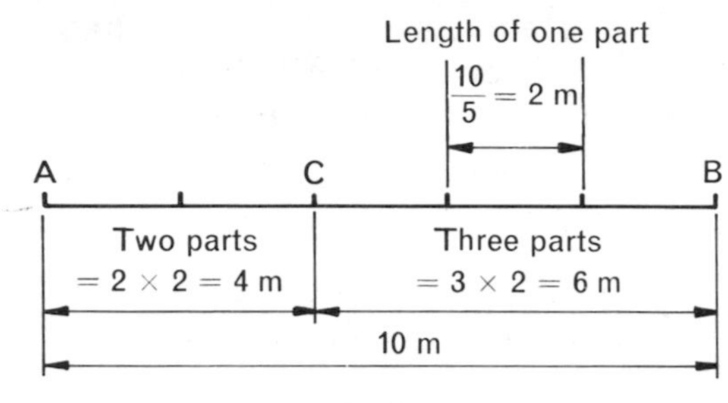

Fig. 4.1

We could tackle the problem as follows:

$$\text{Total number of parts} = 2+3 = 5$$

$$\text{Length of each part} = \frac{10}{5} = 2\,\text{m}$$

$$\text{Length of AC} = 2\times 2 = 4\,\text{m}$$

$$\text{Length of BC} = 3\times 2 = 6\,\text{m}$$

EXAMPLE 4.4

a) A certain brass is made by alloying copper and zinc in the ratio of 7:3. How much copper must be mixed with 30 g of zinc?

3 parts have a mass of 30 g

1 part has a mass of 10 g

7 parts have a mass of 70 g

Therefore, Mass of copper needed = 70 g

b) To make a certain solder, tin and lead are mixed in the ratio 6:2. How much of each metal will be required to make 20 kg of solder?

$$\text{Total number of parts} = 6+2 = 8$$

$$\text{Each part has a mass of } \frac{20}{8} = 2.5\,\text{kg}$$

Therefore, Mass of tin required $= 6 \times 2.5 = 15$ kg

and the Mass of lead required $= 2 \times 2.5 = 5$ kg

Exercise 4.1

1) Express the following ratios as fractions reduced to their lowest terms:

(a) 15 g to 2 kg (b) 30p to £5
(c) 20 cm^2 to 100 mm^2 (d) 400 m to 3 km

2) The length of a ship and the length of its model are in the ratio of 200:1. If the ship is 300 m long how long is its model?

3) A general arrangement drawing is made $\frac{1}{5}$ full size. If a dimension of 740 mm is to be represented on the drawing, what size will it be?

4) A copper wire 8 m long has a resistance of 0.22 ohm. If the resistance is directly proportional to the length find:

(a) the resistance of a wire 14 m long,
(b) the length of a wire which has a resistance of 0.17 ohms.

5) Two shafts are to rotate at 150 and 250 rev/min respectively. A 120 mm diameter pulley is fitted to the slower shaft and by means of a belt it drives a pulley on the faster shaft. What diameter pulley is required on the faster shaft?

6) A motor running at 400 rev/min has a pulley of 125 mm diameter attached to its shaft. It drives a parallel shaft which has a 1000 mm diameter pulley attached to it. Find the speed of this shaft.

7) A gear wheel having 40 teeth revolves at 120 rev/min. It meshes with a wheel having 25 teeth. Find the speed of the 25 tooth wheel.

8) Divide a line 140 mm long in the ratio 4:3.

9) A white metal suitable for high speed bearings is made from tin and lead in the ratio 8.6:1.4 by mass. Find the mass of each metal in a sample of the metal which has a mass of 15 kg.

10) A bar of metal 10.5 m long is to be cut into three parts in the ratio of $\frac{1}{2}:1\frac{3}{4}:3$. Find the length of each part.

11) A mass is composed of 3 parts copper to 2 parts zinc. Find the mass of copper and zinc in a casting which has a mass of 80 kg.

12) How much copper is required to be melted with 40 kg zinc to make a brass so that the ratio of copper to zinc is 7 : 3.

13) A right-angled triangle has sides in the ratio of 3 : 4 : 5. If the hypotenuse (the longest side) is 70 mm long, how long are the other sides?

PERCENTAGES

When comparing fractions it is often convenient to express them with a denominator of 100. Fractions expressed with a denominator of 100 are called percentages.

$$\frac{1}{2} = \frac{50}{100} = 50 \text{ per cent}$$

$$\frac{2}{5} = \frac{40}{100} = 40 \text{ per cent}$$

The sign % is often used instead of the words per cent.

EXAMPLE 4.5

a) $\frac{3}{4} = \frac{3}{4} \times 100 = 75\%$

b) $0.3 = 0.3 \times 100 = 30\%$

c) $0.245 = 0.245 \times 100 = 24.5\%$

To convert a fraction into a percentage we multiply the fraction by 100

EXAMPLE 4.6

a) $45\% = \frac{45}{100} = 0.45$

b) $3.9\% = \frac{3.9}{100} = 0.039$

To convert a percentage into a fraction we divide the percentage by 100

Exercise 4.2

Convert the following to percentages:

1) $\frac{3}{10}$ 2) $\frac{11}{20}$ 3) $\frac{9}{25}$ 4) $\frac{4}{5}$ 5) $\frac{31}{50}$

6) 0.63 7) 0.813 8) 0.667 9) 0.723 10) 0.027

Convert the following into decimal figures:

11) 32% 12) 78% 13) 6% 14) 24% 15) 31.5%

16) 48.2% 17) 2.5% 18) 1.25% 19) 3.95% 20) 20.1%

PERCENTAGE OF A QUANTITY

It is easy to find the percentage of a quantity if we express the percentage as a vulgar fraction.

EXAMPLE 4.7

a) What is 10% of 40?

Expressing 10% as a fraction it is $\frac{10}{100}$ and the problem then becomes what is $\frac{10}{100}$ of 40?

$$10\% \text{ of } 40 = \frac{10}{100} \text{ of } 40 = \frac{10}{100} \times 40 = 4$$

b) The current flowing through a circuit is 5 amperes. If it is increased by 6%, what is the final value of the current?

$$\text{Increase in the current} = \frac{6}{100} \times 5 = 0.3 \text{ amperes}$$

$$\text{Final value of the current} = 5 + 0.3 = 5.3 \text{ amperes}$$

c) A bronze is made by alloying 140 kg of copper, 20 kg of lead and 40 kg of tin. Find the percentage composition of the bronze.

$$\text{Total mass of bronze} = 140 + 20 + 40 = 200 \text{ kg}$$

$$\therefore \quad \text{Percentage of copper} = \frac{140}{200} \times 100 = 70\%$$

$$\text{Percentage of lead} = \frac{20}{200} \times 100 = 10\%$$

$$\text{Percentage of tin} = \frac{40}{200} \times 100 = 20\%$$

Exercise 4.3

1) What is:

(a) 12% of 80 (b) 20.3% of 105 (c) 3.7% of 68

2) (a) What percentage of 150 is 24?
(b) What percentage of 178 is 29?
(c) What percentage of 33 is 15?

3) If 20% of the length of a bar is 230 mm, what is the complete length?

4) The composition of an alloy is 44 parts copper, 14 parts tin and 2 parts antimony. What is the percentage of each metal in the alloy?

5) In a sample of iron ore 20% is iron. How much ore is needed to produce 15 000 kg of iron?

6) An alloy contains 7 kg of copper, 2 kg of zinc and 1 kg of lead. Calculate the percentage composition of the alloy.

7) The alloy called Elektron contains 10% of aluminium, 3.5% of zinc, 0.5% of manganese, the remainder being magnesium. Calculate the amount of each element in a sample containing 2000 kg.

8) An alloy is composed of 40 kg zinc and 10 kg of copper; another alloy contains 80 kg copper and 20 kg of tin. If three parts by mass of the first alloy are fused with one part by mass of the second alloy to make a third alloy, what percentage of copper and tin will be found in the third alloy?

9) The current flowing through a circuit is increased from 8 amperes to 8.56 amperes. What is the percentage increase in current?

10) Copper wire 0.063 mm diameter has a resistance of 5.53 ohms per metre whilst copper wire 0.050 mm diameter has a resistance of 8.79 ohms per metre. If a wire 0.050 mm diameter is used instead of one 0.063 mm diameter, what is the percentage increase in resistance?

SUMMARY

a) A ratio is a comparison between two similar quantities. A ratio may be expressed as a fraction. Thus the ratio $5:7 = \frac{5}{7}$.

b) Before a ratio can be stated the units must be the same.

c) If a quantity is divided in the ratio of $3:4:5$ then it has been divided into $3+4+5=12$ parts altogether. The first part is $\frac{3}{12}$th of the total, the second part is $\frac{4}{12}$th of the total and the third part is $\frac{5}{12}$th of the total.

d) Two quantities are in direct proportion if they increase or decrease at the same rate. In solving problems on direct proportion either the unitary method or the fractional method may be used.

e) Two quantities vary inversely if when one is doubled the other is halved, and so on.

f) Percentages are fractions with a denominator of 100.

g) To convert a fraction into a percentage multiply it by 100.

h) To convert a percentage into a fraction divide it by 100.

i) To find the percentage of a quantity first convert the percentage into a fraction and then multiply the quantity by the fraction.

Self-Test 4

In questions 1 to 20 the answer is either true or false. State which.

1) The ratio $6:3$ is the same as the ratio $2:1$

2) The ratio $5:10$ is the same as the ratio $2:1$

3) The ratios $20:100:300$ are the same as the ratios $1:5:15$

4) The fraction $\frac{2}{3}$ means the same as the ratio $2:3$

5) The ratio $18:24$ is the same as $\frac{3}{4}$

6) The ratio $18:30$ is the same as $\frac{5}{3}$

7) The ratio $9:2$ may be written $4\frac{1}{2}:1$

8) The ratio 8 pence: £4 is the same as $\frac{2}{1}$

9) The ratio 20 g to 0.4 kg is the same as $50:1$

10) The ratio 200 m to 5 km is the same as $\frac{1}{25}$.

11) When 600 kg is divided in the ratio 3 : 2 the two amounts are 360 kg and 240 kg.

12) When 900 m is divided in the ratios 2 : 3 : 5. the three lengths are 200 m, 300 m and 400 m.

13) A fraction expressed with a denominator of 100 is called a percentage.

14) $\frac{13}{25}$ is the same as 42%.

15) 0.725 is the same as 72.5%.

16) 3.5% is the same as $\frac{7}{20}$.

17) 20.45% is the same as 2.045.

18) 20% of 80 is 16.

19) If 15% of a complete length is 45 mm the complete length is 300 mm.

20) An alloy consists of 80% copper and 20% tin. The amount of tin in 50 kg of the alloy is 20 kg.

In questions 21 to 30 state the letter corresponding to the correct answer.

21) When £1200 is divided in the ratio 7 : 5 the smallest amount is:

a £700 **b** £500 **c** £240 **d** £480

22) When a length 3.6 m is divided in the ratio 5 : 4 : 3 the smallest amount is:

a 0.3 m **b** 1.5 m **c** 1.20 m **d** 0.9 m

23) An alloy contains copper, lead and tin in the ratio of 15 : 3 : 2. The amount of lead in 400 kg of the alloy is:

a 300 kg **b** 60 kg **c** 40 kg **d** 200 kg

24) A line 920 mm long is divided into four parts in the ratio 15 : 13 : 10 : 8. The longest part is:

a 260 mm **b** 200 mm **c** 300 mm **d** 160 mm

25) 40 men working in a factory produce 6000 articles in 12 working days. The length of time required for 15 men to produce the 6000 articles is:

a 32 days **b** $4\frac{1}{2}$ days **c** 64 days **d** 9 days

26) 35% is not the same as:

a $\frac{35}{100}$ b $\frac{7}{20}$ c $\frac{35}{10}$ d 0.35

27) $\frac{11}{25}$ is the same as:

a 4.4%   b 44%   c 22%   d 440%

28) 30% of a certain length is 600 mm. The complete length is:

a 20 mm   b 200 mm   c 2000 mm   d 2 m

29) What percentage of 150 is 48?

a 48%   b 36%   c 32%   d 72%

30) The composition of an alloy is 36 parts of copper, 12 parts of tin and 2 parts of antimony. The percentage of tin in the alloy is:

a 12%   b 30%   c 24%   d 6%

5. DIRECTED NUMBERS

After reaching the end of this chapter you should be able to:

1. *Recognise positive and negative numbers.*
2. *Add, subtract, multiply and divide numb having different signs.*

INTRODUCTION

Directed numbers are numbers which have either a plus or a minus sign attached to them such as $+7$ and -5. In this chapter we shall study the rules for the addition, subtraction, multiplication and division of directed numbers.

POSITIVE AND NEGATIVE NUMBERS

Fig. 5.1 shows part of a Celsius thermometer. The freezing point of water is 0°C (zero degrees Celsius). Temperatures above freezing point may be read off the scale directly and so may those below freezing.

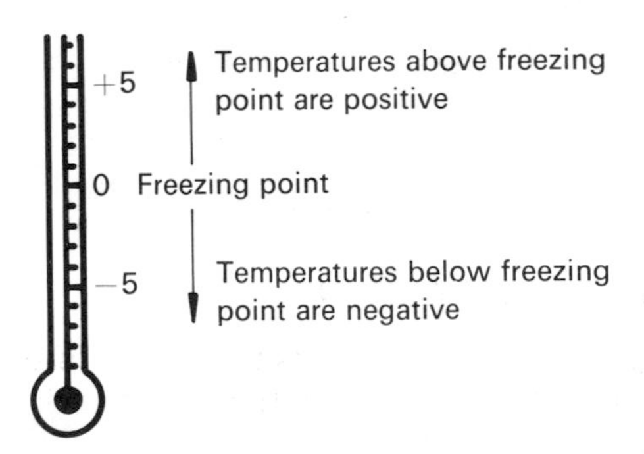

Fig. 5.1

We now have to decide on a method for showing whether a temperature is above or below zero. We may say that a temperature is 6 degrees above zero or 5 degrees below zero but these statements are not compact enough for calculations. Therefore we say that a temperature of $+6^\circ$ is a temperature which is 6° above zero and a temperature of 5 degrees below zero would be written -5°. We have thus used the signs + and − to indicate a change of direction.

Again if starting from a given point, distances measured to the right are regarded as being positive then distances measured to the left are regarded as being negative. As stated in the introduction, numbers which have a sign attached to them are called directed numbers. Thus $+7$ is a positive number and -7 is a negative number.

THE ADDITION OF DIRECTED NUMBERS

In Fig. 5.2, a movement from left to right (i.e. in the direction $0A$) is regarded as positive, whilst a movement from right to left (i.e. in the direction $0B$) is regarded as negative.

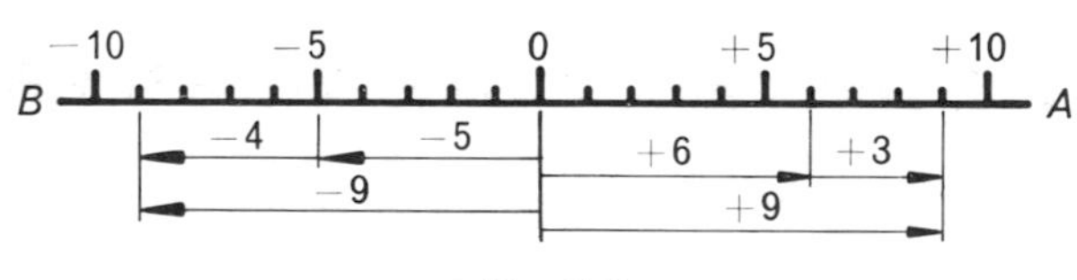

Fig. 5.2

To find the value of $+6+3$

Measure 6 units to the right of 0 (Fig. 5.2) and then measure a further 3 units to the right. The final position is 9 units to the right of 0. Hence,

$$+6+3 = +9$$

To find the value of $-5+(-4)$

Again in Fig. 5.2, measure 5 units to the left of 0 and then measure a further 4 units to the left. The final position is 9 units to the left of 0. Hence,

$$-5+(-4) = -9$$

From these results we obtain the rule:

To add several numbers together whose signs are the same add the numbers together. The sign of the sum is the same as the sign of each of the numbers.

Positive signs are frequently omitted as shown in the following examples.

a) $+5+9=+14$

More often this is written $5+9=14$

b) $-7+(-9)=-16$

More often this is written $-7-9=-16$

c) $-7-6-4=-17$

Exercise 5.1

Find the values of the following:

1) $+8+7$ 2) $-7-5$ 3) $-15-17$

4) $8+6$ 5) $-9-6-5-4$ 6) $3+6+8+9$

7) $-2-5-8-3$ 8) $9+6+5+3$

THE ADDITION OF NUMBERS HAVING DIFFERENT SIGNS

To find the value of $-4+11$

Measure 4 units to the left of 0 (Fig. 5.3) and from this point measure 11 units to the right. The final position is 7 units to the right of 0. Hence,

$$-4+11 = 7$$

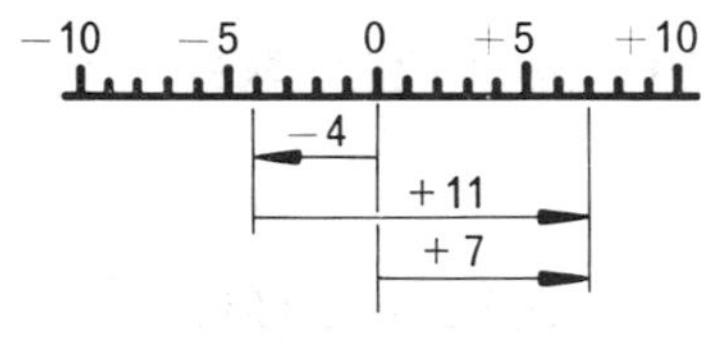

Fig. 5.3

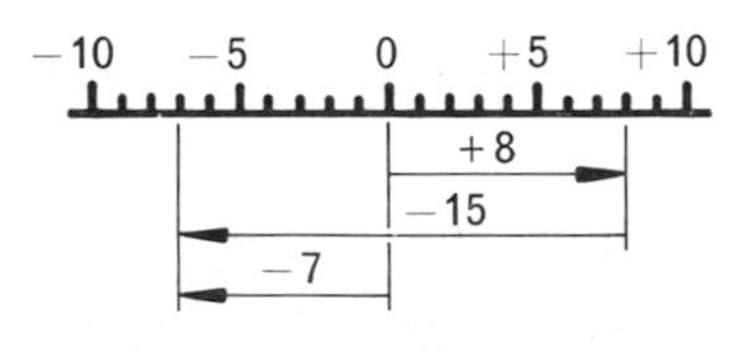

Fig. 5.4

To find the value of $8-15$

Measure 8 units to the right of 0 (Fig. 5.4) and from this point measure 15 units to the left. The final position is 7 units to the left of 0. Hence,

$$8-15 = -7$$

From these results we obtain the rule:

To add two numbers together whose signs are different, subtract the numerically smaller from the numerically larger. The sign of the result will be the same as the sign of the numerically larger number.

EXAMPLE 5.1

a) $-12+6=-6$

b) $11-16=-5$

When dealing with several numbers having mixed signs add the positive and negative numbers together separately. The set of numbers is then reduced to two numbers, one positive and the other negative, which are added in the way shown above.

EXAMPLE 5.2

$-16+11-7+3+8=-23+22=-1$

Exercise 5.2

Find values for the following:

1) $6-11$ 2) $7-16$ 3) $-5+10$

4) $12-7$ 5) $-8+9-2$ 6) $15-7-8$

7) $23-21-8+2$ 8) $-7+11-9-3+15$

SUBTRACTION OF DIRECTED NUMBERS

To find the value of $-4-(+7)$

To represent $+7$ we measure 7 units to the right of 0 (Fig. 5.5). Therefore to represent $-(+7)$ we must reverse direction and measure 7 units to the left of 0 and hence $-(+7)$ is the same as -7. Hence,

$$-4-(+7) = -4-7 = -11$$

To find the value of $+3-(-10)$

To represent -10 we measure 10 units to the left of 0 (Fig. 5.5).

Therefore to represent $-(-10)$ we measure 10 units to the right of 0 and hence $-(-10)$ is the same as $+10$. Hence,

$$+3-(-10) = 3+10 = 13$$

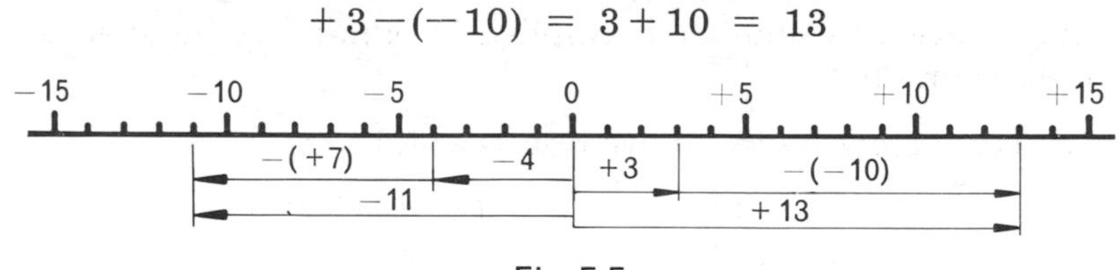

Fig. 5.5

The rule is:

To subtract a directed number change its sign and add the resulting number.

EXAMPLE 5.3

a) $-10-(-6) = -10+6 = -4$

b) $7-(+8) = 7-8 = -1$

c) $8-(-3) = 8+3 = 11$

Exercise 5.3

Find values for the following:

1) $8-(+6)$ 2) $-5-(-8)$ 3) $8-(-6)$

4) $-3-(-7)$ 5) $-4-(-5)$ 6) $-2-(+3)$

7) $-10-(-5)$ 8) $7-(-9)$

MULTIPLICATION OF DIRECTED NUMBERS

Now $5+5+5 = 15$

That is $3\times 5 = 15$

Thus two positive numbers multiplied together give a positive product.

Now $(-5)+(-5)+(-5) = -15$

That is $3\times(-5) = -15$

Thus a positive number multiplied by a negative number gives a negative product. Suppose, now, that we wish to find the value of

$(-3)\times(-5)$. We can write (-3) as $-(+3)$ and hence,

$$(-3)\times(-5) = -(+3)\times(-5) = -(-15) = +15$$

Thus a negative number multiplied by a negative number gives a positive product.

We may summarise the above results as follows:

$$(+)\times(+) = (+) \qquad (-)\times(+) = (-)$$
$$(+)\times(-) = (-) \qquad (-)\times(-) = (+)$$

and the rule is:

The product of two numbers with like signs is positive whilst the product of two numbers with unlike signs is negative.

EXAMPLE 5.4

a) $7\times 4 = 28$

b) $7\times(-4) = -28$

c) $(-7)\times 4 = -28$

d) $(-7)\times(-4) = 28$

Exercise 5.4

Find the values of the following:

1) $7\times(-6)$

2) $(-6)\times 7$

3) 7×6

4) $(-7)\times(-6)$

5) $(-2)\times(-4)\times(-6)$

6) $(-2)^2$

7) $3\times(-4)\times(-2)\times 5$

8) $(-3)^2$

DIVISION OF DIRECTED NUMBERS

The rules for division must be very similar to those used for multiplication, since if

$$3\times(-5) = -15$$

then

$$\frac{-15}{3} = -5$$

Also

$$\frac{-15}{-5} = 3$$

The rule is:

When dividing, numbers with like signs give a positive quotient and numbers with unlike signs give a negative quotient.

The rule may be summarised as follows:

$$(+)\div(+) = (+) \qquad (+)\div(-) = (-)$$
$$(-)\div(+) = (-) \qquad (-)\div(-) = (+)$$

EXAMPLE 5.5

a) $\dfrac{20}{4} = 5$

b) $\dfrac{20}{-4} = -5$

c) $\dfrac{-20}{4} = -5$

d) $\dfrac{-20}{-4} = 5$

e) $\dfrac{(-9)\times(-4)\times 5}{3\times(-2)} = \dfrac{36\times 5}{-6} = \dfrac{180}{-6} = -30$

Exercise 5.5

1) $6\div(-2)$

2) $(-6)\div 2$

3) $(-6)\div(-2)$

4) $6\div 2$

5) $(-10)\div 5$

6) $1\div(-1)$

7) $(-4)\div(-2)$

8) $(-3)\div 3$

9) $8\div(-4)$

10) $\dfrac{(-6)\times 4}{(-2)}$

11) $\dfrac{(-8)}{(-4)\times(-2)}$

12) $\dfrac{(-3)\times(-4)\times(-2)}{3\times 4}$

13) $\dfrac{4\times(-6)\times(-8)}{(-3)\times(-2)\times(-4)}$

14) $\dfrac{5\times(-3)\times 6}{10\times 3}$

SUMMARY

a) Directed numbers are numbers with a sign attached to them. $+7$ is a positive number and -3 is a negative number.

b) To add several numbers together whose signs are the same, add the numbers together. The sign of the sum is the same as the sign of each of the numbers.

c) To add two numbers together whose signs are different, subtract the numerically smaller number from the numerically larger. The sign of the result is the same as the sign of the numerically larger number.

d) To subtract a directed number change its sign and add the resulting number.

e) The product of two numbers having like signs is positive. The product of two numbers having unlike signs is negative.

f) The quotient of two numbers having like signs is positive. The quotient of two numbers having unlike signs is negative.

Self-Test 5

The answers to the following are either true or false. Write down the appropriate word for each problem.

1) $-5-6=11$

2) $-8+3=-5$

3) $-7-(+5)=12$

4) $(-5)-(-8)=3$

5) $(-6)\times(-7)=42$

6) $(-8)\times 5=40$

7) $3\times(-4)=12$

8) $8\div(-2)=-4$

9) $(-9)\div(-3)=3$

10) $(-6)^2=36$

INDICES AND LOGARITHMS

After reaching the end of this chapter you should be able to:

1. *Define the terms base, index and power.*
2. *Apply the rules*

$$a^m a^n = a^{m+n}$$

$$\frac{a^m}{a^n} = a^{m-n}$$

$$(a^m)^n = a^{mn}$$

3. *Deduce that* $a^0 = 1$ *for all values of* a.
4. *Deduce that* $a^{-n} = \dfrac{1}{a^n}$.
5. *Deduce that* $a^{1/n} = \sqrt[n]{a}$.
6. *Evaluate expressions which combine positive, negative and fractional indices.*
7. *Define the inverse* $a^x = y$ *as* $x = \log_a y$.
8. *Recognise that only positive numbers have real logarithms.*
9. *Convert numbers to standard form.*
10. *Add, subtract, multiply and divide numbers in standard form.*
11. *Recognise and use SI units – preferred standard form.*

BASE, INDEX AND POWER

The quantity $2 \times 2 \times 2 \times 2$ may be written as 2^4

Now 2^4 is called the fourth power of the base 2

The figure 4, which gives the number of 2s to be multiplied together is called the index (plural: indices).

Similarly $$a \times a \times a = a^3$$

Here a^3 is the third power of the base a, and the index is 3

Thus in the expression x^n

x^n is called the nth power of x,

x is called the base, and

n is called the index.

Remember that, in algebra, letters such as a in the above expression merely represent numbers. Hence the laws of arithmetic apply strictly to algebraic terms as well as numbers.

RECIPROCAL

The expression $\frac{1}{2}$ is called the reciprocal of 2

Similarly the expression $\frac{1}{p}$ is called the reciprocal of p

Likewise the expression $\frac{1}{x^n}$ is called the reciprocal of x^n

LAWS OF INDICES

(i) Multiplication

Let us see what happens when we multiply powers of the same base together.

Now $5^2 \times 5^4 = (5 \times 5) \times (5 \times 5 \times 5 \times 5)$

$= 5 \times 5 \times 5 \times 5 \times 5 \times 5 = 5^6$

and $c^3 \times c^5 = (c \times c \times c) \times (c \times c \times c \times c \times c)$

$= c \times c \times c \times c \times c \times c \times c \times c = c^8$

In both the examples above we see that we could have obtained the result by adding the indices together.

Thus $5^2 \times 5^4 = 5^{2+4} = 5^6$

and $c^3 \times c^5 = c^{3+5} = c^8$

In general terms the law is

$$a^m \times a^n = a^{m+n}$$

We may apply this idea when multiplying more than two powers of the same base together.

Thus $7^2 \times 7^5 \times 7^9 = 7^{2+5} \times 7^9 = 7^7 \times 7^9 = 7^{7+9} = 7^{16}$

We see that the same result would have been obtained by adding the indices, hence

$$7^2 \times 7^5 \times 7^9 = 7^{2+5+9} = 7^{16}$$

The law is:

When multiplying powers of the same base together, add the indices.

EXAMPLE 6.1

Simplify $m^5 \times m^4 \times m^6 \times m^2$

$$m^5 \times m^4 \times m^6 \times m^2 = m^{5+4+6+2} = m^{17}$$

(ii) Division of Powers

Now let us see what happens when we divide powers of the same base

$$\frac{3^5}{3^2} = \frac{3 \times 3 \times 3 \times 3 \times 3}{3 \times 3} = 3 \times 3 \times 3 = 3^3$$

We see that the same result could have been obtained by subtracting the indices.

Thus $$\frac{3^5}{3^2} = 3^{5-2} = 3^3$$

In general terms the law is

$$\boxed{\frac{a^m}{a^n} = a^{m-n}}$$

or

When dividing powers of the same base subtract the index of the denominator from the index of the numerator.

EXAMPLE 6.2

a) $$\frac{4^8}{4^5} = 4^{8-5} = 4^3$$

b) $$\frac{z^3 \times z^4 \times z^8}{z^5 \times z^6} = \frac{z^{3+4+8}}{z^{5+6}} = \frac{z^{15}}{z^{11}} = z^{15-11} = z^4$$

(iii) Powers of Powers

How do we simplify $(5^3)^2$? One way is to proceed as follows:

$$(5^3)^2 = 5^3 \times 5^3 = 5^{3+3} = 5^6$$

We see that the same result would have been obtained if we multiplied the two indices together.

Thus $$(5^3)^2 = 5^{3 \times 2} = 5^6$$

In general terms the law is

$$(a^m)^n = a^{mn}$$

or

When raising the power of a base to a power, multiply the indices together.

EXAMPLE 6.3

a) $(8^4)^3 = 8^{4\times 3} = 8^{12}$

b) $(p^2 \times q^4)^3 = (p^2)^3 \times (q^4)^3 = p^{2\times 3} \times q^{4\times 3} = p^6 \times q^{12}$

c) $\left(\frac{a^7}{b^5}\right)^6 = \frac{(a^7)^6}{(b^5)^6} = \frac{a^{7\times 6}}{b^{5\times 6}} = \frac{a^{42}}{b^{30}}$

(iv) Zero Index

Now $$\frac{2^5}{2^5} = \frac{2\times 2\times 2\times 2\times 2}{2\times 2\times 2\times 2\times 2} = 1$$

but using the laws of indices

$$\frac{2^5}{2^5} = 2^{5-5} = 2^0$$

Thus $$2^0 = 1$$

Also $$\frac{c^4}{c^4} = \frac{c\times c\times c\times c}{c\times c\times c\times c} = 1$$

But using the laws of indices

$$\frac{c^4}{c^4} = c^{4-4} = c^0$$

Thus $$c^0 = 1$$

In general terms the law is

$$x^0 = 1$$

or

Any base raised to the index of zero is equal to 1

EXAMPLE 6.4

a) $25^0 = 1$ b) $(0.56)^0 = 1$ c) $(\frac{1}{4})^0 = 1$

(v) Negative Indices

Now $$\frac{2^3}{2^7} = \frac{2\times2\times2}{2\times2\times2\times2\times2\times2\times2} = \frac{1}{2\times2\times2\times2} = \frac{1}{2^4}$$

but using the laws of indices

$$\frac{2^3}{2^7} = 2^{3-7} = 2^{-4}$$

It follows that

$$2^{-4} = \frac{1}{2^4}$$

Also $$\frac{d}{d^2} = \frac{d}{d\times d} = \frac{1}{d}$$

but using the laws of indices

$$\frac{d}{d^2} = \frac{d^1}{d^2} = d^{1-2} = d^{-1}$$

It follows that

$$d^{-1} = \frac{1}{d}$$

In general terms the law is

$$\boxed{x^{-n} = \frac{1}{x^n}}$$

or

> The power of a base which has a negative index is the reciprocal of the power of the base having the same, but positive, index.

EXAMPLE 6.5

a) $3^{-1} = \frac{1}{3^1} = \frac{1}{3}$ b) $5x^{-3} = 5\times x^{-3} = 5\times\frac{1}{x^3} = \frac{5}{x^3}$

c) $(2a)^{-4} = \frac{1}{(2a)^4} = \frac{1}{2^4\times a^4} = \frac{1}{16a^4}$

d) $\frac{1}{z^{-5}} = \frac{1}{\frac{1}{z^5}} = 1\div\frac{1}{z^5} = 1\times\frac{z^5}{1} = z^5$

Summary of the meaning of positive, zero, and negative indices.

$$x^4 = x \times x \times x \times x$$

$$x^3 = x \times x \times x$$

$$x^2 = x \times x$$

$$x^1 = x$$

$$x^0 = 1$$

$$x^{-1} = \frac{1}{x}$$

$$x^{-2} = \frac{1}{x \times x} = \frac{1}{x^2}$$

$$x^{-3} = \frac{1}{x \times x \times x} = \frac{1}{x^3}$$

Each line of the above sequence is obtained by dividing the previous line by x.

The above sequence may help you to appreciate the meaning of positive and negative indices, and especially the zero index. Remember: $(\text{elephants})^0 = 1$

Exercise 6.1

Simplify the following, giving each answer as a power:

1) $2^5 \times 2^6$
2) $a \times a^2 \times a^5$
3) $n^8 \div n^5$
4) $3^4 \times 3^7$
5) $b^2 \div b^5$
6) $10^5 \times 10^3 \div 10^4$
7) $z^4 \times z^2 \times z^{-3}$
8) $3^2 \times 3^{-3} \div 3^3$
9) $\dfrac{m^5 \times m^6}{m^4 \times m^3}$
10) $\dfrac{x^2 \times x}{x^6}$
11) $(9^3)^4$
12) $(y^2)^{-3}$
13) $(t \times t^3)^2$
14) $(c^{-7})^{-2}$
15) $\left(\dfrac{a^2}{a^5}\right)^3$
16) $\left(\dfrac{1}{7^3}\right)^4$
17) $\left(\dfrac{b^2}{b^7}\right)^{-2}$
18) $\dfrac{1}{(s^3)^3}$

Without using tables or calculating machines find the values of the following:

19) $\dfrac{8^3 \times 8^2}{8^4}$ 20) $\dfrac{7^2 \times 7^5}{7^3 \times 7^4}$ 21) $\dfrac{2^2}{2^2 \times 2}$

22) $2^4 \times 2^{-1}$ 23) 2^{-2} 24) $\dfrac{1}{(10)^{-2}}$

25) $\dfrac{2^{-1}}{2}$ 26) $\dfrac{24^0}{7}$ 27) $(5^{-1})^2$

28) $3^{-3} \div 3^{-4}$ 29) $\dfrac{7}{24^0}$ 30) $\left(\dfrac{1}{5}\right)^{-2}$

31) 7×24^0 32) $\left(\dfrac{2}{3}\right)^{-3}$ 33) $\left(\dfrac{2}{2^{-3}}\right)^{-2}$

Fractional Indices

The cube root of 5 (written as $\sqrt[3]{5}$) is the number which, when multiplied by itself three times, gives 5.

Thus $$\sqrt[3]{5} \times \sqrt[3]{5} \times \sqrt[3]{5} = 5$$

But we also know that

$$5^{1/3} \times 5^{1/3} \times 5^{1/3} = 5^{1/3+1/3+1/3} = 5$$

Comparing these expressions

$$\sqrt[3]{5} = 5^{1/3}$$

Similarly the fourth root of base d (written as $\sqrt[4]{d}$) is the number which, when multiplied by itself four times, gives d.

Thus $$\sqrt[4]{d} \times \sqrt[4]{d} \times \sqrt[4]{d} \times \sqrt[4]{d} = d$$

But we also know that

$$d^{1/4} \times d^{1/4} \times d^{1/4} \times d^{1/4} = d^{1/4+1/4+1/4+1/4} = d$$

Comparing these expressions $\sqrt[4]{d} = d^{1/4}$

In general terms the law is

$$\boxed{\sqrt[n]{x} = x^{1/n}}$$

Thus a fractional index represents a root—the denominator of the index denotes the root to be taken.

EXAMPLE 6.6

a) $7^{1/2} = \sqrt{7}$ (note that for square roots the figure 2 indicating the root is usually omitted)

b) Find the value of $81^{1/4}$

$$81^{1/4} = \sqrt[4]{81} = 3$$

c) Find the value of $8^{2/3}$

$$8^{2/3} = 8^{(1/3)\times 2} = (8^{1/3})^2 = (\sqrt[3]{8})^2 = (2)^2 = 4$$

d) Find the value of $16^{-3/4}$

$$16^{-3/4} = \frac{1}{16^{3/4}} = \frac{1}{16^{(1/4)\times 3}} = \frac{1}{(16^{1/4})^3} = \frac{1}{(\sqrt[4]{16})^3} = \frac{1}{(2)^3}$$

$$= \frac{1}{8} = 0.125$$

e) Find the value of $(\tfrac{3}{2})^{-3}$

$$\left(\frac{3}{2}\right)^{-3} = \frac{(3)^{-3}}{(2)^{-3}} = (3)^{-3} \times \frac{1}{(2)^{-3}} = \frac{1}{(3)^3} \times (2)^3 = \frac{1}{27} \times 8$$

$$= \frac{8}{27} = 0.296$$

f) Find the value of $9^{2.5}$

$$9^{2.5} = 9^{5/2} = 9^{(1/2)\times 5} = (9^{1/2})^5 = (\sqrt{9})^5 = (3)^5 = 243$$

g) Find the value of $\dfrac{1}{(\sqrt{5})^{-2}}$

$$\frac{1}{(\sqrt{5})^{-2}} = (\sqrt{5})^2 = (5^{1/2})^2 = 5^{(1/2)\times 2} = 5^1 = 5$$

EXAMPLE 6.7

Write the following as powers:

a) $\sqrt{x^3} = (x^3)^{1/2} = x^{3\times(1/2)} = x^{3/2} = x^{1.5}$

b) $\dfrac{1}{\sqrt[4]{a^5}} = \dfrac{1}{(a^5)^{1/4}} = \dfrac{1}{a^{5\times(1/4)}} = \dfrac{1}{a^{5/4}} = \dfrac{1}{a^{1.25}} = a^{-1.25}$

c) $\dfrac{\sqrt{b}}{\sqrt[3]{b^{-2}}} = \dfrac{b^{1/2}}{(b^{-2})^{1/3}} = \dfrac{b^{1/2}}{b^{-2\times(1/3)}} = \dfrac{b^{1/2}}{b^{-2/3}} = b^{1/2} \times b^{2/3} = b^{(1/2)+(2/3)}$

$= b^{7/6} = b^{1.167}$

EXAMPLE 6.8

Simplify the following:

a) $\dfrac{m^4}{m^{-3}} = m^{4-(-3)} = m^{4+3} = m^7$

b) $\dfrac{r^{-3}}{r^{2/3}} = r^{-3-(2/3)} = r^{-3.667}$

c) $\dfrac{t^{3/4}(\sqrt{t})}{t^{1/4}} = \dfrac{t^{3/4}(t^{1/2})}{t^{1/4}} = \dfrac{t^{3/4} \times t^{1/2}}{t^{1/4}} = \dfrac{t^{(3/4)+(1/2)}}{t^{1/4}} = \dfrac{t^{1.25}}{t^{0.25}} = t^{1.25-(0.25)}$

$= t^1 = t$

d) $\dfrac{z^{2.7} \times z^3}{z^{0.12}} = \dfrac{z^{2.7+3}}{z^{0.12}} = \dfrac{z^{5.7}}{z^{0.12}} = z^{5.7-0.12} = z^{5.58}$

Exercise 6.2

Write each of the following as a single power:

1) $\sqrt{x}$
2) $\sqrt[5]{x^4}$
3) $\dfrac{1}{\sqrt{x}}$
4) $\dfrac{1}{\sqrt[3]{x}}$
5) $\dfrac{1}{\sqrt[3]{x^4}}$
6) $\sqrt{x^{-3}}$
7) $\dfrac{1}{\sqrt[3]{x^{-2}}}$
8) $\dfrac{1}{\sqrt[4]{x^{-0.3}}}$
9) $(\sqrt[3]{-x})^2$
10) $\sqrt{x^{2/3}}$
11) $(\sqrt{x})^{2/3}$
12) $\left(\dfrac{1}{\sqrt[3]{x^4}}\right)^{-3/4}$

Without using tables or calculating machines find the values of the following:

13) $5^2 \times 5^{1/2} \times 5^{-3/2}$
14) $4 \div 4^{1/2}$
15) $8^{1/3}$
16) $64^{1/6}$
17) $8^{2/3}$
18) $25^{3/2}$
19) $(16^{1/4})^3$
20) $\dfrac{1}{9^{-3/2}}$
21) $\left(\dfrac{1}{4}\right)^{-1/2}$
22) $16^{0.5}$
23) $36^{-0.5}$
24) $(4^{-3})^{1/2}$
25) $\left(\dfrac{1}{4}\right)^{5/2}$
26) $\left(\dfrac{1}{16^{0.5}}\right)^{-3}$
27) $\dfrac{1}{(\sqrt{3})^{-2}}$

Simplify the following, expressing each answer as a power:

28) $\dfrac{\sqrt[3]{a}}{a^2 \times \sqrt{a}}$ 29) $\dfrac{a^{-3}}{a^{2/3}}$ 30) $\dfrac{x^3}{x^{-1.5}}$

31) $\dfrac{b^{5/2} \times b^{-3/2}}{b^{1/2}}$ 32) $\dfrac{m^{-3/4}}{m^{-5/2}}$ 33) $\dfrac{z^{2.3} \times z^{-1.5}}{z^{-3.5} \times z^2}$

34) $\dfrac{(x^{1/2})^3}{(x^3)^{1/2}}$ 35) $\dfrac{\sqrt{u}}{u^3}$ 36) $\dfrac{\sqrt[4]{y^3}}{\sqrt{y}}$

37) $\dfrac{(\sqrt[4]{n})^3}{\sqrt{n}}$ 38) $\dfrac{\sqrt[4]{x^2}}{\sqrt[7]{x^{-2}}}$ 39) $\dfrac{\sqrt[3]{t} \times \sqrt{t^3}}{t^{5/2}}$

LOGARITHMS

If N is a number such that $N = b^x$

we may write this in the alternative form $\log_b N = x$

which, in words, is

the logarithm of N, to the base b, is x

or

x is the logarithm of N to the base b

The word 'logarithm' is often abbreviated to just 'log'.

It is helpful to remember that

$$\text{Number} = \text{base}^{\text{logarithm}}$$

Alternatively, in words:

> The log of a number is the power to which the base must be raised to give that number.

Thus:

We may write $8 = 2^3$ in log form as $\log_2 8 = 3$	We may write $81 = 3^4$ in log form as $\log_3 81 = 4$
We may write $2 = \sqrt{4}$ or $2 = 4^{1/2}$ or $2 = 4^{0.5}$ in log form as $\log_4 2 = 0.5$	We may write $\dfrac{1}{4} = \dfrac{1}{2^2}$ or $0.25 = 2^{-2}$ in log form as $\log_2 0.25 = -2$

EXAMPLE 6.9

If $\log_7 49 = x$, find the value of x.

Writing the equation in index form we have $49 = 7^x$

or $7^2 = 7^x$

Since the bases are the same on both sides of the equation the indices must be the same. Thus $x = 2$

EXAMPLE 6.10

If $\log_x 8 = 3$, find the value of x.

Writing this equation in index form we have $8 = x^3$

or $2^3 = x^3$

Since the indices on both sides of the equation are the same the bases must be the same. Thus $x = 2$

EXAMPLE 6.11

Find the value of $\log_3(-9)$.

Let $\log_3(-9) = x$

Thus in index form $-9 = 3^x$

Now whatever value we give to the index x, whether positive or negative, it is *not* possible to obtain a power of 3 which is a negative number. Thus there is *no* value of x which will satisfy the equation $-9 = 3^x$ or the alternative form $\log_3(-9) = x$.

It follows that there is no value for $\log_3(-9)$.

LOGARITHM OF A NEGATIVE NUMBER

If we examine Example 6.11, we can see that whatever the value of the negative number, or whatever the value of the base, it is not possible to find a value for the index, or logarithm, x. Therefore the logarithm of a negative number does not have a real value. In other words:

> Only positive numbers have real logarithms.

Exercise 6.3

Express in logarithmic form:

1) $n = a^x$ 2) $2^3 = 8$ 3) $5^{-2} = 0.04$

4) $10^{-3} = 0.001$ 5) $x^0 = 1$ 6) $10^1 = 10$

7) $a^1 = a$ 8) $e^2 = 7.39$ 9) $10^0 = 1$

Find the value of x in each of the following:

10) $\log_x 9 = 2$ 11) $\log_x 81 = 4$ 12) $\log_2 16 = x$

13) $\log_5 125 = x$ 14) $\log_3 x = 2$ 15) $\log_4 x = 3$

16) $\log_{10} x = 2$ 17) $\log_7 x = 0$ 18) $\log_x 8 = 3$

19) $\log_x 27 = 3$ 20) $\log_9 3 = x$ 21) $\log_n n = x$

NUMBERS IN STANDARD FORM

Any number can be expressed as a value between 1 and 10 multiplied by a power of 10. A number expressed in this way is said to be in standard form. The repeating of zeros in very large and very small numbers often leads to errors. Stating the number in standard form helps to avoid these errors.

EXAMPLE 6.12

a) $49.4 = 4.94 \times 10$

b) $385.3 = 3.853 \times 100 = 3.853 \times 10^2$

c) $20\,000\,000 = 2 \times 10\,000\,000 = 2 \times 10^7$

d) $0.596 = \dfrac{5.96}{10} = 5.96 \times 10^{-1}$

e) $0.000\,478 = \dfrac{4.78}{10\,000} = \dfrac{4.78}{10^4} = 4.78 \times 10^{-4}$

We sometimes need to convert a number stated in standard form into a number in ordinary decimal form.

(1) When the power of 10 is positive the index, which gives the power of 10, shows the number of places that the decimal point has to be moved to the *right*. Thus,

$6.187 \times 10^2 = 618.7$ (decimal point moved 2 places to the right)

$8.463 \times 10^5 = 846\,300$ (decimal point moved 5 places to the right)

(2) When the power of 10 is negative the numerical value of the index gives the number of places that the decimal point has to be moved to the *left*. Thus,

$3.167 \times 10^{-2} = 0.031\,67$ (decimal point moved 2 places to the left)

$8.463 \times 10^{-4} = 0.000\,846\,3$ (decimal point moved 4 places to the left)

Exercise 6.4

1) Write the following in standard form:

(a) 19.6 (b) 385 (c) 59 876 (d) 1 500 000
(e) 0.013 (f) 0.003 85 (g) 0.000 698 (h) 0.023 85

2) Write down the following in ordinary decimal form:

(a) 1.5×10^2 (b) 4.7×10^4 (c) 3.6×10^6 (d) 9.45×10^3
(e) 2.5×10^{-1} (f) 4.0×10^{-3} (g) 8.0×10^{-5} (h) 4.0×10^{-2}

3) State which of the following pairs of numbers is the larger:

(a) 5.8×10^2 and 2.1×10^3 (b) 9.4×10^3 and 9.95×10^3
(c) 8.58×10^4 and 9.87×10^3

4) State which of the following pairs of numbers is the smaller:

(a) 2.1×10^{-2} and 3.8×10^{-2} (b) 8.72×10^{-3} and 9.7×10^{-2}
(c) 3.83×10^{-2} and 2.11×10^{-4}

Adding and Subtracting Numbers in Standard Form

(1) If the numbers to be added or subtracted have the same power of 10 then the numbers are added or subtracted directly.

EXAMPLE 6.13

a) $(1.859 \times 10^2) + (2.387 \times 10^2) + (9.163 \times 10^2)$

$= (1.859 + 2.387 + 9.163) \times 10^2$

$= 13.409 \times 10^2$ or 1.3409×10^3

b) $(8.768 \times 10^{-3}) - (4.381 \times 10^{-3})$

$= (8.768 - 4.381) \times 10^{-3}$

$= 4.387 \times 10^{-3}$

(2) If the numbers have different powers of 10 the easiest way is first to convert them to numbers in ordinary decimal form and then add or subtract them.

EXAMPLE 6.14

$$(3.475\times10^3)+(4.826\times10^2) = 3475+482.6$$
$$= 3957.6 \quad \text{or} \quad 3.9576\times10^3$$

Multiplying and Dividing Numbers Stated in Standard Form

By using the laws of indices, numbers stated in standard form can be easily multiplied or divided.

EXAMPLE 6.15

a) $$(8.463\times10^2)\times(1.768\times10^3) = (8.463\times1.768)\times(10^2\times10^3)$$
$$= 14.96\times10^5 \quad \text{or} \quad 1.496\times10^6$$

b) $$\frac{3.258\times10^2}{7.197\times10^4} = \frac{3.258}{7.197}\times10^{2-4} = 0.4527\times10^{-2}$$
$$\text{or} \quad 4.527\times10^{-3}$$

Exercise 6.5

State the answers to the following in standard form:

1) $(3.582\times10^3)+(8.907\times10^3)$
2) $(7.81\times10^{-2})+(1.88\times10^{-2})+(8.89\times10^{-2})$
3) $(1.809\times10^2)-(1.705\times10^2)$
4) $(8.89\times10^{-3})-(8.85\times10^{-3})$
5) $(1.78\times10^2)+(2.58\times10^3)$
6) $(5.987\times10^3)+(8.91\times10^2)+(7.635\times10^4)$
7) $(8.902\times10^{-2})-(7.652\times10^{-3})$
8) $(1.832\times10^{-1})-(9.998\times10^{-2})$
9) $(7.58\times10^2)\times(6\times10^3)$ (to 3 significant figures)
10) $(6\times10^{-1})\times(2.58\times10^{-2})$ (to 2 significant figures)
11) $(5\times10^3)\times(2.11\times10^4)\times(4\times10^2)$ (to 3 significant figures)

12) $\dfrac{2.68\times10^2}{8\times10^3}$ (to 2 significant figures)

13) $\dfrac{1.78\times10^{-1}}{3\times10^{-2}}$ (to 3 significant figures)

14) $\dfrac{(7\times10^2)\times(6.58\times10^3)}{8\times10^4}$ (to 2 significant figures)

SI UNITS: PREFERRED STANDARD FORM

The Systèmе Internationale d'Unités (the international system of units) usually abbreviated to SI, is based upon six fundamental units as follows:

Length—the metre (abbreviation: m)
Mass—the kilogram (kg)
Time—the second (s)
Electric current—the ampere (A)
Luminous intensity—the candela (cd)
Temperature—the kelvin (K)

For many applications some of the above units are too small or too large and hence multiples and sub-multiples of these units are often needed. These multiples and sub-multiples are given special names as follows:

Multiplication Factor		Prefix	Symbol
1 000 000 000 000	10^{12}	tera	T
1 000 000 000	10^{9}	giga	G
1 000 000	10^{6}	mega	M
1 000	10^{3}	kilo	k
100	10^{2}	hecto	h
10	10^{1}	deca	da
0.1	10^{-1}	deci	d
0.01	10^{-2}	centi	c
0.001	10^{-3}	milli	m
0.000 001	10^{-6}	micro	μ
0.000 000 001	10^{-9}	nano	n
0.000 000 000 001	10^{-12}	pico	p
0.000 000 000 000 001	10^{-15}	femto	f
0.000 000 000 000 000 001	10^{-18}	atto	a

Where possible multiples and sub-multiples should use powers of ten which are multiples of three. Thus 5000 metres should be written as 5 kilometres and not as 50 hectometres. These powers of ten which are multiples of three are called *preferred multiples.*

EXAMPLE 6.16

A measurement is 18 350 000 metres. Express this as preferred multiples of the metre.

Now $18\,350\,000\,\text{m} = 18\,350 \times 10^3\,\text{m}$

$= 18\,350$ kilometres or km

or $18\,350\,000\,\text{m} = 18.35 \times 10^6\,\text{m}$

$= 18.35$ megametres or Mm

Thus $18\,350\,000\,\text{m} = 18\,350\,\text{km}$ or $18.35\,\text{Mm}$

EXAMPLE 6.17

A measurement is taken as 0.000 000 082 metres. Express this as preferred sub-multiples of the metre.

Now $0.000\,000\,082\,\text{m} = 0.000\,082 \times 10^{-3}\,\text{m}$

$= 0.000\,082$ millimetres or mm

or $0.000\,000\,082\,\text{m} = 0.082 \times 10^{-6}\,\text{m}$

$= 0.082$ micrometres or μm

or $0.000\,000\,082\,\text{m} = 82 \times 10^{-9}\,\text{m}$

$= 82$ nanometres or nm

Thus $0.000\,000\,082\,\text{m} = 0.000\,082\,\text{mm}$ or $0.082\,\mu\text{m}$ or $82\,\text{nm}$

Exercise 6.6

Express each of the following as a standard multiple or sub-multiple.

1) 8000 m
2) 15 000 kg
3) 3800 km
4) 1 891 000 kg
5) 0.007 m
6) 0.000 001 3 m
7) 0.028 kg
8) 0.000 36 km
9) 0.000 064 kg
10) 0.0036 A

7. THE SCIENTIFIC ELECTRONIC CALCULATOR

After reaching the end of this chapter you should be able to:

1. *Understand the need for a rough check answer.*
2. *Obtain a rough check answer for any calculation.*
3. *Use the calculator to find the values of arithmetical expressions involving addition, subtraction, multiplication, division and use of memory.*
4. *Extend operations to include reciprocals, square roots, and numbers in standard form.*
5. *Evaluate arithmetical expressions involving whole number, negative, and fractional indices.*
6. *Use all the techniques in 1-5 to evaluate algebraic expressions and formulae which arise from relevant technology.*

Trigonometrical functions are included in the chapter on trigonometry.

KEYBOARD LAYOUT

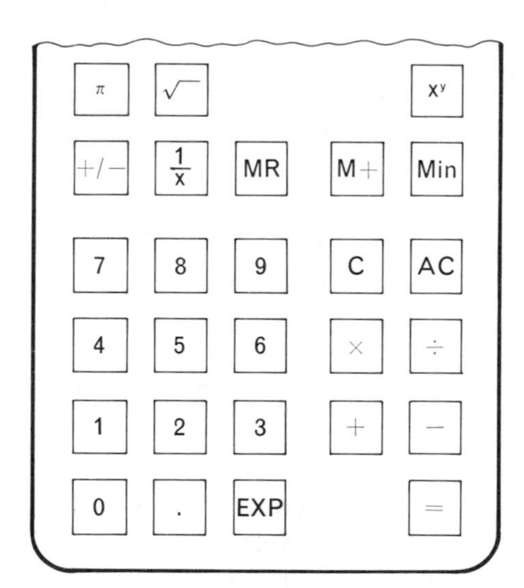

Fig. 7.1

The following are figure keys:

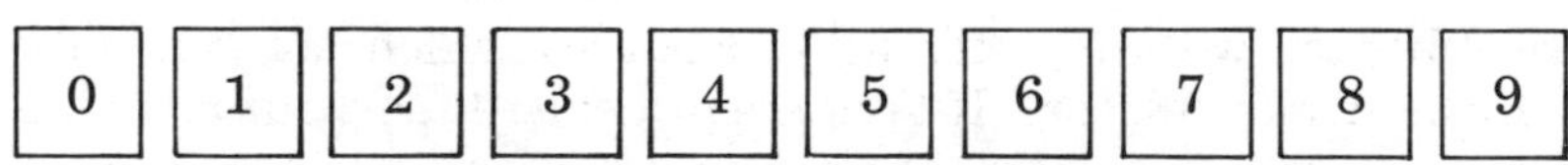

The other keys are summarised as follows:

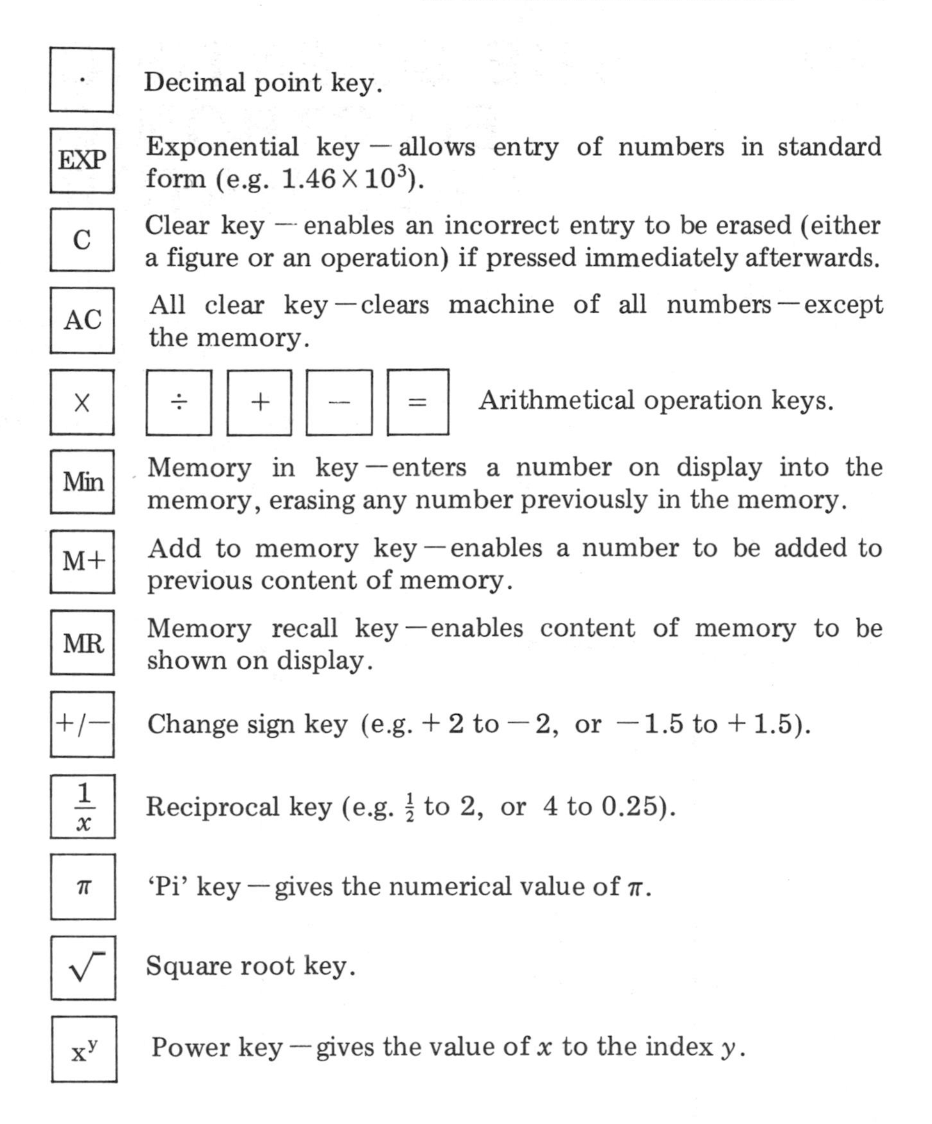

[.] Decimal point key.

[EXP] Exponential key — allows entry of numbers in standard form (e.g. 1.46×10^3).

[C] Clear key — enables an incorrect entry to be erased (either a figure or an operation) if pressed immediately afterwards.

[AC] All clear key — clears machine of all numbers — except the memory.

[×] [÷] [+] [−] [=] Arithmetical operation keys.

[Min] Memory in key — enters a number on display into the memory, erasing any number previously in the memory.

[M+] Add to memory key — enables a number to be added to previous content of memory.

[MR] Memory recall key — enables content of memory to be shown on display.

[+/−] Change sign key (e.g. $+2$ to -2, or -1.5 to $+1.5$).

[$\frac{1}{x}$] Reciprocal key (e.g. $\frac{1}{2}$ to 2, or 4 to 0.25).

[π] 'Pi' key — gives the numerical value of π.

[$\sqrt{\ }$] Square root key.

[x^y] Power key — gives the value of x to the index y.

Part of a typical keyboard layout of an electronic calculator is shown in Fig. 7.1. Calculators vary in layout and operation, just as motor cars do from different manufacturers, but the methods of using each type are similar. Each calculator is supplied with an instruction booklet and you should work through this carrying out any worked examples which are given.

In this chapter we are outlining procedures which are generally common to all calculators. If they are not exactly as your machine requires you will have to make allowance according to the instruction booklet.

ROUGH CHECKS

When using a calculator it is essential for you to do a rough check in order to obtain an approximate result. Any error in carrying out a sequence of operations, however small it may seem, will result in a wrong answer.

Suppose, for instance, that you had £1000 in the bank and then withdrew £97.82. The bank staff then used a calculator to find how much money you had left in your account—they calculated that £1000 less £978.2 left only £21.80 credited to you. You would be extremely annoyed and probably point out to them that a rouch check of £1000 less £100 would leave £900, and that if this had been done much embarrassment would have been avoided.

The 'small' mistake was to get the decimal point in the wrong place when recording the money withdrawn, which is typical of errors we all make from time to time. You should get in the habit of doing a rough check on any calculation *before* using your machine. The advantage of a rough check answer before the actual calculation avoids the possibility of forgetting it in the excitement of obtaining a machine result. Also your rough check will not be influenced by the result obtained on your calculator.

WORKED EXAMPLES

After first switching on the calculator, or commencing a fresh problem, you should press the [AC] key. This ensures that all figures entered previously have been erased and will not interfere with new data to be entered.

The memory is not cleared but this is done automatically when a new number is entered in the memory using the [Min] key.

EXAMPLE 7.1

Evaluate $18.24 + 4.39 - 9.72$

A rough check gives $18 + 5 - 10 = 13$

The sequence used on the calculator is similar to the order in which the problem is given. This is shown by:

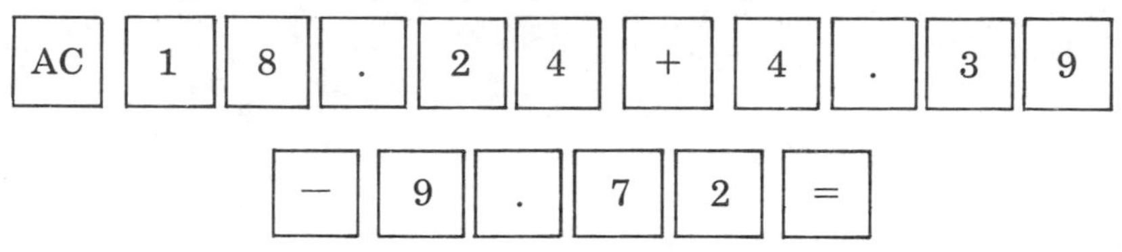

The display gives 12.91 and, since all the data are given to two decimal places, the answer, correct to two decimal places, is 12.91

It should be noted that the order of operations is not important. Try for yourself the sequence $18.24-9.72+4.39$ and you will obtain the same answer.

EXAMPLE 7.2

Evaluate $\dfrac{20.3 \times 3092}{1.563}$

A rough check gives: $\dfrac{20 \times 3000}{2} = 30\,000$

The sequence of operations is:

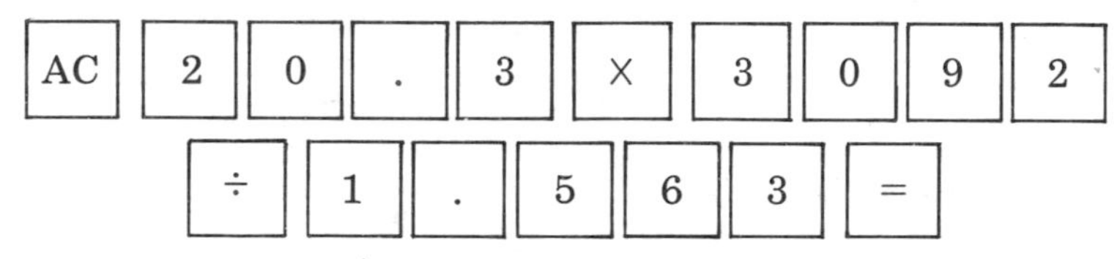

The display gives 40 158.413 which seems rather way out when compared with the approximate answer 30 000 obtained from the rough check. This is not unexpected, however, since we did allow for 2 in the denominator instead of 1.563, which would make the rough check answer smaller. It does confirm that the true answer is of the correct 'order' (i.e. it is *not* 4015.8413 or 401 584.13).

It would not be correct to state the answer as 40 158.413 given on the display.

In general, it is significant figures of the data (not the decimal places) which are important. An answer is not correct if given to an accuracy greater than that of the least accurate of the given numbers.

We see that 20.3 has three significant figures, whilst 3092 and 1.563 both have four significant figures. (Remember that significant figures are counted from left to right, starting with the first non-zero figure.)

Hence the least accurate given number has three significant figures and this gives us the accuracy to which we should state the answer.

Thus the answer is 40 200 correct to three significant figures.

EXAMPLE 7.3

Find the value of $6\,857\,000 \times 119\,000 \times 85.3$

For the rough check numbers which contain as many figures as these are better considered in standard form:

$$(6.857 \times 10^6) \times (1.19 \times 10^5) \times (8.53 \times 10)$$

or approximately:

$$\begin{aligned}(7 \times 10^6) \times (1 \times 10^5) \times (10 \times 10) &= 7 \times 1 \times 10 \times 10^{12} \\ &= 70 \times 10^{12} \\ &= 7 \times 10^{13}\end{aligned}$$

The sequence of operation is:

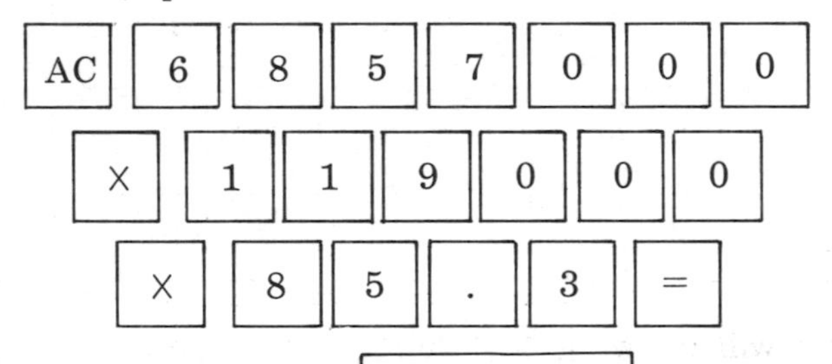

The display will show 6.96033 13 which represents $6.960\,33 \times 10^{13}$.

The least number of significant figures in the given numbers is three, thus the answer is 6.96×10^{13}.

An alternative method is to enter the numbers in standard form using the EXP (exponential) key. The sequence would then be:

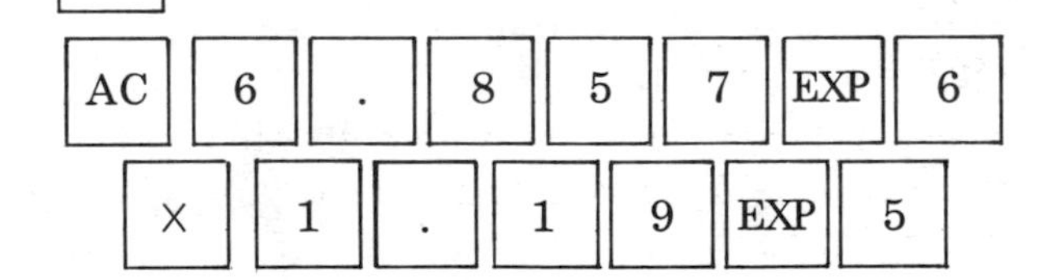

×	8	.	5	3	EXP	1	=

giving the same result.

The sequence used in a problem such as this would be personal choice, but if the problem includes numbers with powers of 10, the latter sequence is better.

EXAMPLE 7.4

Evaluate $\dfrac{5.745 \times 10^3 \times 56.7 \times 10^{-4}}{0.0343 \times 10^6}$

The rough check gives:

$$\frac{6 \times 10^3 \times 6 \times 10^{-3}}{3 \times 10^4} = \frac{6 \times 6}{3} \times 10^{3-3-4} = 12 \times 10^{-4}$$

The following sequence includes use of the +/− (change sign) key to enter the negative index:

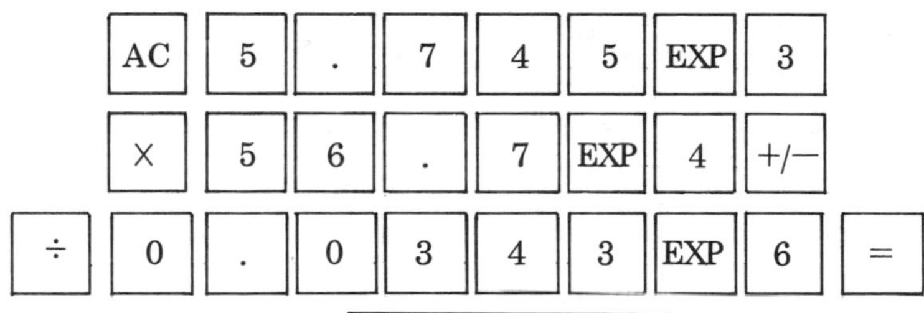

The display will show 9.499684 −04 and since the least number of significant figures in the given numbers is three, then the answer is 9.50×10^{-4}.

Note that we give the answer stating 9.50 and not merely 9.5 which would imply only two significant figure accuracy.

EXAMPLE 7.5

Evaluate $\dfrac{0.674}{1.239} - \dfrac{0.564 \times 1.89}{0.379}$

The rough check will need a little more care when approximating numbers. It is not possible to give rules but as you gain experience you will have no difficulty.

Thus the rough check gives:

$$\frac{0.5}{1}-\frac{0.5\times 2}{0.5} = 0.5-2 = -1.5$$

The sequence of operations includes use of the memory in which intermediate results may be kept for use later in the sequence. You should note also how a number may be subtracted from the contents of the memory by using the M+ (add to memory) key after changing the sign of the number.

The sequence of operations is:

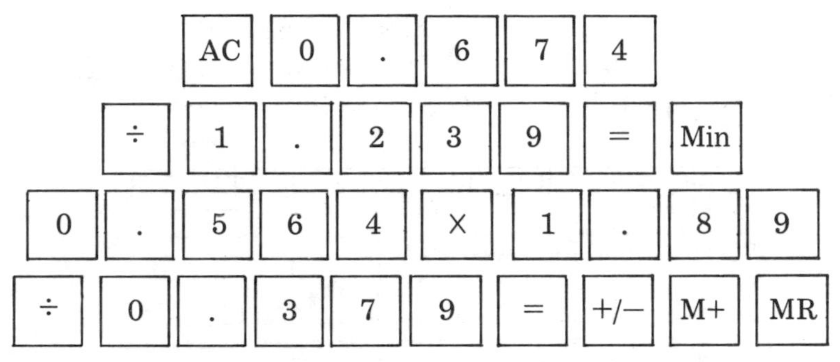

The display shows − 2.268 572 3

This answer is considerably higher than that obtained by the rough check, but it is of the correct order, i.e. *not* −22.7 *or* −0.227

Hence the required answer is −2.27 correct to three significant figures.

EXAMPLE 7.6

Find the value of $9.7+\dfrac{55.15}{29.6-8.64}$

The rough check gives:

$$10+\frac{60}{30-9} = 10+\frac{60}{21} \simeq 13$$

It is possible to work this problem out on the calculator by rearranging and using the memory. However, no rearrangement is necessary if we make use of the $\boxed{\frac{1}{x}}$ (the reciprocal) key.

This key enables us to find the reciprocal of a number—for example the reciprocal of 2 is $\frac{1}{2}$ or 0.5

Let us consider $\dfrac{1}{\left(\dfrac{29.6-8.64}{55.15}\right)}$

This may be written as:

$$1 \div \left(\frac{29.6-8.64}{55.15}\right) = 1 \times \left(\frac{55.15}{29.6-8.64}\right)$$

$$= \frac{55.15}{29.6-8.64}$$

If we make use of this knowledge then the sequence of operations is:

[AC] [9] [.] [7] [Min] [2] [9] [.] [6]

[−] [8] [.] [6] [4]

[=] [÷] [5] [5] [.] [1] [5]

[=] [$\frac{1}{x}$] [+] [MR] [=]

The display gives 12.331 202

It is always difficult to assess accuracy of the answer to a calculation which involves addition and subtraction. Although the 9.7 has only two significant figures the addition of the other portion of the calculation will increase the figures before the decimal point to two.

Thus the answer may be given as 12.3

In most engineering problems you will not often be wrong if you give the answers to three significant figures—this is consistent with the accuracy of much of the data (such as ultimate tensile strengths of materials). There are exceptions, of course, such as certain machine shop problems which may require a much greater degree of accuracy.

'WHOLE NUMBER' POWERS

The obvious method, but not necessarily the quickest, to find the fifth power of 5.6 (for example) is to multiply 5.6 by itself four times. To avoid entering the number fives times use may be made of the memory as the following sequence shows:

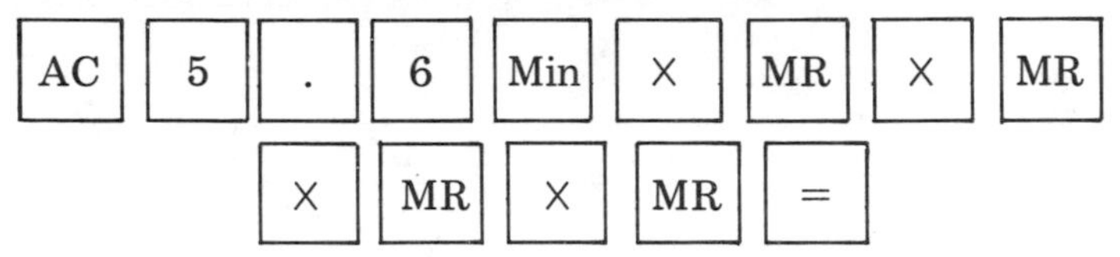

The display shows 5507.3178

Thus $(5.6)^5 = 5510$ correct to three significant figures.

Use of the Constant Multiplier Facility

You should check with the calculator booklet that this is possible with your machine.

Suppose that we have the following to evaluate:

3.1×7.89, 3.1×6.2, 3.1×3.45, 3.1×9.8, 3.1×10.9

If each calculation is carried out individually, then the number 3.1 will have to be entered for each calculation. This may be avoided by two consecutive pressings of the [×] key. Thus the sequence would be:

[AC] [3] [.] [1] [×] [×] [7] [.] [8] [9] [=]

giving 24.459

[6] [.] [2] [=]

giving 19.22

[3] [.] [4] [5] [=]

giving 10.695

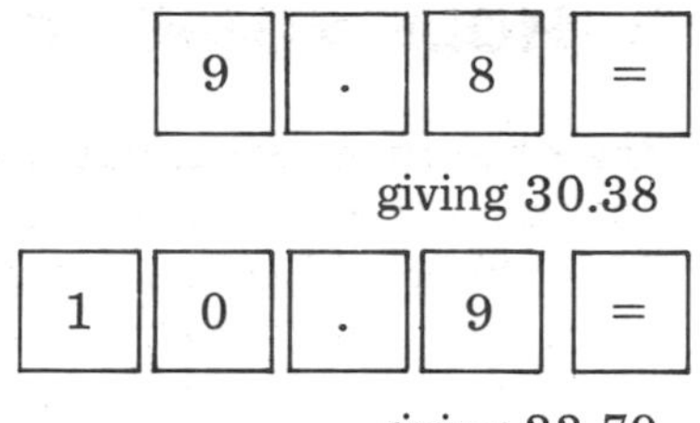

giving 33.79

If in the above sequence entry of each of the numbers preceding the equals operation were omitted, then the calculator would automatically substitute 3.1 for each of them.

Thus $(3.1)^4 = 3.1 \times 3.1 \times 3.1 \times 3.1$ may be found by the sequence:

AC	3	.	1	×	×	=	=	=

giving 92.3521

Exercise 7.1

Evaluate:

1) $45.6 + 3.563 - 21.42 - 14.6$

2) $-23.94 - 6.93 + 1.92 + 17.6$

3) $\dfrac{40.72 \times 3.86}{5.73}$

4) $\dfrac{4.86 \times 0.008\,34 \times 0.64}{0.86 \times 0.934 \times 21.7}$

5) $\dfrac{57.3 + 64.29 + 3.17}{64.2}$

6) $\dfrac{32.2}{6.45 + 7.29 - 21.3}$

7) $\dfrac{1}{\frac{1}{3} + \frac{1}{4} + \frac{1}{5}}$

8) $\dfrac{3.76 + 42.4}{1.6 + 0.86}$

9) $\dfrac{4.82 + 7.93}{-0.73 \times 6.92}$

10) $9.38(4.86 + 7.6 \times 1.89^3)$

11) $4.93^2 - 6.86^2$

12) $(4.93 + 6.86)(4.93 - 6.86)$

13) $\dfrac{1}{6.3^2 + 9.6^2}$

14) $\dfrac{3.864^2 + 9.62}{3.74 - 8.62^2}$

15) $\dfrac{9.5}{(6.4 \times 3.2) - (6.7 \times 0.9)}$

16) $1 - \dfrac{5}{3.6 + 7.49}$

17) $\dfrac{1}{6} - \dfrac{1}{5}(4.6)^2$

18) $\dfrac{6.4}{20.2}\left(3.94^2 - \dfrac{5.7 + 4.9}{6.7 - 3.2}\right)$

19) $\dfrac{3.64^3 + 5.6^2 - \dfrac{1}{0.085}}{9.76 + 3.4 - 2.9}$

20) $\dfrac{0.000\,076\,9 \times 6.54}{0.643^2 - 79.3 \times 0.000\,321}$

SQUARE ROOT AND 'PI' KEYS

| $\sqrt{\ }$ | gives the square root of any number in the display.

| π | gives the numerical value of π to whatever accuracy the machine is designed.

EXAMPLE 7.7

The period, T seconds (the time for a complete swing), of a simple pendulum is given by the formula $T = 2\pi\sqrt{\dfrac{l}{g}}$ where l m is its length and g m/s^2 is the acceleration due to gravity.

Find the value of T if $l = 1.37$ m and $g = 9.81$ m/s^2.

Substituting the given values into the formula we have

$$T = 2\pi\sqrt{\frac{1.37}{9.81}}$$

The rough check gives:

$$T = 2 \times 3\sqrt{\frac{1}{10}} \simeq 2 \times 3 \times \frac{1}{3} = 2$$

The sequence of operations would be:

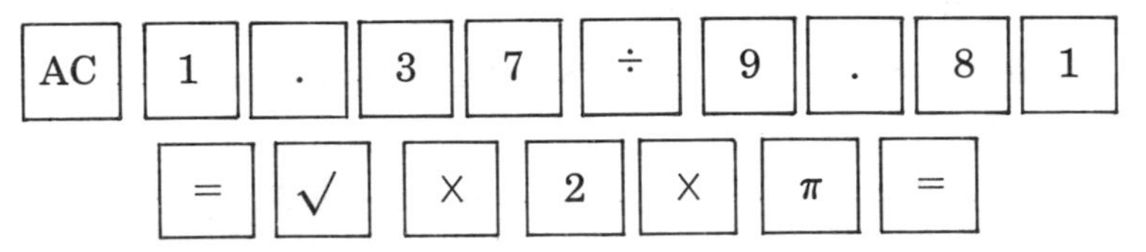

The display gives 2.348 040 8

Thus the value of T is 2.35 seconds, correct to three significant figures.

THE POWER KEY

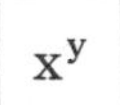 gives the value of x to the index y.

EXAMPLE 7.8

The relationship between the luminosity, I, of a metal filament lamp and the voltage, V, is given by the equation $I = aV^4$ where a is a constant. Find the value of I if $a = 9 \times 10^{-7}$ and $V = 60$.

Substituting the given values into the equations we have:

$$I = (9 \times 10^{-7})60^4$$

The rough check gives:

$$I = (10 \times 10^{-7})(6 \times 10)^4 = 10^{-6} \times 6^4 \times 10^4 = 10^{-2} \times 36 \times 36$$

and if we approximate by putting 30×40 instead of 36×36:

$$I = 10^{-2} \times 30 \times 40 = 10^{-2} \times 1200 = 12$$

The sequence of operations would be:

AC | 6 | 0 | x^y | 4

= | × | 9 | EXP | 7 | +/− | =

The display gives 11.664

Thus the value of I is 11.7 correct to three significant figures.

EXAMPLE 7.9

The law of expansion of a gas is given by the expression $pV^{1.2} = k$ where p is the pressure, V is the volume, and k is a constant. Find the value of k if $p = 0.8 \times 10^6$ and $V = 0.2$.

Substituting the given values into the formula we have:

$$k = (0.8 \times 10^6)0.2^{1.2}$$

The rough check gives:

$$k = 1 \times 10^6 \times \left(\frac{2}{10}\right)^{1.2} = 10^6 \times \frac{2^{1.2}}{10^{1.2}} \simeq 10^6 \times \frac{3}{30} = 1 \times 10^5$$

Since it is difficult to assess the approximate value of a decimal to an index, it becomes simpler to express the decimal number as a fraction using whole numbers. In this case it is convenient to express 0.2 as $\frac{2}{10}$. We can guess the rough value of $2^{1.2}$, since we know that $2^1 = 2$ and $2^2 = 4$. Similarly we judge the value of $10^{1.2}$ as being between $10^1 = 10$ and $10^2 = 100$. The more practice you have in doing calculations of this type, the more accurate your guess will be.

The sequence of operations would be:

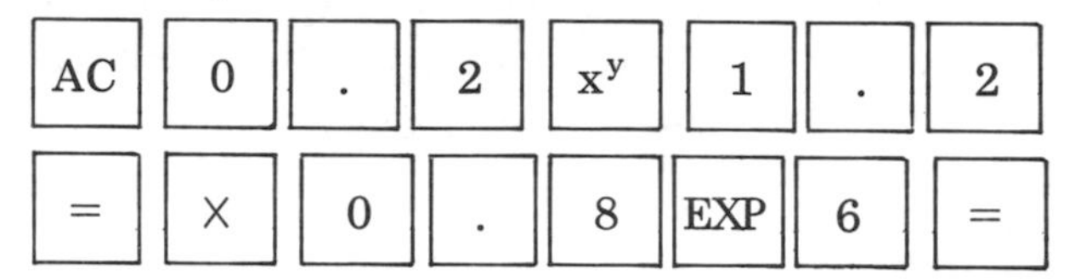

The display gives 115 964.74

Thus the required value of k is 116 000 or 1.16×10^5 correct to three significant figures.

EXAMPLE 7.10

The intensity of radiation, R, from certain radioactive materials at a particular time, t, follows the law $R = 95t^{-1.8}$. If $t = 5$ find the value of R.

Substituting $t = 5$ into the given equation gives $R = 95 \times 5^{-1.8}$.

For a rough check on the value of an expression containing a negative index it helps to rearrange so that the index becomes positive.

The rough check gives:

$$R = 100 \times 5^{-2} = 100 \times \frac{1}{5^2} = 100 \times \frac{1}{25} = 4$$

The sequence of operations would be:

[AC] [5] [x^y] [1] [.] [8] [+/−]

[=] [×] [9] [5] [=]

The display gives 5.242 972 8

Hence the value of R is 5.24 correct to three significant figures.

EXAMPLE 7.11

In the formula $Q = 2.37H^{5/2}$ find Q when $H = 2.81$

Substituting $H = 2.81$ into the given formula gives:

$$Q = 2.37(2.81)^{5/2}$$

The fractional index should be expressed as a decimal so that we may use the [x^y] key. Since $\frac{5}{2} = 2.5$ then the equation may be stated as:

$$Q = 2.37(2.81)^{2.5}$$

The rough check gives:

$$Q = 2 \times 3^{2.5} \simeq 2 \times 15 = 30$$

We estimated the value of $3^{2.5}$ since it lies between $3^2 = 9$ and $3^3 = 27$

The sequence of operations is:

[AC] [2] [.] [8] [1] [x^y] [2] [.] [5]

[=] [×] [2] [.] [3] [7] [=]

The display shows 31.369 974

Hence $Q = 31.4$ correct to three significant figures.

Exercise 7.2

Evaluate the expressions in questions 1 to 16 giving the answers correct to three significant figures:

1) 2.32^4	2) 1.52^6	3) 0.523^5	4) 7.9^{-2}
5) 4.59^{-3}	6) 0.321^{-4}	7) $12.1^{1.5}$	8) $6.83^{2.32}$
9) $0.879^{3.1}$	10) $5.56^{0.62}$	11) $14.7^{0.347}$	12) $3.9^{-0.5}$
13) $6.64^{3/2}$	14) $13.6^{2/5}$	15) $1.23^{7/3}$	16) $0.334^{3/5}$

17) Evaluate $4\pi r^2$ when $r = 6.1$

18) Evaluate $5\pi(R^2 - r^2)$ when $R = 1.32$ and $r = 1.24$

19) In a beam the stress, σ, due to bending is given by the expression $\sigma = \frac{My}{I}$. Find σ if $M = 12 \times 10^6$, $y = 60$, and $I = 11.5 \times 10^6$

20) The polar second moment of area, J, of a hollow shaft is given by the equation $J = \frac{\pi}{32}(D^4 - d^4)$. Find J if $D = 220$ and $d = 140$

21) The velocity, v, of a body performing simple harmonic motion is given by the expression $v = \omega\sqrt{A^2 - x^2}$. Find v if $\omega = 20.9$, $A = 0.060$, and $x = 0.02$

22) The natural frequency of oscillation, f, of a mass, m, supported by a spring of stiffness, λ, is given by the formula $f = \frac{1}{2\pi}\sqrt{\frac{\lambda}{m}}$. Find f if $\lambda = 5000$ and $m = 1.5$

23) The volume rate of flow, $\dot{Q}$, of water through a venturimeter is given by:

$$\dot{Q} = A_2\sqrt{\frac{2gH}{1 - \left(\frac{A_2}{A_1}\right)^2}}$$

Find $\dot{Q}$ if $A_1 = 0.0201$, $A_2 = 0.005\,03$, $g = 9.81$ and $H = 0.554$

TABLES AND CHARTS

After reaching the end of this chapter you should be able to:

1. *Use tables requiring the use of positive and negative mean differences.*
2. *Use conversion tables and charts.*
3. *Use nomographs.*
4. *Use network diagrams.*

TABLES AND CHARTS

As a technician it is inevitable that, from time to time, you will need to obtain information from tables or charts. In this chapter you will be introduced to various types of tables and charts, and the methods of using them.

It is only possible to include typical examples of tables and charts. For convenience, these have been listed under the following headings:

(1) Tables requiring the use of positive mean differences.
(2) Tables requiring the use of negative mean differences.
(3) Conversion tables and charts.
(4) Nomographs.
(5) Network diagrams.

Tables Requiring the Use of Positive Mean Differences

These include tables of the squares, square roots and logarithms of numbers, and degrees to radians for angles.

EXAMPLE 8.1

Use the table of squares to find:

a) 1.54^2 b) 2.463^2 c) 232^2 d) 0.2078^2

An extract from the table of squares is given overleaf:

TABLE OF SQUARES

											Mean differences								
x	0	1	2	3	4	5	6	7	8	9	1	2	3	4	5	6	7	8	9
1.0	1.000	1.020	1.040	1.061	1.082	1.103	1.124	1.145	1.166	1.188	2	4	6	8	10	13	15	17	19
1.1	1.210	1.232	1.254	1.277	1.300	1.323	1.346	1.369	1.392	1.416	2	5	7	9	11	14	16	18	21
1.2	1.440	1.464	1.488	1.513	1.538	1.563	1.588	1.613	1.638	1.664	2	5	7	10	12	15	17	20	22
1.3	1.690	1.716	1.742	1.769	1.796	1.823	1.850	1.877	1.904	1.932	3	5	8	11	13	16	19	22	24
1.4	1.960	1.988	2.016	2.045	2.074	2.103	2.132	2.161	2.190	2.220	3	6	9	12	14	17	20	23	26
1.5	2.250	2.280	2.310	2.341	2.372	2.403	2.434	2.465	2.496	2.528	3	6	9	12	15	19	22	25	28
1.6	2.560	2.592	2.624	2.657	2.690	2.723	2.756	2.789	2.822	2.856	3	7	10	13	16	20	23	26	30
1.7	2.890	2.924	2.958	2.993	3.028	3.063	3.098	3.133	3.168	3.204	3	7	10	14	17	21	24	28	31
1.8	3.240	3.276	3.312	3.349	3.386	3.423	3.460	3.497	3.534	3.572	4	7	11	15	18	22	26	30	33
1.9	3.610	3.648	3.686	3.725	3.764	3.803	3.842	3.881	3.920	3.960	4	8	12	16	19	23	27	31	35
2.0	4.000	4.040	4.080	4.121	4.162	4.203	4.244	4.285	4.326	4.368	4	8	12	16	20	25	29	33	37
2.1	4.410	4.452	4.494	4.537	4.580	4.623	4.666	4.709	4.752	4.796	4	9	13	17	21	26	30	34	39
2.2	4.840	4.884	4.928	4.973	5.018	5.063	5.108	5.153	5.198	5.244	4	9	13	18	22	27	31	36	40
2.3	5.290	5.336	5.382	5.429	5.476	5.523	5.570	5.617	5.664	5.712	5	9	14	19	23	28	33	38	42
2.4	5.760	5.808	5.856	5.905	5.954	6.003	6.052	6.101	6.150	6.200	5	10	15	20	24	29	34	39	44
2.5	6.250	6.300	6.350	6.401	6.452	6.503	6.554	6.605	6.656	6.708	5	10	15	20	25	31	36	41	46
2.6	6.760	6.812	6.864	6.917	6.970	7.023	7.076	7.129	7.182	7.236	5	11	16	21	26	32	37	42	48
2.7	7.290	7.344	7.398	7.453	7.508	7.563	7.618	7.673	7.728	7.784	5	11	16	22	27	33	38	44	49
2.8	7.840	7.896	7.952	8.009	8.066	8.123	8.180	8.237	8.294	8.352	6	11	17	23	28	34	40	46	51
2.9	8.410	8.468	8.526	8.585	8.644	8.703	8.762	8.821	8.880	8.940	6	12	18	24	29	35	41	47	53

a) The value of 1.54^2 may be found from the main body of the tables. Find 1.5 in the extreme left-hand column and proceed along this row until you reach the column headed by the figure 4 (i.e. the sixth column).

The number here, 2.372, is the value of 1.54^2.

b) The value of 2.463^2 requires use of both the main body of the tables and also the mean differences.

Using the method in part **a)** find 2.46^2, which is 6.052 in the main body.

The effect of the fourth significant figure 3 is allowed for by following the row which includes 6.052 along until the column in the mean differences headed by 3 is reached. The number here, 15 (representing 0.015), must now be added to 6.052 giving

$$\begin{array}{r} 6.052 \\ + \quad 15 \\ \hline 6.067 \end{array}$$

Thus $$2.463^2 = 6.067$$

c) The number 232 is outside the range of the tables. This may be remedied by stating it in standard form.

Thus $232^2 = (2.32 \times 10^2)^2 = 2.32^2 \times (10^2)^2 = 2.32^2 \times (10^2)^2$

From the body of the tables we find that $2.32^3 = 5.382$

Hence $232^2 = 5.382 \times 10^4 = 53\,820$

d) Again we have a number, 0.2078, which is outside the range of the tables and we proceed as in part **c)**.

Thus

$$0.2078^2 = (2.078 \times 10^{-1})^2 = 2.078^2 \times (10^{-1})^2 = 2.078^2 \times 10^{-2}$$

Using the main body of the tables, and also the mean differences we obtain:

$$\begin{aligned} 2.078^2 = \quad & 4.285 \\ + \quad & \underline{\quad 33} \\ & 4.318 \end{aligned}$$

Hence $0.2078^2 = 4.318 \times 10^{-2} = 0.043\,18$

Tables Requiring the Use of Negative Mean Differences

These include tables of reciprocal of numbers, and cosines of angles.

EXAMPLE 8.2

Find the reciprocals of: **a)** 1.932 **b)** 24.51 **c)** 0.022 28

An extract from the table of reciprocals is given below:

TABLE OF RECIPROCALS OF NUMBERS

Mean differences (must be subtracted NOT added)

	0	1	2	3	4	5	6	7	8	9	1	2	3	4	5	6	7	8	9
1.0	1.0000	0.9901	0.9804	0.9709	0.9615	0.9524	0.9434	0.9346	0.9259	0.9174									
1.1	0.9091	0.9009	0.8929	0.8850	0.8772	0.8696	0.8621	0.8547	0.8475	0.8403									
1.2	0.8333	0.8264	0.8197	0.8130	0.8065	0.8000	0.7937	0.7874	0.7813	0.7752									
1.3	0.7692	0.7634	0.7576	0.7519	0.7463	0.7407	0.7353	0.7299	0.7246	0.7194									
1.4	0.7143	0.7092	0.7042	0.6993	0.6944	0.6897	0.6849	0.6803	0.6757	0.6711									
1.5	0.6667	0.6623	0.6579	0.6536	0.6494	0.6452	0.6410	0.6369	0.6329	0.6289	4	8	12	17	21	25	29	33	37
1.6	0.6250	0.6211	0.6173	0.6135	0.6098	0.6061	0.6024	0.5988	0.5952	0.5917	4	7	11	15	18	22	26	29	33
1.7	0.5882	0.5848	0.5814	0.5780	0.5747	0.5714	0.5682	0.5650	0.5618	0.5587	3	7	10	13	16	20	23	26	29
1.8	0.5556	0.5525	0.5495	0.5464	0.5435	0.5405	0.5376	0.5348	0.5319	0.5291	3	6	9	12	15	18	20	23	26
1.9	0.5263	0.5236	0.5208	0.5181	0.5155	0.5128	0.5102	0.5076	0.5051	0.5025	3	5	8	11	13	16	18	21	24
2.0	0.5000	0.4975	0.4950	0.4926	0.4902	0.4878	0.4854	0.4831	0.4808	0.4785	2	5	7	10	12	14	17	19	21
2.1	0.4762	0.4739	0.4717	0.4695	0.4673	0.4651	0.4630	0.4608	0.4587	0.4566	2	4	6	9	11	13	15	17	19
2.2	0.4545	0.4525	0.4505	0.4484	0.4464	0.4444	0.4425	0.4405	0.4386	0.4367	2	4	6	8	10	12	14	16	18
2.3	0.4348	0.4329	0.4310	0.4292	0.4274	0.4255	0.4237	0.4219	0.4202	0.4184	2	4	5	7	9	11	13	14	16
2.4	0.4167	0.4149	0.4132	0.4115	0.4098	0.4082	0.4065	0.4049	0.4032	0.4016	2	3	5	7	8	10	12	13	15
2.5	0.4000	0.3984	0.3968	0.3953	0.3937	0.3922	0.3906	0.3891	0.3876	0.3861	2	3	5	6	8	9	11	12	14
2.6	0.3846	0.3831	0.3817	0.3802	0.3788	0.3774	0.3759	0.3745	0.3731	0.3717	1	3	4	6	7	9	10	11	13
2.7	0.3704	0.3690	0.3676	0.3663	0.3650	0.3636	0.3623	0.3610	0.3597	0.3584	1	3	4	5	7	8	9	11	12
2.8	0.3571	0.3559	0.3546	0.3534	0.3521	0.3509	0.3497	0.3484	0.3472	0.3460	1	2	4	5	6	7	9	10	11
2.9	0.3448	0.3436	0.3425	0.3413	0.3401	0.3390	0.3378	0.3367	0.3356	0.3344	1	2	3	5	6	7	8	9	10

a) The procedure is similar to that used with the table of squares, except that the mean difference numbers must be subtracted and *not* added.

Thus $\dfrac{1}{1.932}$ is found as

$$\begin{array}{r} 0.5181 \\ -\quad 5 \\ \hline 0.5176 \end{array}$$

Hence

$$\text{the reciprocal of } 1.932 = \frac{1}{1.932} = 0.5176$$

b) The number 24.51 is outside the range of the tables and, as before, we express it in standard form.

Thus
$$\frac{1}{24.51} = \frac{1}{2.451 \times 10} = \frac{1}{2.451} \times 10^{-1}$$

Now $\dfrac{1}{2.451}$ is found as

$$\begin{array}{r} 0.4082 \\ -\quad 2 \\ \hline 0.4080 \end{array}$$

Hence

$$\text{the reciprocal of } 24.51 = \frac{1}{24.51} = 0.4080 \times 10^{-1} = 0.040\,80$$

c) Again the number 0.022 28 is outside the range of the tables.

Thus
$$\frac{1}{0.022\,28} = \frac{1}{2.228 \times 10^{-2}} = \frac{1}{2.228} \times 10^{2}$$

Now $\dfrac{1}{2.228}$ is found as

$$\begin{array}{r} 0.4505 \\ -\quad 16 \\ \hline 0.4489 \end{array}$$

Hence

$$\text{the reciprocal of } 0.022\,28 = \frac{1}{0.022\,28} = 0.4489 \times 10^{2} = 44.89$$

CONVERSION TABLES — ILLUSTRATING USE OF INTERPOLATION

An example of the above is a table connecting temperature measurement in degrees Fahrenheit (°F) with degrees Celsius (°C).

EXAMPLE 8.3

Convert: a) 122.7°F to °C b) 43.3°F to °C

c) 62.7°C to °F d) −5.3°C to °F

An extract from the temperature conversion table is given below:

TEMPERATURE Degrees Fahrenheit to degrees Celsius (Centigrade)

°F	0	1	2	3	4	5	6	7	8	9
0	−17.8	−17.2	−16.7	−16.1	−15.6	−15.0	−14.4	−13.9	−13.3	−12.8
10	−12.2	−11.7	−11.1	−10.6	−10.0	−9.4	−8.9	−8.3	−7.8	−7.2
20	−6.7	−6.1	−5.6	−5.0	−4.4	−3.9	−3.3	−2.8	−2.2	−1.7
30	−1.1	−0.6	0	0.6	1.1	1.7	2.2	2.8	3.3	3.9
40	4.4	5.0	5.6	6.1	6.7	7.2	7.8	8.3	8.9	9.4
50	10.0	10.6	11.1	11.7	12.2	12.8	13.3	13.9	14.4	15.0
60	15.6	16.1	16.7	17.2	17.8	18.3	18.9	19.4	20.0	20.6
70	21.1	21.7	22.2	22.8	23.3	23.9	24.4	25.0	25.6	26.1
80	26.7	27.2	27.8	28.3	28.9	29.4	30.0	30.6	31.1	31.7
90	32.2	32.8	33.3	33.9	34.4	35.0	35.6	36.1	36.7	37.2
100	37.8	38.3	38.9	39.4	40.0	40.6	41.1	41.7	42.2	42.8
110	43.3	43.9	44.4	45.0	45.6	46.1	46.7	47.2	47.8	48.3
120	48.9	49.4	50.0	50.6	51.1	51.7	52.2	52.8	53.3	53.9
130	54.4	55.0	55.6	56.1	56.7	57.2	57.8	58.3	58.9	59.4
140	60.0	60.6	61.1	61.7	62.2	62.8	63.3	63.9	64.4	65.0
150	65.6	66.1	66.7	67.2	67.8	68.3	68.9	69.4	70.0	70.6
160	71.1	71.7	72.2	72.8	73.3	73.9	74.4	75.0	75.6	76.1
170	76.7	77.2	77.8	78.3	78.9	79.4	80.0	80.6	81.1	81.7
180	82.2	82.8	83.3	83.9	84.4	85.0	85.6	86.1	86.7	87.2
190	87.8	88.3	88.9	89.4	90.0	90.6	91.1	91.7	92.2	92.8

Interpolation:	0.1	0.2	0.3	0.4	0.5	0.6	0.7	0.8	0.9
deg F:	0.1	0.2	0.3	0.4	0.5	0.6	0.7	0.8	0.9
deg C:	0.1	0.1	0.2	0.2	0.3	0.3	0.4	0.4	0.5

a) In this table the mean differences are listed under the heading 'interpolation' and are given at the bottom of the table.

If we consider 122.7°F we can find values in the body of the tables for the whole number values immediately above and below, i.e. 122°F and 123°F.

Thus 122°F = 50.0°C and 123°F = 50.6°C

Now we can deduce that the required value for 122.7°F will lie between 50.0°C and 50.6°C. and being a little over half way may well be 50.4°C.

This process is called *interpolation* and is simplifed by use of the figures given at the foot of the table.

If we look along the top row of numbers, we can find 0.7 °F and this is equivalent to 0.4 °C which is shown immediately below in the second row.

Thus 122.7 °F is found as

$$\begin{array}{r} 50.0 \\ +\ \ 0.4 \\ \hline 50.4 \end{array}$$

Hence $122.7\,^\circ\text{F} = 50.4\,^\circ\text{C}$

b) Now for 18.2 °F we can find that $18\,^\circ\text{F} = -7.8$ from the body of the tables. Also the 'interpolation' numbers give 0.2 °F as being equivalent to 0.1 °C.

Thus 18.2 °F is found as

$$\begin{array}{r} -7.8 \\ +0.1 \\ \hline -7.7 \end{array}$$

This result should be checked very carefully, bearing in mind the negative numbers. This check involves looking at the values in the table for the whole numbers immediately above and below 18.2 °F. Now $18\,^\circ\text{F} = -7.8$ and $19\,^\circ\text{F} = -7.2$, and so the value -7.7 is reasonable.

Hence $18.2\,^\circ\text{F} = -7.7\,^\circ\text{C}$

c) Conversion from °C to °F may be carried out using the tables 'in reverse'. We must look in the body of the table in order to find numbers as close to the given 62.7 as possible. They are 62.2 corresponding to 144 °F and 62.8 corresponding to 145 °F. Now the interpolation numbers show that 0.5 °C is equivalent to 0.9 °F.

Thus

$$62.7\,^\circ\text{C} = 62.2\,^\circ\text{C} + 0.5\,^\circ\text{C} = 144\,^\circ\text{F} + 0.9\,^\circ\text{F} = 144.9\,^\circ\text{F}$$

A rough check shows this is reasonable since it lies between 144 °F and 145 °F.

d) We again use the method in part **c)** using extreme caution with the negative numbers. Now $-5.3\,^\circ\text{C}$ lies between -5.6 corresponding to 22 °F, and -5.0 corresponding to 23 °F. The interpolation numbers show that 0.3 °C is equivalent to 0.5 °F or 0.6 °F.

Thus

$$-5.3\,^\circ\text{C} = -5.6\,^\circ\text{C} + 0.3\,^\circ\text{C} = 22\,^\circ\text{F} + 0.5\,^\circ\text{F} = 22.5\,^\circ\text{F}$$

A rough check shows this answer is reasonable.

PARALLEL SCALE CONVERSION CHARTS

A system of parallel scales may be used when we wish to convert from one set of units to another related set.

An example of the above is a chart relating British Imperial pound (lb) mass units with SI kilogram (kg) mass units.

EXAMPLE 8.4

Convert: **a)** 3.4 lb to kg **b)** 568 lb to kg
c) 1.8 kg to lb **d)** 0.23 kg to lb.

The conversion chart is shown below:

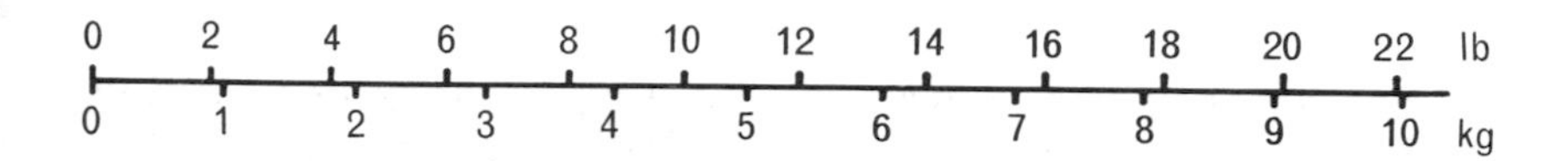

Observation of the scales shows that we cannot expect accuracy greater than 0.1 on either of them.

a) Direct reading from the adjacent scale shows that

$$3.4\,\text{lb} = 1.6\,\text{kg}$$

b) Since 568 lb is outside the range of the scales we may express it in standard form. Thus $568\,\text{lb} = 5.68 \times 10^2\,\text{lb}$.

Bearing in mind the accuracy limitations we will have to consider 5.68 as 5.7 and obtain

$$568\,\text{lb} = 5.7 \times 10^2\,\text{lb} = 2.6 \times 10^2\,\text{kg} = 260\,\text{kg}$$

correct to two significant figures only.

c) Direct reading from the adjacent scales shows that

$$1.8\,\text{kg} = 4.0\,\text{lb}$$

d) Although 0.23 kg may be found directly on the scale, accuracy will be improved if we express it as a standard number, and proceed as in part **c)**.

Thus $0.23\,\text{kg} = 2.3 \times 10^{-1}\,\text{kg} = 5.1 \times 10^{-1}\,\text{lb} = 0.51\,\text{lb}$

NOMOGRAPHS

These are charts which represent graphically mathematical laws or relationships.

EXAMPLE 8.5

The data given on the chart shown in Fig. 8.1 refers to wires of circular cross-section supporting dead loads. The chart connects stress (MN/m^2) at failure of the wire, the load (kg) which causes the wire to break, and the diameter (mm) of the wire. Use the chart to find:

a) the stress at failure if 600 kg breaks a wire of 4 mm diameter,

b) the diameter of a wire which breaks under a load of 850 kg with a stress at failure of 1100 MN/m^2.

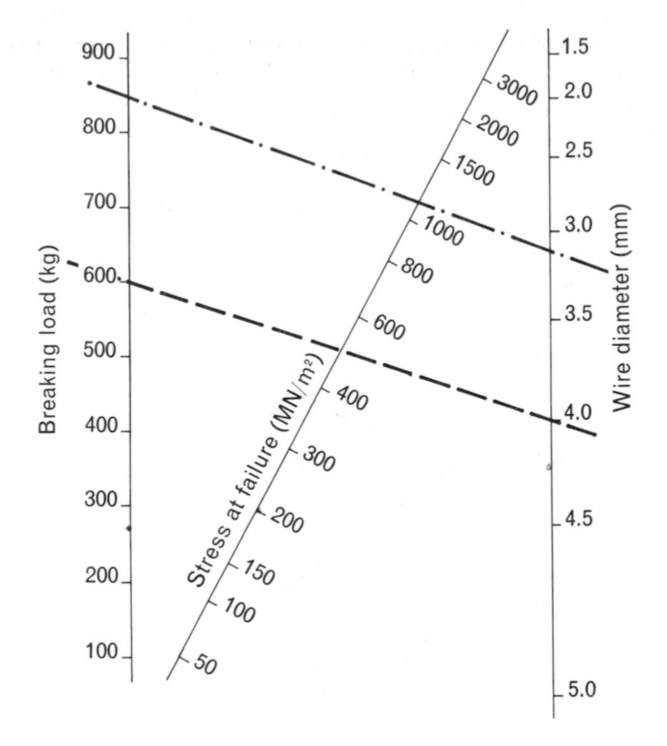

Fig. 8.1

The chart is used by placing a straight edge, such as the edge of a rule, across the three scales.

a) The dashed line shows the straight edge passing through 600 kg on the load scale and 4 mm on the wire diameter scale. The edge cuts the sloping scale at 500 MN/m^2, and this is the required stress at failure.

Careful judgement is needed in reading off the scales inbetween the marked points—especially the non-linear scales (those with unequal intervals).

b) The chain-dotted line shows the straight edge passing through 850 kg on the load scale and 1100 MN/m^2 on the stress at failure scale. The edge then cuts the right-hand vertical scale at 3.1 mm and this is the required diameter of the wire.

EXAMPLE 8.6

In calculations on electrical circuits we often need to find the single resistance, R_E ohms, which is equivalent to two resistances, R_A and R_B ohms connected in 'parallel' arrangement. The chart below in Fig. 8.2 relates the values of R_A, R_B, and R_E to each other. Use the chart to find:

a) R_E when $R_A = 5$ ohms and $R_B = 7$ ohms,

b) R_E when $R_A = 3.7$ ohms and $R_B = 2.4$ ohms.

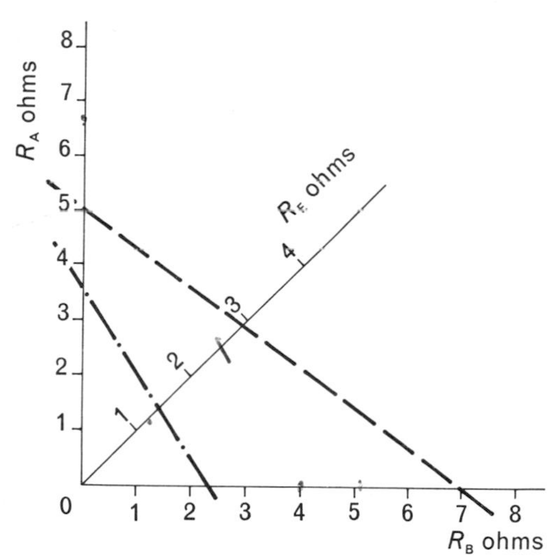

Fig. 8.2

The chart is used in a similar manner to that of the previous example by placing a straight edge across the scales.

a) The dashed line shows the straight edge passing through 5 ohms on the R_A scale and 7 ohms on the R_B scale. The intersection on the R_E scale gives 2.9 ohms and this is the required value of R_E.

b) The chain-dotted line shows the position of the straight edge for $R_A = 3.7$ ohms and $R_B = 2.4$ ohms cutting the inclined axis at 1.4 ohms, which is the required value of R_E.

EXAMPLE 8.7

The specific speed is a number used in the design of hydraulic turbines. It depends on the speed of the turbine (rev/min), the head of water available (metres), and the power output from the machine (kilowatts). The chart shown in Fig. 8.3 enables the specific speed to be found for any given speed, head, and power output. Use the chart to find the specific speed of a turbine running at 1400 rev/min, under a head of 120 metres and developing 20 megawatts of power.

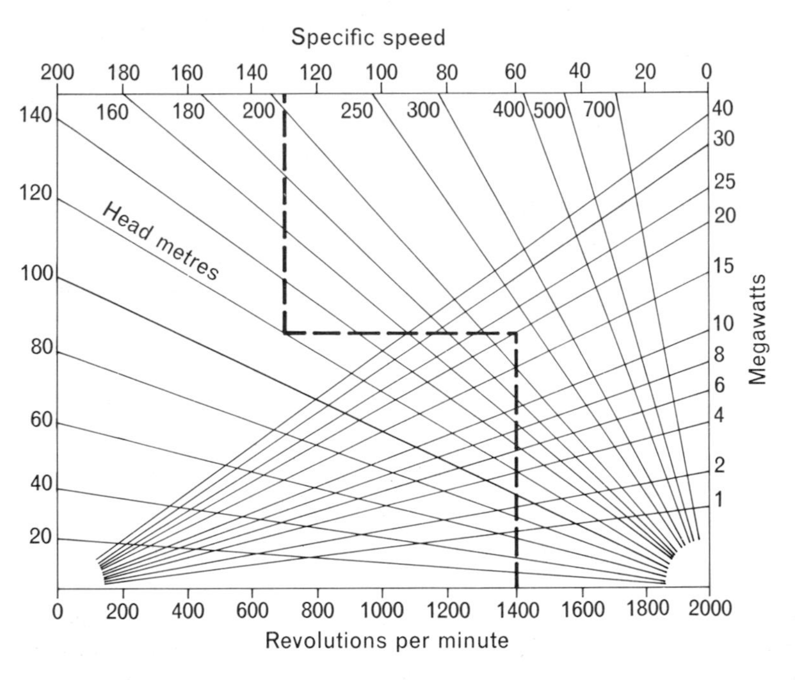

Fig. 8.3

The dashed line follows the path for solving our problem. We first locate 1400 rev/min on the bottom horizontal axis line. We then go vertically until meeting the line indicating 20 megawatts (this is one of the series of lines which radiate upwards to the right from the left-hand lower corner). We then move horizontally until we meet the line indicating 120 metres (this is one of the series of lines which start from the lower right corner and radiate upwards to the left). We then proceed vertically upwards until meeting the top horizontal line on which the required specific speeds are marked. This is 128. Accuracy is limited on this type of chart and it would be more reasonable to give an answer of 130.

NETWORK DIAGRAMS

A network diagram is a method of showing sequences of activities, and the corresponding times taken, on a chart.

Fig. 8.4 represents an activity A which takes 2 days to complete.

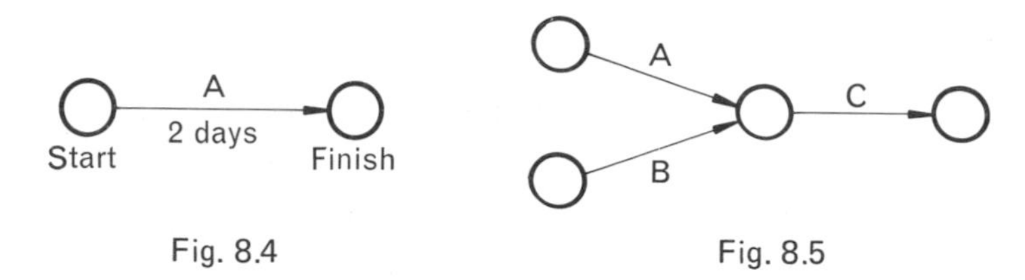

Fig. 8.4 Fig. 8.5

Fig. 8.5 shows that activity C cannot start until both A and B have been completed.

Fig. 8.6 shows that none of activities B, C, and D may commence until A has been completed.

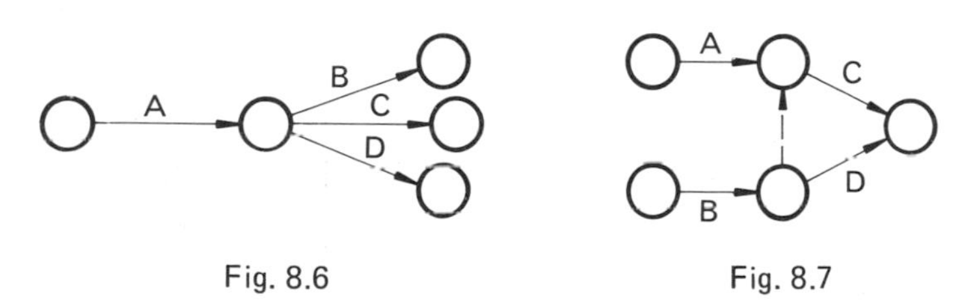

Fig. 8.6 Fig. 8.7

In Fig. 8.7 the broken line is called a 'dummy', which has *no* time meaning. It is put in to indicate that activity C cannot commence until both A and B have been completed. Activity D may commence, however, when B is completed (i.e. it is independent of activity A).

EXAMPLE 8.8

In the network diagram shown in Fig. 8.8 determine the critical path and hence find the least time to complete the whole project.

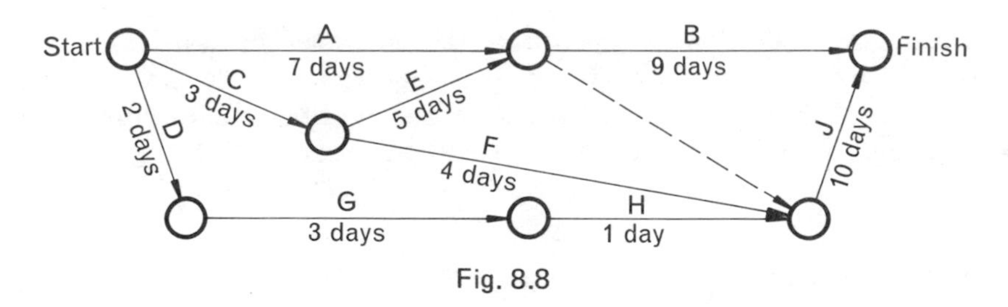

Fig. 8.8

You should note that in a network diagram the lengths of the lines are *not* proportional to the times taken.

One of the most useful pieces of information that we can obtain is the critical path.

The *critical path* is the path through the network on which any delay will lengthen the completion time.

If it were not for the dummy (the dotted line) activity J could commence after both F and H ended. F would take the longer time as it depends on C giving a total of $3+4=7$ days from the start. However, the dummy means that J cannot start until both A and E have finished. The longer of these is E, which depends on C, and will take $3+5=8$ days from the start. Thus the critical path comprises activities C, E and J which give a total time of $3+5+10=18$ days from start to finish.

You can see that all the activity sequences except C, E and J have time to spare and delays in their completion may not affect the 18 day start to finish time.

However any delay in the critical path will prevent the completion of the whole project in 18 days.

Many other facts may be obtained such as earliest and latest starts and finishes of activities but these are beyond the scope of our work here, the object of this being to introduce you to the overall idea of network analysis from a diagram.

Exercise 8.1

1) Use the table of squares on p. 110 to find the values of:

(a) 1.306^2 (b) 0.2871^2 (c) 12.34^2 (d) 2111^2

2) Use the table of reciprocals on p. 111 to find the values of:

(a) $\dfrac{1}{2.822}$ (b) $\dfrac{1}{17.72}$ (c) $\dfrac{1}{0.2941}$ (d) $\dfrac{1}{0.0209}$

3) Use the temperature table on p. 113 to convert:

(a) 76.0°F to °C (b) 161.5°F to °C

(c) 68.7°C to °F (d) −11.3°C to °F

4) Use the conversion chart on p. 115 to convert:

(a) 7.7 kg to lb (b) 332 kg to lb

(c) 18.3 lb to kg (d) 0.42 lb to kg

5) Use the chart on p. 116 to find:

(a) the stress at failure if a 3 mm diameter wire breaks under a load of 400 kg,

(b) the diameter of a wire which breaks under a load of 728 kg if the stress at failure is 1760 MN/m^2,

(c) the breaking load of a 4.43 mm diameter wire which has a stress of 273 MN/m^2 at failure.

6) Using the chart on p. 117 find the value of:

(a) R_E when $R_A = 6.7$ ohms and $R_B = 4$ ohms

(b) R_A when $R_B = 5.1$ ohms and $R_E = 3.1$ ohms

(c) R_B when $R_A = 2$ ohms when $R_E = 1.21$ ohms.

7) Use the chart on p. 118 to find the specific speed of a turbine having:

(a) a speed of 1200 rev/min, a head of 120 m and developing 20 megawatts of power,

(b) a head of 60 m and developing 4 megawatts of power at a speed of 600 rev/min.

8) Find the critical path in each of the networks shown and hence the least time to complete each of the projects:

(a)

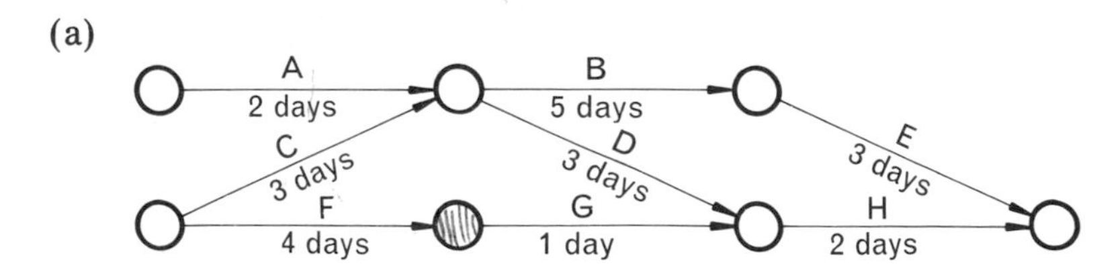

(b)

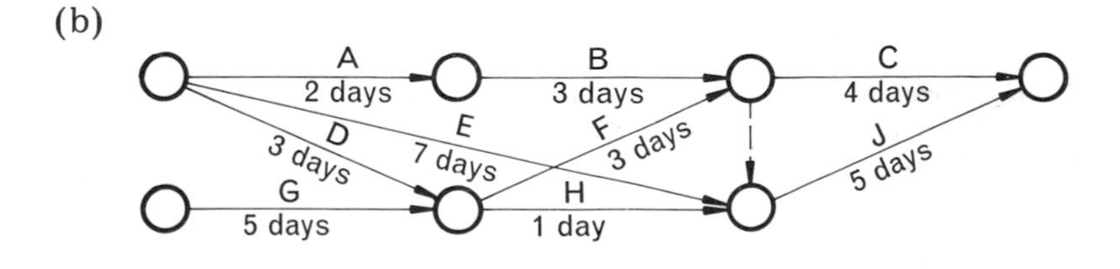

INTRODUCTION TO ALGEBRA

After reaching the end of this chapter you should be able to:

1. *Represent quantities by numbers and letters.*
2. *Simplify expressions involving symbols and numbers using (i) the four arithmetic operations of addition, subtraction, multiplication and division (ii) the commutative, associative and distributive laws (iii) the precedence in the use of brackets and arithmetic operations.*
3. *Multiply an expression in brackets by number, a symbol or by another expression in a bracket.*
4. *Factorise expressions by (i) extraction of common factor, (ii) grouping.*
5. *Multiply, divide, add and subtract algebraic fractions.*

USE OF SYMBOLS

A technician often has to indicate that certain quantities or measurements have to be added, subtracted, multiplied or divided. Frequently this has to be done without using actual numbers. The statement:

$$\text{Area of a rectangle} = \text{length} \times \text{breadth}$$

is a perfectly *general* statement which applies to all rectangles. If we use symbols we obtain a much shorter statement. Thus,

if A = the area of the rectangle,

l = the length of the rectangle

and b = the breadth of the rectangle

then the statement becomes:

$$A = l \times b$$

Knowing what the symbols A, l and b stand for, this statement conveys as much information as the first statement. To find the area of a particular rectangle we replace the symbols l and b by the actual dimensions of the rectangle, first making sure that l and b have the same units. Thus, to find the area of a rectangle whose length is 50 mm and whose breadth is 30 mm we put $l = 50$ mm and $b = 30$ mm.

$$A = l \times b = 50 \times 30 = 1500\,\text{mm}^2$$

Many verbal statements can be translated into symbols as the following statements show:

The difference of two numbers $= x - y$

Two numbers multiplied together $= a \times b$

One number divided by another $= p \div q$

ADDITION AND SUBTRACTION OF ALGEBRAIC TERMS

Like terms are numerical multiples of the same algebraic quantity. Thus,

$$7x,\ 5x \text{ and } -3x$$

are three like terms.

An expression consisting of like terms can be reduced to a single term by adding or subtracting the numerical coefficients. Thus,

$$7x - 5x + 3x = (7 - 5 + 3)x = 5x$$

$$3b^2 + 7b^2 = (3 + 7)b^2 = 10b^2$$

$$-3y - 5y = (-3 - 5)y = -8y$$

$$q - 3q = (1 - 3)q = -2q$$

Only like terms can be added or subtracted. Thus $7a + 3b - 2c$ is an expression containing three unlike terms and it cannot be simplified any further. Similarly with $8a^2b + 7ab^3 - 6a^2b^2$ which are all unlike terms. It is possible to have several sets of like terms in an expression and each set can then be simplified:

$$\begin{aligned} &8x + 3y - 4z - 5x + 7z - 2y + 2z \\ &= (8 - 5)x + (3 - 2)y + (-4 + 7 + 2)z \\ &= 3x + y + 5z \end{aligned}$$

MULTIPLICATION AND DIVISION SIGNS

When using symbols multiplication signs are nearly always omitted and $l \times b$ becomes lb. Of course the same scheme cannot apply to numbers and we cannot write 9×6 as 96. The multiplication sign can, however, be omitted when a symbol and a number are to be multiplied together. Thus $5 \times m$ is written $5m$. The system may be extended to three or more quantities and hence $P \times L \times A \times N$ is written $PLAN$. The symbols need not be written in any special order because the order in which numbers are multiplied together is unimportant. Thus $PLAN$ is the same as $LANP$ or $NAPL$. It is usual, however, to write numbers before symbols, that is, it is better to write $8xy$ than $xy8$ or $x8y$. In algebraic expressions the number in front of the symbols is called the *coefficient*. Thus in the expression $8x$ the coefficient of x is 8.

The division sign ÷ is seldom used in algebra and it is more convenient to write $p \div q$ in the fractional form $\frac{p}{q}$. Thus,

$$\frac{lp}{2\pi R} \quad \text{means} \quad lp \div 2\pi R$$

MULTIPLICATION AND DIVISION OF ALGEBRAIC QUANTITIES

The rules are exactly the same as those used with directed numbers.

$$(+x)(+y) = +(xy) = +xy = xy$$

$$5x \times 3y = 5 \times 3 \times x \times y = 15xy$$

$$(x)(-y) = -(xy) = -xy$$

$$(2x)(-3y) = -(2x)(3y) = -6xy$$

$$(-4x)(2y) = -(4x)(2y) = -8xy$$

$$(-3x)(-2y) = +(3x)(2y) = 6xy$$

$$\frac{+x}{+y} = +\frac{x}{y} = \frac{x}{y} \qquad \frac{-3x}{2y} = -\frac{3x}{2y}$$

$$\frac{-5x}{-6y} = +\frac{5x}{6y} = \frac{5x}{6y} \qquad \frac{4x}{-3y} = -\frac{4x}{3y}$$

When *multiplying* expressions containing the same symbols, indices are used:

$$m \times m = m^2$$

$$3m \times 5m = 3 \times m \times 5 \times m = 15m^2$$

$$(-m) \times m^2 = (-m) \times m \times m = -m^3$$

$$5m^2n \times 3mn^3 = 5 \times m \times m \times n \times 3 \times m \times n \times n \times n$$
$$= 15m^3n^4$$
$$3mn \times (-2n^2) = 3 \times m \times n \times (-2) \times n \times n = -6mn^3$$

When *dividing* algebraic expressions, cancellation between numerator and denominator is often possible. Cancelling is equivalent to dividing both numerator and denominator by the same quantity:

$$\frac{pq}{p} = \frac{\not{p} \times q}{\not{p}} = q$$

$$\frac{3p^2q}{6pq^2} = \frac{3 \times \not{p} \times p \times \not{q}}{6 \times \not{p} \times \not{q} \times q} = \frac{3p}{6q} = \frac{p}{2q}$$

$$\frac{18x^2y^2z}{6xyz} = \frac{18 \times \not{x} \times x \times \not{y} \times y \times \not{z}}{6 \times \not{x} \times \not{y} \times \not{z}} = 3xy$$

Exercise 9.1

Simplify the following:

1) $7x + 11x$ 2) $7x - 5x$ 3) $3x - 6x$

4) $-2x - 4x$ 5) $-8x + 3x$ 6) $-2x + 7x$

7) $8a - 6a + 7a$ 8) $5m + 13m - 6m$ 9) $6b^2 - 4b^2 + 3b^2$

10) $6ab - 3ab - 2ab$ 11) $14xy + 5xy - 7xy + 2xy$

12) $-5x + 7x - 3x - 2x$

13) $-4x^2 - 3x^2 + 2x^2 - x^2$

14) $3x - 2y + 4z - 2x - 3y + 5z + 6x + 2y - 3z$

15) $3a^2b + 2ab^3 + 4a^2b^2 - 5ab^3 + 11b^4 + 6a^2b$

16) $1.2x^3 - 3.4x^2 + 2.6x + 3.7x^2 + 3.6x - 2.8$

17) $pq + 2.1qr - 2.2rq + 8qp$

18) $2.6a^2b^2 - 3.4b^3 - 2.7a^3 - 3a^2b^2 - 2.1b^3 + 1.5a^3$

19) $2x \times 5y$ 20) $3a \times 4b$

21) $3 \times 4m$ 22) $\frac{1}{4}q \times 16p$

23) $x \times (-y)$ 24) $(-3a) \times (-2b)$

25) $8m \times (-3n)$

26) $(-4a) \times 3b$

27) $8p \times (-q) \times (-3r)$

28) $3a \times (-4b) \times (-c) \times 5d$

29) $12x \div 6$

30) $4a \div (-7b)$

31) $(-5a) \div 8b$

32) $(-3a) \div (-3b)$

33) $4a \div 2b$

34) $4ab \div 2a$

35) $12x^2yz^2 \div 4xz^2$

36) $(-12a^2b) \div 6a$

37) $8a^2bc^2 \div 4ac^2$

38) $7a^2b^2 \div 3ab$

39) $a \times a$

40) $b \times (-b)$

41) $(-m) \times m$

42) $(-p) \times (-p)$

43) $3a \times 2a$

44) $5X \times X$

45) $5q \times (-3q)$

46) $3m \times (-3m)$

47) $(-3pq) \times (-3q)$

48) $8mn \times (-3m^2n^3)$

49) $7ab \times (-3a^2)$

50) $2q^3r^4 \times 5qr^2$

51) $(-3m) \times 2n \times (-5p)$

52) $5a^2 \times (-3b) \times 5ab$

53) $m^2n \times (-mn) \times 5m^2n^2$

SEQUENCE OF MIXED OPERATIONS ON ALGEBRAIC QUANTITIES

Since algebraic quantities contain symbols (or letters) which represent numbers the sequence of operations is exactly the same as used with numbers.

Remember the word BODMAS which gives the initial letters of the correct sequence, i.e. Brackets, Of, Divide, Multiply, Add, Subtract.

Thus

$$\begin{aligned} 2x^2 + (12x^4 - 3x^4) \div 3x^2 - x^2 &= 2x^2 + 9x^4 \div 3x^2 - x^2 \\ &= 2x^2 + 3x^2 - x^2 \\ &= 5x^2 - x^2 \\ &= 4x^2 \end{aligned}$$

BRACKETS

Brackets are used for convenience in grouping terms together. When removing brackets each *term* within the bracket is multiplied by the quantity outside the bracket:

$$3(x+y) = 3x+3y$$
$$5(2x+3y) = 5\times 2x+5\times 3y = 10x+15y$$
$$4(a-2b) = 4\times a-4\times 2b = 4a-8b$$
$$m(a+b) = ma+mb$$
$$3x(2p+3q) = 3x\times 2p+3x\times 3q = 6px+9qx$$
$$4a(2a+b) = 4a\times 2a+4a\times b = 8a^2+4ab$$

When a bracket has a minus sign in front of it, the signs of all the terms inside the bracket are changed when the bracket is removed. The reason for this rule may be seen from the following examples:

$$-3(2x-5y) = (-3)\times 2x+(-3)\times(-5y) = -6x+15y$$
$$-(m+n) = -m-n$$
$$-(p-q) = -p+q$$
$$-2(p+3q) = -2p-6q$$

When simplifying expressions containing brackets first remove the brackets and then add the like terms together:

$$(3x+7y)-(4x+3y) = 3x+7y-4x-3y = -x+4y$$
$$3(2x+3y)-(x+5y) = 6x+9y-x-5y = 5x+4y$$
$$x(a+b)-x(a+3b) = ax+bx-ax-3bx = -2bx$$
$$2(5a+3b)+3(a-2b) = 10a+6b+3a-6b = 13a$$

Exercise 9.2

Remove the brackets in the following:

1) $3(x+4)$

2) $2(a+b)$

3) $3(3x+2y)$

4) $\frac{1}{2}(x-1)$

5) $5(2p-3q)$

6) $7(a-3m)$

7) $-(a+b)$

8) $-(a-2b)$

9) $-(3p-3q)$

10) $-(7m-6)$

11) $-4(x+3)$
12) $-2(2x-5)$
13) $-5(4-3x)$
14) $2k(k-5)$
15) $-3y(3x+4)$
16) $a(p-q-r)$
17) $4xy(ab-ac+d)$
18) $3x^2(x^2-2xy+y^2)$
19) $-7P(2P^2-P+1)$
20) $-2m(-1+3m-2n)$

Remove the brackets and simplify the following:

21) $3(x+1)+2(x+4)$
22) $5(2a+4)-3(4a+2)$
23) $3(x+4)-(2x+5)$
24) $4(1-2x)-3(3x-4)$
25) $5(2x-y)-3(x+2y)$
26) $\frac{1}{2}(y-1)+\frac{1}{3}(2y-3)$
27) $-(4a+5b-3c)-2(2a+3b-4c)$
28) $2x(x-5)-x(x-2)-3x(x-5)$
29) $3(a-b)-2(2a-3b)+4(a-3b)$
30) $3x(x^2+7x-1)-2x(2x^2+3)-3(x^2+5)$

THE PRODUCT OF TWO BINOMIAL EXPRESSIONS

A binomial expression consists of two terms. Thus $3x+5$, $a+b$, $2x+3y$ and $4p-q$ are all binomial expressions.

To find the product of $(a+b)(c+d)$ consider the diagram (Fig. 9.1).

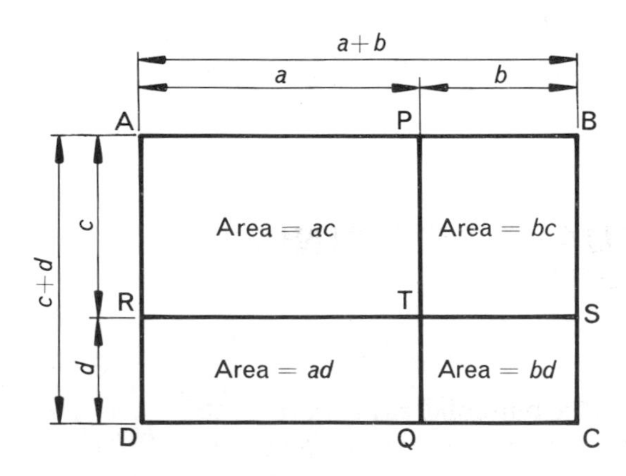

Fig. 9.1

In Fig. 9.1 the rectangular area ABCD is made up as follows:

$$\text{ABCD} = \text{APTR} + \text{TQDR} + \text{PBST} + \text{STQC}$$

$$\text{i.e. } (a+b)(c+d) = ac + ad + bc + bd$$

It will be noticed that the expression on the right hand side is obtained by multiplying each term in the one bracket by each term in the other bracket. The process is illustrated below:

$$(a+b)(c+d) = ac + ad + bc + bd$$

EXAMPLE 9.1

a) $(3x+2)(4x+5) = 3x \times 4x + 3x \times 5 + 2 \times 4x + 2 \times 5$
$= 12x^2 + 15x + 8x + 10$
$= 12x^2 + 23x + 10$

b) $(2p-3)(4p+7) = 2p \times 4p + 2p \times 7 - 3 \times 4p - 3 \times 7$
$= 8p^2 + 14p - 12p - 21$
$= 8p^2 + 2p - 21$

c) $(z-5)(3z-2) = z \times 3z + z \times (-2) - 5 \times 3z - 5 \times (-2)$
$= 3z^2 - 2z - 15z + 10$
$= 3z^2 - 17z + 10$

d) $(2x+3y)(3x-2y) = 2x \times 3x + 2x \times (-2y) + 3y \times 3x$
$+ 3y \times (-2y)$
$= 6x^2 - 4xy + 9xy - 6y^2$
$= 6x^2 + 5xy - 6y^2$

THE SQUARE OF A BINOMIAL EXPRESSION

$$(a+b)^2 = (a+b)(a+b) = a^2 + ab + ba + b^2 = a^2 + 2ab + b^2$$

The square of a binomial expression is the sum of the squares of the two terms and twice their product.

$$(a-b)^2 = (a-b)(a-b) = a^2 - ab - ba + b^2 = a^2 - 2ab + b^2$$

EXAMPLE 9.2

a) $(2x+5)^2 = (2x)^2 + 2\times 2x \times 5 + 5^2$
$= 4x^2 + 20x + 25$

b) $(3x-2)^2 = (3x)^2 + 2\times 3x \times (-2) + (-2)^2$
$= 9x^2 - 12x + 4$

c) $(2x+3y)^2 = (2x)^2 + 2\times 2x \times 3y + (3y)^2$
$= 4x^2 + 12xy + 9y^2$

THE PRODUCT OF THE SUM AND DIFFERENCE OF TWO TERMS

$(a+b)(a-b) = a^2 - ab + ba - b^2 = a^2 - b^2$

This result is the difference of the squares of the two terms.

EXAMPLE 9.3

a) $(8x+3)(8x-3) = (8x)^2 - 3^2 = 64x^2 - 9$

b) $(2x+5y)(2x-5y) = (2x)^2 - (5y)^2 = 4x^2 - 25y^2$

Exercise 9.3

Find the products of the following:

1) $(x+1)(x+2)$
2) $(x+3)(x+1)$
3) $(x+4)(x+5)$
4) $(2x+5)(x+3)$
5) $(3x+7)(x+6)$
6) $(5x+1)(x+4)$
7) $(2x+4)(3x+2)$
8) $(5x+1)(2x+3)$
9) $(7x+2)(3x+5)$
10) $(x-1)(x-3)$
11) $(x-4)(x-2)$
12) $(x-6)(x-3)$
13) $(2x-1)(x-4)$
14) $(x-2)(3x-5)$
15) $(x-8)(4x-1)$
16) $(2x-4)(3x-2)$
17) $(3x-1)(2x-5)$
18) $(7x-5)(3x-2)$
19) $(x+3)(x-1)$
20) $(x-2)(x+7)$
21) $(x-5)(x+3)$
22) $(2x+5)(x-2)$

23) $(3x-5)(x+6)$

24) $(3x+5)(x+6)$

25) $(3x+5)(2x-3)$

26) $(6x-7)(2x+3)$

27) $(3x-5)(2x+3)$

28) $(3x+2y)(x+y)$

29) $(2p-q)(p-3q)$

30) $(3v+2u)(2v-3u)$

31) $(2a+b)(3a-b)$

32) $(5a-7)(a-6)$

33) $(3x+4y)(2x-3y)$

34) $(x+1)^2$

35) $(2x+3)^2$

36) $(3x+7)^2$

37) $(x-1)^2$

38) $(3x-5)^2$

39) $(2x-3)^2$

40) $(2a+3b)^2$

41) $(x+y)^2$

42) $(P+3Q)^2$

43) $(a-b)^2$

44) $(3x-4y)^2$

45) $(2x+y)(2x-y)$

46) $(a-3b)(a+3b)$

47) $(2m-3n)(2m+3n)$

48) $(x^2+y)(x^2-y)$

HIGHEST COMMON FACTOR (HCF)

The HCF of a set of algebraic expressions is the highest expression which is a factor of each of the given expressions.

The method used is similar to that for finding the HCF of a set of numbers (p. 17).

EXAMPLE 9.4

Find the HCF of $ab^2c^2, a^2b^3c^3, a^2b^4c^4$.

We express each expression as the product of its factors.

Thus $\quad ab^2c^2 = a \times b \times b \times c \times c$

and $\quad a^2b^3c^3 = a \times a \times b \times b \times b \times c \times c \times c$

and $\quad a^2b^4c^4 = a \times a \times b \times b \times b \times b \times c \times c \times c \times c$

We now note the factors which are common to each of the lines. Factor a is common once, factor b twice, and factor c twice. The product of these factors gives the required HCF.

Thus

$$\text{HCF} = a \times b \times b \times c \times c$$
$$= ab^2c^2$$

EXAMPLE 9.5

Find the HCF of $\dfrac{x^3y}{m^2n^4}, \dfrac{x^2y^3}{m^2n^2}, \dfrac{x^4y^2}{mn^3}$.

Now $$\frac{x^3y}{m^2n^4} = x \times x \times x \times y \times \frac{1}{m} \times \frac{1}{m} \times \frac{1}{n} \times \frac{1}{n} \times \frac{1}{n} \times \frac{1}{n}$$

and $$\frac{x^2y^3}{m^2n^2} = x \times x \times y \times y \times y \times \frac{1}{m} \times \frac{1}{m} \times \frac{1}{n} \times \frac{1}{n}$$

and $$\frac{x^4y^2}{mn^3} = x \times x \times x \times x \times y \times y \times \frac{1}{m} \times \frac{1}{n} \times \frac{1}{n} \times \frac{1}{n}$$

Factor x is common twice, factor y once, factor $\dfrac{1}{m}$ once, and factor $\dfrac{1}{n}$ twice.

Thus $$\text{HCF} = x \times x \times y \times \frac{1}{m} \times \frac{1}{n} \times \frac{1}{n}$$
$$= \frac{x^2y}{mn^2}$$

An alternative method is to select the lowest power of each of the quantities which occur in **all** of the expressions, and then multiply them together.

EXAMPLE 9.6

Find the HCF of $3m^2np^3$, $6m^3n^2p^2$, $24m^3p^4$.

Dealing with the numerical coefficients 3, 6 and 24 we note that 3 is a factor of each of them. The quantities m and p occur in all three expressions, their lowest powers being m^2 and p^2. Hence,

$$\text{HCF} = 3m^2p^2$$

(Note that n does not occur in each of the three expressions and hence it does not appear in the HCF.)

FACTORISING

A factor is a common part of two or more terms which make up an algebraic expression. Thus the expression $3x + 3y$ has two terms which have the number 3 common to both of them. Thus $3x + 3y = 3(x + y)$. We say that 3 and $(x + y)$ are the factors of

$3x+3y$. To factorise algebraic expressions of this kind, we first find the HCF of all the terms making up the expression. The HCF then appears outside the bracket. To find the terms inside the bracket divide each of the terms making up the expression by the HCF.

EXAMPLE 9.7

a) Find the factors of $ax+bx$.

The HCF of ax and bx is x.

$$\therefore \quad ax+bx = x(a+b) \quad \left(\text{since } \frac{ax}{x} = a \text{ and } \frac{bx}{x} = b\right)$$

b) Find the factors of $m^2n - 2mn^2$.

The HCF of m^2n and $2mn^2$ is mn.

$$\therefore \quad m^2n - 2mn^2 = mn(m-2n)$$

$$\left(\text{since } \frac{m^2n}{mn} = m \text{ and } \frac{2mn^2}{mn} = 2n\right)$$

c) Find the factors of $3x^4y + 9x^3y^2 - 6x^2y^3$.

The HCF of $3x^4y$, $9x^3y^2$ and $6x^2y^3$ is $3x^2y$.

$$\therefore \quad 3x^4y + 9x^3y^2 - 6x^2y^3 = 3x^2y(x^2+3xy-2y^2)$$

$$\left(\text{since } \frac{3x^4y}{3x^2y} = x^2,\ \frac{9x^3y^2}{3x^2y} = 3xy \text{ and } \frac{6x^2y^3}{3x^2y} = 2y^2\right)$$

d) Find the factors of $\dfrac{ac}{x} + \dfrac{bc}{x^2} - \dfrac{cd}{x^3}$

The HCF of $\dfrac{ac}{x}$, $\dfrac{bc}{x^2}$ and $\dfrac{cd}{x^3}$ is $\dfrac{c}{x}$.

$$\therefore \quad \frac{ac}{x} + \frac{bc}{x^2} - \frac{cd}{x^3} = \frac{c}{x}\left(a + \frac{b}{x} - \frac{d}{x^2}\right)$$

$$\left(\text{since } \frac{ac}{x} \div \frac{c}{x} = a,\ \frac{bc}{x^2} \div \frac{c}{x} = \frac{b}{x} \text{ and } \frac{cd}{x^3} \div \frac{c}{x} = \frac{d}{x^2}\right)$$

Exercise 9.4

Find the HCF of the following:

1) p^3q^2, p^2q^3, p^2q
2) $a^2b^3c^3, a^3b^3, ab^2c^2$
3) $3mn^2, 6mnp, 12m^2np^2$
4) $2ab, 5b, 7ab^2$
5) $3x^2yz, 12x^2yz, 6xy^2z^3, 3xyz^2$

Factorise the following:

6) $2x+6$
7) $4x-4y$
8) $5x-5$
9) $4x-8xy$
10) $mx-my$
11) $ax+bx+cx$
12) $\frac{x}{2}-\frac{y}{8}$
13) $5a-10b+15c$
14) ax^2+ax
15) $2\pi r^2+\pi rh$
16) $3y-9y^2$
17) ab^3-a^2b
18) $x^2y^2-axy+bxy^2$
19) $5x^3-10x^2y+15xy^2$
20) $9x^3y-6x^2y^2+3xy^5$
21) $I_0+I_0\alpha t$
22) $\frac{x}{3}-\frac{y}{6}+\frac{z}{9}$
23) $2a^2-3ab+b^2$
24) x^3-x^2+7x
25) $\frac{m^2}{pn}-\frac{m^3}{pn^2}+\frac{m^4}{p^2n^2}$

FACTORISING BY GROUPING

To factorise the expression $ax+ay+bx+by$ first group the terms in pairs so that each pair of terms has a common factor. Thus,

$$ax+ay+bx+by = (ax+ay)+(bx+by) = a(x+y)+b(x+y)$$

Now notice that in the two terms $a(x+y)$ and $b(x+y)$, $(x+y)$ is a common factor. Hence,

$$a(x+y)+b(x+y) = (x+y)(a+b)$$

$$\therefore \quad ax+ay+bx+by = (x+y)(a+b)$$

Similarly,

$$\begin{aligned} np+mp-qn-qm &= (np+mp)-(qn+qm) \\ &= p(n+m)-q(n+m) \\ &= (n+m)(p-q) \end{aligned}$$

Exercise 9.5

Factorise the following:

1) $ax + by + bx + ay$
2) $mp + np - mq - nq$
3) $a^2c^2 + acd + acd + d^2$
4) $2pr - 4ps + qr - 2qs$
5) $4ax + 6ay - 4bx - 6by$
6) $ab(x^2 + y^2) - cd(x^2 + y^2)$
7) $mn(3x - 1) - pq(3x - 1)$
8) $k^2l^2 - mnl - k^2l + mn$

LOWEST COMMON MULTIPLE (LCM)

The LCM of a set of algebraic terms is the simplest expression of which each of the given terms is a factor.

The method used is similar to that for finding the LCM of a set of numbers (p. 16).

EXAMPLE 9.8

Find the LCM of $2a$, $3ab$, and a^2b.

We express each term as a product of its factors.

Thus $\quad 2a = 2 \times a$

and $\quad 3ab = 3 \times a \times b$

and $\quad a^2b = a \times a \times b$

We now note the greatest number of times each factor occurs in any one particular line.

Now factor 2 occurs once in the line for $2a$,

and factor 3 occurs once in the line for $3ab$,

and factor a occurs twice in the line for a^2b,

and factor b occurs once in either of the lines for $3ab$ or a^2b.

The product of these factors gives the required LCM.

Thus $\quad \text{LCM} = 2 \times 3 \times a \times a \times b$

$\quad = 6a^2b$

EXAMPLE 9.9

Find the LCM of $4x$, $8yz$, $2x^2y$ and yz^2.

With practice the LCM may be found by inspection, by finding the product of the highest powers of **all** factors which occur in **any** of the terms.

Thus
$$\text{LCM} = 8 \times x^2 \times y \times z^2$$
$$= 8x^2yz^2$$

EXAMPLE 9.10

Find the LCM of $(a-1)$, $n(m+n)$, $(m+n)^2$.

Brackets must be treated as single factors—**not** the individual terms inside each bracket.

Hence
$$\text{LCM} = (a-1) \times n \times (m+n)^2$$
$$= n(a-1)(m+n)^2$$

Exercise 9.6

Find the LCM for the terms in each of the following examples:

1) $2a, 3a^2, a, a^2$
2) $xy, x^2y, 2x, 2y$
3) m^2n, mn^2, mn, m^2n^2
4) $2ab, abc, bc^2$
5) $2(x+1), (x+1)$
6) $(a+b), x(a+b)^2, x^2$
7) $(a+b), (a-b)$
8) $x, (1-x), (x+1)$
9) $(x-2), (x+2), (x-2)^2$
10) $(x+2), (x^2-4)$
11) $2(a-b), 3(a+b), (a^2-b^2)$
12) $2(a+b)^2, 3(a+b), (a^2+b^2)$

HANDLING ALGEBRAIC FRACTIONS

Since algebraic expressions contain symbols (or letters) which represent numbers all the rule of operations with numbers also apply to algebraic terms, including fractions.

Thus

$$\frac{1}{\frac{1}{a}} = 1 \div \frac{1}{a} = 1 \times \frac{a}{1} = \frac{1 \times a}{1} = a$$

and

$$\frac{\frac{a}{b}}{\frac{c}{d}} = \frac{a}{b} \div \frac{c}{d} = \frac{a}{b} \times \frac{d}{c} = \frac{a \times d}{b \times c} = \frac{ad}{bc}$$

and

$$\frac{\frac{x+y}{1}}{x-y} = \frac{\frac{(x+y)}{1}}{(x-y)} = (x+y) \div \frac{1}{(x-y)} = (x+y) \times \frac{(x-y)}{1}$$

$$= (x+y)(x-y)$$

You should note in the last example how we put brackets round the $x+y$ and $x-y$ to remind us that they must be treated as single expressions—otherwise we may have been tempted to handle the terms x and y on their own.

ADDING AND SUBTRACTING ALGEBRAIC FRACTIONS

Consider the expression $\frac{a}{b} + \frac{c}{d}$ which is the addition of two fractional terms. These are called partial fractions.

If we wish to express the sum of these partial fractions as one single fraction then we proceed as follows (see p. 28 for the similar method used when adding or subtracting number fractions).

First find the lowest common denominator. This is the LCM of b and d which is bd. Each fraction is then expressed with bd as the denominator.

Now $\frac{a}{b} = \frac{a \times d}{b \times d} = \frac{ad}{bd}$ and $\frac{c}{d} = \frac{c \times b}{d \times b} = \frac{cb}{bd}$

and adding these new fractions we have:

$$\frac{a}{b} + \frac{c}{d} = \frac{ad}{bd} + \frac{cb}{bd} = \frac{ad + cb}{bd}$$

EXAMPLE 9.11

Express each of the following as a single fraction:

a) $\frac{1}{x}-\frac{1}{y}$ b) $a-\frac{1}{b}$ c) $\frac{1}{m}+n-\frac{a}{b}$

d) $\frac{a}{b^2}-\frac{1}{bc}$ e) $\frac{2}{x}+\frac{3}{x-1}$ f) $\frac{x}{x+1}-\frac{2}{x+3}$

a) The lowest common denominator is the LCM of x and y which is xy.

Therefore $$\frac{1}{x}-\frac{1}{y} = \frac{y}{xy}-\frac{x}{xy} = \frac{y-x}{xy}$$

b) $$a-\frac{1}{b} = \frac{a}{1}-\frac{1}{b} = \frac{ab}{b}-\frac{1}{b} = \frac{ab-1}{b}$$

c) $$\frac{1}{m}+n-\frac{a}{b} = \frac{1}{m}+\frac{n}{1}-\frac{a}{b} = \frac{b}{mb}+\frac{nmb}{mb}-\frac{am}{mb} = \frac{b+nmb-am}{mb}$$

d) $$\frac{a}{b^2}-\frac{1}{bc} = \frac{ac}{b^2c}-\frac{b}{b^2c} = \frac{ac-b}{b^2c}$$

e) $$\frac{2}{x}+\frac{3}{(x-1)} = \frac{2(x-1)}{x(x-1)}+\frac{3x}{x(x-1)} = \frac{2(x-1)+3x}{x(x-1)}$$
$$= \frac{2x-2+3x}{x(x-1)} = \frac{5x-2}{x(x-1)}$$

f) $$\frac{x}{(x+1)}-\frac{2}{(x+3)} = \frac{x(x+3)}{(x+1)(x+3)}-\frac{2(x+1)}{(x+3)(x+1)}$$
$$= \frac{x(x+3)-2(x+1)}{(x+1)(x+3)}$$
$$= \frac{x^2+3x-2x-2}{(x+1)(x+3)}$$
$$= \frac{x^2+x-2}{(x+1)(x+3)}$$

You should note that we have not attempted to multiply out the brackets in the denominator. They should be left as they are, for multiplying them out would complicate the expression rather than simplify it.

Exercise 9.7

Rearrange the following and thus express in a simplified form:

1) $\dfrac{\frac{1}{b}}{a}$ 2) $\dfrac{\frac{1}{a}}{\frac{1}{b}}$ 3) $\dfrac{\frac{x}{y}}{\frac{y}{x}}$ 4) $\dfrac{1}{\frac{2}{xy}}$

5) $\dfrac{\frac{a}{b}}{a^2}$ 6) $\dfrac{(a+b)}{\frac{1}{c}}$ 7) $\dfrac{1-x}{\frac{1}{1+x}}$ 8) $\dfrac{\frac{1}{a-b}}{\frac{a-b}{c}}$

Express with a common denominator:

9) $\dfrac{1}{x}+\dfrac{1}{y}$ 10) $1+\dfrac{1}{a}$ 11) $\dfrac{m}{n}-1$

12) $\dfrac{b}{c}-c$ 13) $\dfrac{a}{b}-\dfrac{c}{d}$ 14) $\dfrac{a}{b}-\dfrac{1}{bc}$

15) $\dfrac{1}{\frac{1}{a}}-\dfrac{1}{a}$ 16) $\dfrac{1}{xy}+\dfrac{1}{x}+1$ 17) $\dfrac{3}{x}+\dfrac{x}{4}$

18) $\dfrac{3}{c}+\dfrac{2}{d}-\dfrac{5}{e}$ 19) $\dfrac{a}{b}+\dfrac{c}{d}+1$ 20) $\dfrac{1}{3fg}-\dfrac{5}{6gh}-\dfrac{1}{2fh}$

21) $\dfrac{5}{y+3}+\dfrac{3}{y-5}$ 22) $\dfrac{2}{x}-\dfrac{4}{x+2}$ 23) $\dfrac{1}{x^2}+\dfrac{1}{x-1}$

24) $\dfrac{x}{1-x}+\dfrac{1}{1+x}$ 25) $1-\dfrac{x}{x-2}$ 26) $\dfrac{x+2}{x-2}+1$

EXPRESSING A SINGLE FRACTION AS PARTIAL FRACTIONS

When considering $\dfrac{x-y}{x}$ many students are tempted to cancel the x terms and obtain $\dfrac{\cancel{x}-y}{\cancel{x}}=\dfrac{1-y}{1}=1-y$

This is completely wrong!

The correct method is to reverse the procedure used for adding algebraic terms.

Thus $$\frac{x-y}{x} = \frac{x}{x} - \frac{y}{x} = 1 - \frac{y}{x}$$

Alternatively we may consider the numerator as enclosed in a bracket giving:

$$\frac{(x-y)}{x} = \frac{1}{x}(x-y) = \frac{1}{x}\times x - \frac{1}{x}\times y = 1 - \frac{y}{x}$$

EXAMPLE 9.12

Express as partial fractions:

a) $\dfrac{ab+bc-1}{abc}$ b) $\dfrac{a-b}{x+y}$ c) $\dfrac{(x-1)+y}{a(x-1)}$

a) $$\frac{ab+bc-1}{abc} = \frac{ab}{abc} + \frac{bc}{abc} - \frac{1}{abc} = \frac{1}{c} + \frac{1}{a} - \frac{1}{abc}$$

b) $$\frac{a-b}{x+y} = \frac{a}{x+y} - \frac{b}{x+y}$$

c) $$\frac{(x-1)+y}{a(x-1)} = \frac{(x-1)}{a(x-1)} + \frac{y}{a(x-1)} = \frac{1}{a} + \frac{y}{a(x-1)}$$

MIXED OPERATIONS WITH FRACTIONS

We will now combine all the ideas already used. It helps to work methodically and avoid taking short cuts by leaving out stages of simplification.

EXAMPLE 9.13

Simplify: a) $\dfrac{\frac{1}{x}+x}{\frac{1}{x}}$ b) $\dfrac{\frac{a}{b}}{\frac{1}{2ab}-\frac{3}{b}}$ c) $\dfrac{x-\frac{1}{1-x}}{\frac{1-x}{1+x}}$

In all these solutions the numerators and denominators are first expressed as single fractions. It is then possible to divide the numerator by the denominator.

a) $$\frac{\frac{1}{x}+x}{\frac{1}{x}} = \frac{\frac{1}{x}+\frac{x^2}{x}}{\frac{1}{x}} = \frac{\frac{1+x^2}{x}}{\frac{1}{x}} = \frac{(1+x^2)}{x} \div \frac{1}{x}$$

$$= \frac{(1+x^2)}{x} \times \frac{x}{1} = 1+x^2$$

b) $$\frac{\frac{a}{b}}{\frac{1}{2ab}-\frac{3}{b}} = \frac{\frac{a}{b}}{\frac{1}{2ab}-\frac{6a}{2ab}} = \frac{\frac{a}{b}}{\frac{(1-6a)}{2ab}} = \frac{a}{b} \div \frac{(1-6a)}{2ab}$$

$$= \frac{a}{b} \times \frac{2ab}{(1-6a)}$$

$$= \frac{2a^2}{1-6a}$$

c) $$\frac{x-\frac{1}{(1+x)}}{\frac{(1-x)}{(1+x)}} = \frac{\frac{x(1+x)}{(1+x)}-\frac{1}{(1+x)}}{\frac{(1-x)}{(1+x)}} = \frac{\frac{x(1+x)-1}{(1+x)}}{\frac{(1-x)}{(1+x)}}$$

$$= \frac{x(1+x)-1}{(1+x)} \div \frac{(1-x)}{(1+x)}$$

$$= \frac{x(1+x)-1}{(1+x)} \times \frac{(1+x)}{(1-x)}$$

$$= \frac{x(1+x)-1}{(1-x)}$$

Exercise 9.8

Express as partial fractions:

1) $\frac{a+b}{a}$ 2) $\frac{a-b}{ab}$ 3) $\frac{1+c}{c}$

4) $\frac{x^2+y}{2x}$ 5) $\frac{a^2-ab+ac}{abc}$ 6) $\frac{(x-1)+(x+1)}{(x+1)}$

7) $\dfrac{xy(1-a)+y^2}{x(1-a)}$ 8) $\dfrac{x+(x-y)}{x(x-y)}$ 9) $\dfrac{(a+b)-(a-b)}{(a-b)(a+b)}$

Simplify:

10) $\dfrac{\frac{7}{c}-\frac{3}{d}}{\frac{4}{d}}$ 11) $\dfrac{1}{1+\frac{1}{x}}$ 12) $\dfrac{a}{a-\frac{1}{a}}$

13) $\dfrac{\frac{1}{2a}}{4+\frac{3}{a}}$ 14) $\dfrac{1+\frac{1}{m}}{\frac{1}{m}-1}$ 15) $\dfrac{3+\frac{5}{t}}{\frac{1}{3t}-2}$

16) $\dfrac{1}{\frac{1}{R_1}+\frac{1}{R_2}}$ 17) $\dfrac{a-\frac{b}{c}}{1+\frac{c}{a}}$ 18) $\dfrac{\frac{b}{c}+\frac{x}{y}}{x+\frac{b}{y}}$

19) $\dfrac{ab-a^2}{a^2-\frac{a}{b}}$ 20) $\dfrac{\frac{a}{b}-1}{\frac{a-b}{a+b}}$ 21) $\dfrac{1+\frac{1-x}{1+x}}{\frac{1}{1+x}-1}$

SUMMARY

a) In algebra, multiplication signs are usually omitted. Thus $a \times b \times c$ becomes abc.

b) Division signs are seldom used and $p \div q$ is best written in the fractional form $\dfrac{p}{q}$.

c) The value of an algebraic expression is found by substituting the given values for the symbols into the expression.

d) Formulae are evaluated in the same way as algebraic expressions by substituting numerical values for the symbols.

e) The quantity $a \times a \times a$ is written a^3. The number 3 which gives the number of a's to be multiplied together is called the index.

f) Like terms are numerical multiples of the same algebraic quantity. Only like terms may be added or subtracted by adding or subtracting the numerical coefficients.

g) The rules used when multiplying and dividing are the same as those used with directed numbers. They are: $(+)\times(+)=(+)$;
$(+)\times(-)=(-)$; $(-)\times(+)=(-)$; $(-)\times(-)=(+)$;
$(+)\div(+)=(+)$; $(-)\div(+)=(-)$; $(+)\div(-)=(-)$;
$(-)\div(-)=(+)$.

h) When dividing with algebraic expressions cancelling between the numerator and denominator is often possible. Note that cancelling means dividing the numerator and denominator by the same quantity.

i) When removing a bracket each term within the bracket is multiplied by the quantity outside the bracket.

j) To find the product of two binomial expressions multiply the terms as shown:

$$(a+b)(c+d) = ac+ad+bc+bd$$

k) $(a+b)^2 = a^2+2ab+b^2$

l) $(a-b)^2 = a^2-2ab+b^2$

m) $(a+b)(a-b) = a^2-b^2$

n) To find the Highest Common Factor (HCF) of a set of algebraic quantities first choose the lowest power of each of the symbols which occur in the set. The HCF is then found by multiplying these together.

o) A factor is a common part of two or more terms which go to make up an algebraic expression. The expression $4a+4b$ has two terms which have the number 4 common to both of them. Thus $4a+4b = 4(a+b)$. 4 and $(a+b)$ are the factors of $4a+4b$. To factorise expressions of this type first find the HCF of all the terms making up the expression. This HCF then appears outside the bracket. To find the terms inside the bracket divide each of the terms making up the expression by the HCF.

p) To factorise by grouping, first group the terms in pairs so that each pair of terms has a common factor.

Self-Test 6

In questions 1 to 55 the answer is either 'true' or 'false'. Write down the appropriate word for each question.

1) The sum of two numbers can be represented by the expression $a+b$

2) The expression $a-b$ represents the difference of two numbers a and b

3) The product of 8 and x is $8+x$

4) 3 times a number minus 7 can be written as $3x-7$

5) Two numbers added together minus a third number and the result divided by a fourth number may be written as $(a+b-c)\div d$

6) The value of $3a+7$ when $a=5$ is 36

7) The value of $8x-3$ when $x=3$ is 21

8) The value of $3b-2c$ when $b=4$ and $c=3$ is 6

9) The value of $8ab\div 3c$ when $a=6$, $b=4$ and $c=2$ is 32

10) The quantity $a\times a\times a\times a$ is written a^3

11) The quantity $y\times y\times y$ is written y^3

12) a^3b^2 is equal to $a\times a\times a\times b\times b$

13) The value of a^4 when $a=3$ is 81

14) When $x=2$, $y=3$ and $z=4$ the value of $2x^2y^3z$ is 258

15) $13a-8a-9a$ is equal to $4a$

16) $-8x+11x-20x$ is equal to $-17x$

17) $-11y+3y-(-8y)$ is equal to $-16y$

18) $5p-(-7p)+8p$ is equal to $20p$

19) $(-5x)\times(-7x)$ is equal to $-35x^2$

20) $(-8z)\times(-3z)$ is equal to $24z^2$

21) $(-3a)\times 7a\times(-8a)\times(-2a)$ is equal to $-336a^4$

22) $(-5c)^3$ is equal to $125c^3$

23) $(-8t)^2$ is equal to $-64t^2$

24) $(-2u)^5$ is equal to $-32u^5$

25) $(-8x)\div(-4x)$ is equal to 2

26) $(-6y)\times(-4y)\div(-8y)$ is equal to $-3y^3$

27) $5x+8x$ is equal to $13x^2$

28) $3x+6x$ is equal to $9x$

29) $8x-5x$ is equal to 3

30) $7x-2x$ is equal to $5x$

31) $15xy+7xy-3xy-2xy$ is equal to $17xy$

32) $8a\times 5a$ is equal to $40a$

33) $9x\times 5x$ is equal to $45x^2$

34) $(-5x)\times(-8x)\times 3x$ is equal to $120x^3$

35) a^2b is the same as ba^2

36) $5x^3y^2z$ is the same as $5y^2zx^3$

37) $8a^3b^2c^4$ and $16a^3b^3c^3$ are like terms

38) $6x^2\div(-3x)$ is equal to $3x$

39) $(-5pq^2)\times(-8p^2q)$ is equal to $40p^3q^3$

40) $a^2b^2\times(-a^2b^2)\times 5a^2b^2$ is equal to $5a^2b^2$

41) $3(2x+7)$ is equal to $6x+7$

42) $5(3x+4)$ is equal to $15x+20$

43) $4(x+8)$ is equal to $4x+32$

44) $-(3x+5y)$ is equal to $-3x+5y$

45) $-(2a+3b)$ is equal to $-2a-3b$

46) $4x(3x-2xy)$ is equal to $12x^2-8x^2y$

47) $-8a(a-3b)$ is equal to $-8a^2-24ab$

48) $3(x-y)-5(2x-3y)$ is equal to $12y-7x$

49) $2x(x-2)-3x(x^2-5)$ is equal to $-3x^3+2x^2-19x$

50) $3a(2a^2+3a-1)-2a(3a^2+3)$ is equal to $9a^2+3a$

51) The HCF of a^2bc^3 and ab^2 is ab

52) The HCF of x^2y^3z and x^2yz^2 is $x^2y^3z^2$

53) The HCF of $a^3b^2c^2$, $a^2b^3c^3$ and ab^4c^4 is ab^2c^2

54) The factors of $a^2x^3+bx^2$ are $x^2(a^2x+b)$

55) The factors of $3a^3y+6a^2x+9a^4z$ are $3a(a^2y+2ax+3a^3z)$

10. LINEAR EQUATIONS

After reaching the end of this chapter you should be able to:

1. *Distinguish between an algebraic expression, an equation, and an identity.*
2. *Maintain the equality of a given equation whilst applying arithmetical operations.*
3. *Solve linear equations in one unknown.*
4. *Construct and solve simple equations deriv[ed] from associated relevant technologies.*

INTRODUCTION

An arithmetical quantity has a definite value, such as 93, 3.73 or $\frac{3}{4}$. An algebraic quantity, however, given by algebraic expressions such as $x-3$ or x^2, represents many amounts depending on the value given to x.

In algebra there are two methods we use to show that two quantities are equivalent to each other. One is called an *identity* and the other an *equation*.

IDENTITIES

A statement of the type $x^2 \equiv x \times x$ is called an *identity*.

The sign $\equiv$ means 'is identical to'. Any statement using this sign is true for *all* values of the variable, the variable in this case being x.

Thus when $x = 2$ we have $2^2 \equiv 2 \times 2$

and when $x = 3$ we have $3^2 \equiv 3 \times 3$ and so on.

Another type of identity involves units as, for example, the relationship between kilometres and metres. This may be stated as

$$x\text{ km} \equiv 1000x\text{ m}$$

Thus $7\text{ km} \equiv 7000\text{ m}$

and $9\text{ km} \equiv 9000\text{ m}$ and so on.

In practice the $\equiv$ sign is often replaced by the = (equals) sign and the above identities would be stated as

$$x^2 = x \times x$$

and

$$x \text{ km} = 1000x \text{ m}$$

EQUATIONS

A statement of the type $x - 3 = 5$ is called an *equation.*

This means that the quantity on the left-hand side of the equation is equal to the quantity on the right-hand side. We can see that, unlike an identity, there is only one value of x that will 'satisfy' the equation, or make the left-hand side equal to the right-hand side. The process of finding $x = 8$ is called 'solving' the equation, and the value 8 is known as the 'solution' or 'root' of the equation.

SOLVING LINEAR EQUATIONS

Linear equations contain only the first power of the unknown quantity.

Thus

$$7t - 5 = 4t + 7 \quad \text{and} \quad \frac{5x}{3} = \frac{2x + 5}{2}$$

are both examples of linear equations.

In the process of solving an equation the appearance of the equation may be considerably altered but the values on both sides must remain the same. We must maintain this equality, and hence whatever we do to one side of the equation we must do exactly the same to the other side.

After an equation is solved, the solution should be checked by substituting the result in each side of the equation separately. If each side of the equation then has the same value the solution is correct.

In the detail which follows, LHS means left-hand side and RHS means right-hand side.

Equations Requiring Multiplication and Division

EXAMPLE 10.1

a) Solve the equation $\frac{x}{6} = 3$

Multiplying each side by 6, we get:

$$\frac{x}{6} \times 6 = 3 \times 6$$

$$x = 18$$

Check: when $x = 18$, LHS $= \frac{18}{6} = 3$, RHS $= 3$

Hence the solution is correct.

b) Solve the equation $5x = 10$

Dividing each side by 5, we get:

$$\frac{5x}{5} = \frac{10}{5}$$

$$x = 2$$

Check: when $x = 2$, LHS $= 5 \times 2 = 10$, RHS $= 10$

Hence the solution is correct.

Equations Requiring Addition and Subtraction

EXAMPLE 10.2

a) Solve $x - 4 = 8$

If we add 4 to each side, we get:

$$x - 4 + 4 = 8 + 4$$

$$x = 12$$

The operation of adding 4 to each side is the same as transferring -4 to the RHS but in so doing the sign is changed from a minus to a plus.

Thus
$$x-4 = 8$$
$$x = 8+4$$
$$x = 12$$

Check: when $x = 12$, LHS $= 12-4 = 8$, RHS $= 8$

Hence the solution is correct.

b) Solve $x+5 = 20$

If we subtract 5 from each side, we get:
$$x+5-5 = 20-5$$
$$x = 15$$

Alternatively moving $+5$ to the RHS
$$x = 20-5$$
$$x = 15$$

Check: when $x = 15$, LHS $= 15+5 = 20$, RHS $= 20$

Therefore, the solution is correct.

Equations Containing the Unknown Quantity on Both Sides

In equations of this kind, group all the terms containing the unknown quantity on one side of the equation and the remaining terms on the other side.

EXAMPLE 10.3

a) Solve $7x+3 = 5x+17$

Transferring $5x$ to the LHS and $+3$ to the RHS
$$7x-5x = 17-3$$
$$2x = 14$$
$$x = \frac{14}{2}$$
$$x = 7$$

Check: when $x = 7$, LHS $= 7\times7+3 = 52$, RHS $= 5\times7+17 = 52$

Hence the solution is correct.

b) Solve $3x-2=5x+6$

$$3x-5x = 6+2$$
$$-2x = 8$$
$$x = \frac{8}{-2}$$
$$x = -4$$

Check: when $x=-4$, LHS $= 3\times(-4)-2=-14$
RHS $= 5\times(-4)+6=-14$

Hence the solution is correct.

Equations Containing Brackets

When an equation contains brackets remove these first and then solve as shown previously.

EXAMPLE 10.4

a) Solve $2(3x+7)=16$

Removing the bracket gives $6x+14 = 16$

$$6x = 16-14$$
$$6x = 2$$
$$x = \frac{2}{6}$$
$$x = \frac{1}{3}$$

Check when $x=\frac{1}{3}$,

LHS $= 2\times\left(3\times\frac{1}{3}+7\right) = 2\times(1+7) = 2\times 8 = 16$

RHS $= 16$

Hence the solution is correct.

b) Solve $3(x+4)-5(x-1)=19$

Removing the brackets gives

$$3x+12-5x+5 = 19$$

$$-2x + 17 = 19$$
$$-2x = 19 - 17$$
$$-2x = 2$$
$$x = \frac{2}{-2}$$
$$x = -1$$

Check: when $x = -1$,

LHS $= 3 \times (-1 + 4) - 5 \times (-1 - 1) = 3 \times 3 - 5 \times (-2) = 9 + 10$
$= 19$

RHS $= 19$

Hence the solution is correct.

Equations Containing Fractions

When an equation contains fractions, *multiply each term of the equation* by the LCM of the denominators.

EXAMPLE 10.5

a) Solve $\dfrac{x}{4} + \dfrac{3}{5} = \dfrac{3x}{2} - 2$

The LCM of the denominators 2, 4 and 5 is 20. Multiplying each term by 20 gives:

$$\frac{x}{4} \times 20 + \frac{3}{5} \times 20 = \frac{3x}{2} \times 20 - 2 \times 20$$
$$5x + 12 = 30x - 40$$
$$5x - 30x = -40 - 12$$
$$-25x = -52$$
$$x = \frac{-52}{-25}$$
$$\therefore \quad x = \frac{52}{25} = 2.08$$

The solution may be verified by the check method shown in the previous examples.

b) Solve the equation $\frac{x-4}{3} - \frac{2x-1}{2} = 4$

In solving equations of this type remember that the line separating the numerator and denominator acts as a bracket. The LCM of the denominators 3 and 2 is 6. Multiplying *each term* of the equation by 6 gives

$$\frac{x-4}{3} \times 6 - \frac{2x-1}{2} \times 6 = 4 \times 6$$

$$2(x-4) - 3(2x-1) = 24$$

$$2x - 8 - 6x + 3 = 24$$

$$-4x - 5 = 24$$

$$-4x = 24 + 5$$

$$-4x = 29$$

$$x = \frac{29}{-4}$$

$$x = -\frac{29}{4} = -7.25$$

c) Solve the equation $\frac{5}{2x+5} = \frac{4}{x+2}$

The LCM of the denominators is $(2x+5)(x+2)$. Multiplying each term of the equation by this gives:

$$\frac{5}{2x+5} \times (2x+5)(x+2) = \frac{4}{x+2} \times (2x+5)(x+2)$$

$$\therefore \quad 5(x+2) = 4(2x+5)$$

$$5x + 10 = 8x + 20$$

$$5x - 8x = 20 - 10$$

$$-3x = 10$$

$$x = \frac{10}{-3}$$

$$x = -\frac{10}{3} = -3.33$$

Exercise 10.1

Solve the equations:

1) $x + 2 = 7$
2) $t - 4 = 3$
3) $2q = 4$
4) $x - 8 = 12$
5) $q + 5 = 2$
6) $3x = 9$
7) $\frac{y}{2} = 3$
8) $\frac{m}{3} = 4$
9) $2x + 5 = 9$
10) $5x - 3 = 12$
11) $6p - 7 = 17$
12) $3x + 4 = -2$
13) $7x + 12 = 5$
14) $6x - 3x + 2x = 20$
15) $14 - 3x = 8$
16) $5x - 10 = 3x + 2$
17) $6m + 11 = 25 - m$
18) $3x - 22 = 8x + 18$
19) $0.3d = 1.8$
20) $1.2x - 0.8 = 0.8x + 1.2$
21) $2(x + 1) = 8$
22) $5(m - 2) = 15$
23) $3(x - 1) - 4(2x + 3) = 14$
24) $5(x + 2) - 3(x - 5) = 29$
25) $3x = 5(9 - x)$
26) $4(x - 5) = 7 - 5(3 - 2x)$
27) $\frac{x}{5} - \frac{x}{3} = 2$
28) $\frac{x}{3} + \frac{x}{4} + \frac{x}{5} = \frac{5}{6}$
29) $\frac{m}{2} + \frac{m}{3} + 3 = 2 + \frac{m}{6}$
30) $3x + \frac{3}{4} = 2 + \frac{2x}{3}$
31) $\frac{3}{m} = 3$
32) $\frac{5}{x} = 2$
33) $\frac{4}{t} = \frac{2}{3}$
34) $\frac{7}{x} = \frac{5}{3}$
35) $\frac{4}{7}y - \frac{3}{5}y = 2$
36) $\frac{1}{3x} + \frac{1}{4x} = \frac{7}{20}$
37) $\frac{x + 3}{4} - \frac{x - 3}{5} = 2$
38) $\frac{2x}{15} - \frac{x - 6}{12} - \frac{3x}{20} = \frac{3}{2}$
39) $\frac{2m - 3}{4} = \frac{4 - 5m}{3}$
40) $\frac{3 - y}{4} = \frac{y}{3}$
41) $x - 5 = \frac{3x - 5}{6}$
42) $\frac{x - 2}{x - 3} = 3$

43) $\dfrac{3}{x-2}=\dfrac{4}{x+4}$

44) $\dfrac{3}{x-1}=\dfrac{2}{x-5}$

45) $\dfrac{3}{2x+7}=\dfrac{5}{3(x-2)}$

46) $\dfrac{x}{3}-\dfrac{3x-7}{5}=\dfrac{x-2}{6}$

47) $\dfrac{4p-1}{3}-\dfrac{3p-1}{2}=\dfrac{5-2p}{4}$

48) $\dfrac{3m-5}{4}-\dfrac{9-2m}{3}=0$

49) $\dfrac{x}{3}-\dfrac{2x-5}{2}=0$

50) $\dfrac{4x-5}{2}-\dfrac{2x-1}{6}=x$

MAKING EXPRESSIONS

It is important to be able to translate information into symbols thus making up algebraic expressions. The following examples illustrate how this is done.

EXAMPLE 10.6

a) Find an expression which will give the total mass of a box containing x articles if the box has a mass of 7 kg and each article has a mass of 1.5 kg.

The total mass of x articles is $1.5x$

Therefore total mass of the box of articles is $1.5x + 7$

b) If the price of an article is reduced from x pence to y pence make an expression giving the number of extra articles that can be bought for 80 pence.

At x pence each the number of articles that can be bought for 80 pence is $\dfrac{80}{x}$

At y pence each the number of articles that can be bought for 80 pence is $\dfrac{80}{y}$

The extra articles that can be bought is $\dfrac{80}{y}-\dfrac{80}{x}$

c) If x nails can be bought for 6 pence write down the cost of y nails.

If x nails cost 6 pence

Then 1 nail costs $\frac{6}{x}$ pence

Hence y nails cost $\frac{6}{x} \times y = \frac{6y}{x}$ pence.

Exercise 10.2

1) A boy is x years old now. How old was he 5 years ago?

2) Find the total cost of 3 pencils at a pence each and 8 pens at b pence each.

3) A man works x hours per weekday except Saturday when he works y hours. If he works z hours on Sunday how many hours does he work per week?

4) What is the perimeter of a rectangle l mm long and b mm wide?

5) A man A has £a and a man B has £b. If A gives B £x how much will each have?

6) A factory employs M men, N boys and P women. If a man earns £x per week, a boy £y per week and a woman £z per week what is the total wage bill per week?

7) A man earns £u per week when he is working and he is paid £v per week when he is on holiday. If he is on holiday for 3 weeks per year find his total annual salary.

8) The price of m articles was £M but the price of each article is increased by n pence. How many articles can be bought for £N?

9) A man starts a job at a salary of £u per week. His salary is increased by y pence per week at the end of each year's service. What will be his salary after x years?

10) A number m is divided into two parts. If a is one part what is the product of the two parts?

CONSTRUCTION OF SIMPLE EQUATIONS

It often happens that we are confronted with mathematical problems that are difficult or impossible to solve by arithmetical methods. We then represent the quantity that has to be found by a symbol. Then by constructing an equation which conforms to the data of the problem we can solve it to give us the value of the unknown quantity. It is stressed that both sides of the equation must be in the same units.

EXAMPLE 10.7

a) The perimeter of a rectangle is 56 m. If one of the two adjacent sides is 4 m longer than the other, find the dimensions of the rectangle.

$$\text{Let } x \text{ m} = \text{length of the shorter side.}$$

$$\text{Then } (x+4)\text{ m} = \text{length of the longer side.}$$

$$\text{Total perimeter} = x + x + (x+4) + (x+4) = (4x+8)\text{ m}$$

But

$$\text{the total perimeter} = 56\text{ m}$$

Hence

$$4x + 8 = 56$$

$$4x = 56 - 8$$

$$4x = 48$$

$$x = 12$$

Hence the shorter side is 12 m and the longer side is $12+4 = 16$ m long.

b) A certain type of lathe costs five times as much as a certain make of drilling machine. If two such lathes and five such drilling machines cost £7500, find the cost of each machine.

Let the cost of a drilling machine be £x

Then the cost of a lathe $=$ £$5x$

$$\text{Cost of 2 lathes and 5 drilling machines} = £(2 \times 5x + 5 \times x)$$

$$= £(10x + 5x) = £15x$$

Since the cost of 2 lathes and 5 drilling machines is £7500

$$15x = 7500$$

$$x = 500$$

Hence the cost of a drilling machine is £500

and the cost of a lathe is $5 \times$ £500 $=$ £2500

Exercise 10.3

1) A foreman and 5 men together earn £732 per week. If the foreman earns £12 per week more than each of the men, how much does each earn?

2) One side of a triangle is 4 m shorter, and another 3 m shorter than the longest side. If the perimeter of the triangle is 25 m, find the lengths of the three sides.

3) The three angles of a triangle are x°, $(x+30)^\circ$ and $(x-6)^\circ$. The sum of the three angles is 180°; find each angle.

4) The perimeter of a rectangle is 56 mm. If one of the two adjacent sides is 8 mm longer than the other, find the dimensions of the rectangle.

5) A certain type of motor car costs seven times as much as a certain make of motor cycle. If two such motor cars and three such motor cycles cost £8500, find the cost of each vehicle.

6) The perimeter of a triangle ABC is 26 m. BC is two-thirds of AB and it is also 2 m longer than AC. Find the lengths of the three sides.

7) A transformer compound of rectangular plan is to be constructed using 100 m of chain link fencing for one long and two short sides and an existing wall for the other side. If the length of the compound has to be 10 m longer than twice the breadth, find the dimensions of the compound.

8) Three plugs and five electric light fittings together cost £9.90. If a plug costs twice as much as a light fitting, find the cost of a plug and a light fitting.

9) Two taps are used to fill a cooling tank which has a capacity of 1200 litres. If it takes 16 min to fill the tank and one tap delivers water at twice the rate of the other, find how many litres per minute each tap delivers.

10) A house is fitted with 3 electric radiators and 5 convector heaters at a total cost of £740. If a convector heater costs £20 more than a radiator, find the cost of each.

SUMMARY

a) To solve an equation the same operation must be performed on both sides. Thus the same amount can be added or subtracted from each side or both sides can be multiplied or divided by the same amount.

b) After an equation has been solved the solution may be checked by substituting the result into the original equation. If each side of the equation has the same value the solution is correct.

c) To construct a simple equation the quantity to be found is represented by a symbol. Then using the given data the equation is formed. Note that both sides of the equation must be in the same units.

Self-Test 7

In questions 1 to 25 the answer is either 'true' or 'false'. State which.

1) If $\frac{x}{7} = 3$ then $x = 21$

2) If $\frac{x}{5} = 10$ then $x = 2$

3) If $\frac{x}{4} = 16$ then $x = 64$

4) If $5x = 20$ then $x = 4$

5) If $3x = 6$ then $x = 18$

6) If $x - 5 = 10$ then $x = 5$

7) If $x + 8 = 16$ then $x = 2$

8) If $x + 7 = 14$ then $x = 21$

9) If $x + 3 = 6$ then $x = 3$

10) If $x - 7 = 14$ then $x = 21$

11) If $3x + 5 = 2x + 10$ then $x = 3$

12) If $2x + 4 = x + 8$ then $x = 4$

13) If $5x - 2 = 3x - 8$ then $x = -6$

14) If $3x - 8 = 2 - 2x$ then $x = 10$

15) If $2(3x + 5) = 18$ then $x = 1$

16) If $2(x+4)-5(x-7)=7$ then $x=12$

17) If $6y=10(8-y)$ then $y=15$

18) If $\frac{5}{y}=10$ then $y=2$

19) If $\frac{8}{y}=4$ then $y=2$

20) If $\frac{x}{3}+\frac{x}{4}=\frac{2x}{5}-11$ then $x=60$

21) If $\frac{x}{2}-1=\frac{x}{3}-\frac{1}{2}$ then $x=3$

22) If $\frac{3}{x+5}=\frac{4}{x-2}$ then $x=26$

23) If $\frac{3}{x-6}=\frac{2}{x-4}$ then $x=-2$

24) If $\frac{x-4}{2}-\frac{x-3}{3}=4$ then $x=10$

25) If $\frac{2x-3}{2}-\frac{x-6}{5}=3$ then $x=3$

In questions 26 to 36 state the letter (or letters) corresponding to the correct answer (or answers).

26) If $3(2x-5)-2(x-3)=3$ then x is equal to:

a 3 b 6 c $\frac{5}{4}$ d $\frac{11}{4}$

27) If $2(x+6)-3(x-4)=1$ then x is equal to:

a 25 b 17 c 23 d -7

28) If $\frac{x-5}{3}=\frac{x+2}{2}$ then x is equal to:

a 16 b -16 c 7 d -7

29) If $3(x-2)-5(x-7)=12$ then x is equal to:

a $-8\frac{1}{2}$ b $8\frac{1}{2}$ c -7 d 0

30) If $\frac{3-2y}{4}=\frac{2y}{6}$ then y is equal to:

a $\frac{18}{20}$ b $\frac{9}{10}$ c 3 d -3

31) The cost of electricity is obtained as follows: A fixed charge of £a, rent of a meter £b and a charge of c pence for each unit of electricity supplied. The total cost of using n units of electricity is therefore:

a) £$(a+b+nc)$ b) £$[100(a+b)+nc]$

c) $[100(a+b)+nc]$ pence d) £$\left(a+b+\frac{nc}{100}\right)$

32) At a factory p men earn an average wage of £a, q women earn an average wage of £b and r apprentices earn an average wage of £c. The average wage for all these employees is:

a £$(a+b+c)$ b £$\left(\frac{a}{b}+\frac{b}{q}+\frac{c}{r}\right)$

c £$\frac{(ap+bq+cr)}{a+b+c}$ d £$\left(\frac{ap+bq+cr}{p+q+r}\right)$

33) A dealer ordered N tools from a manufacturer. The manufacturer can produce p tools per day but $x\%$ of these are faulty and unfit for sale. The number of days is takes the manufacturer to complete the order is:

a $\frac{N}{p(100-x)}$ b $\frac{100N}{p(100-x)}$ c $\frac{N}{p(1-x)}$ d $\frac{p(100-x)}{N}$

34) At the beginning of term a student bought x books at a total cost of £22. A few days later he bought three more books for a further expenditure of £4. He found that this purchase had reduced the average cost per book by 20 pence. An equation from which x can be found is:

a $\frac{26}{x}=20$ b $\frac{26}{x+3}=0.20$

c $\frac{22}{x}-\frac{26}{x+3}=0.20$ d $\frac{26}{x+3}-\frac{22}{x}=0.20$

35) The smallest of three consecutive even numbers is m. Twice the square of the larger is greater than the sum of the squares of the other two numbers by 244. Hence:

a $2(m+2)^2=(m+1)^2+m^2+244$
b $2(m+2)^2-(m+1)^2+m^2=244$
c $2(m+4)^2=(m+2)^2+m^2+244$
d $2(m+4)^2-(m+2)^2+m^2=244$

36) A householder can choose to pay for his electricity by one of the two following methods:

a) a basic charge of £5.20 together with a charge of 5.0 pence of each unit of electricity used;

b) a basic charge of £7.20 together with a charge of 4.8 pence for each unit of electricity used.

The number of units N for which the bill would be the same by either method may be found from the equation:

a $5.20 + 5.0N = 7.20 + 4.8N$

b $5.20N + 500 = 7.20N + 480$

c $520 + 5.0N = 720 + 4.8N$

d $520N + 500 = 720N + 480$

11. SIMULTANEOUS LINEAR EQUATIONS

After reaching the end of this chapter you should be able to:

1. *Solve simultaneous linear equations in two unknowns.*
2. *Solve problems involving simultaneous linear equations.*

Consider the equations

$$3x + 2y = 7 \qquad [1]$$

$$4x + y = 6 \qquad [2]$$

The unknown quantities x and y appear in both equations. To solve the equations we have to find values of x and y so that *both* equations are satisfied. Such equations are called *simultaneous equations.*

SOLUTION OF SIMULTANEOUS LINEAR EQUATIONS

The method for solving simultaneous equations is illustrated by the following examples.

EXAMPLE 11.1

a) Solve the equations

$$5x + 3y = 19 \qquad [1]$$

$$3x + 2y = 12 \qquad [2]$$

If we multiply equation [1] by 3 and equation [2] by 5 the coefficient of x will be the same in both equations. Thus,

$$15x + 9y = 57 \qquad [3]$$

$$15x + 10y = 60 \qquad [4]$$

We now eliminate x by subtracting equation [3] from equation [4] which gives:

$$y = 3$$

To find the value of x we substitute for y in either of the original equations. Substituting $y = 3$, in equation [1]

$$5x + 3 \times 3 = 19$$
$$5x + 9 = 19$$
$$5x = 10$$
$$x = 2$$

Therefore the solutions are $y = 3$ and $x = 2$. To check these values substitute them in equation [2]. (There would be no point in substituting them in equation [1] for this was used in finding x.)

$$\text{LHS of equation [2]} = 3 \times 2 + 2 \times 3 = 6 + 6 = 12 = \text{RHS}$$

b) Solve the equations

$$3x + 4y = 29 \qquad [1]$$
$$8x - 2y = 14 \qquad [2]$$

In these equations it is easier to eliminate y because the same coefficient of y can be obtained in both equations simply by multiplying equation [2] by 2. Thus

Multiplying equation [2] by 2 gives

$$16x - 4y = 28 \qquad [3]$$

Adding equations [1] and [3] gives

$$19x = 57$$
$$x = 3$$

Substituting $x = 3$ in equation [1] gives

$$3 \times 3 + 4y = 29$$
$$9 + 4y = 29$$
$$4y = 20$$
$$y = 5$$

Checking in equation [2] shows that

$$\text{LHS} = 8 \times 3 - 2 \times 5 = 14 = \text{RHS}$$

Therefore, the solutions $x = 3$ and $y = 5$ are correct.

Exercise 11.1

Solve the following equations for x and y and check the solutions:

1) $3x + 2y = 7$
   $x + y = 3$

2) $4x - 3y = 1$
   $x + 3y = 19$

3) $x + 3y = 7$
   $2x - 2y = 6$

4) $7x - 4y = 37$
$6x + 3y = 51$

5) $4x - 6y = -2.5$
$7x - 5y = -0.25$

6) $x + y = 17$
$\frac{x}{5} - \frac{y}{7} = 1$

7) $\frac{3x}{2} - 2y = \frac{1}{2}$
$x + \frac{3y}{2} = 6$

8) $2x + \frac{y}{2} = 11$
$\frac{3x}{5} + 3y = 9$

PROBLEMS INVOLVING SIMULTANEOUS EQUATIONS

In problems which involve two unknowns it is necessary to form two separate equations from the given data and then to solve these as shown above.

EXAMPLE 11.2

a) In a certain lifting machine it is found that the effort (E) and the load (W) which is being raised are connected by the equation $E = aW + b$. An effort of 3.7 N raised a load of 10 N whilst an effort of 7.2 N raises a load of 20 N. Find the values of the constants a and b and hence find the effort needed to lift a load of 12 N.

Substituting $E = 3.7$ and $W = 10$ into the given equation we have

$$3.7 = 10a + b \qquad [1]$$

Substituting $E = 7.2$ and $W = 20$ into the given equation we have

$$7.2 = 20a + b \qquad [2]$$

Subtracting equation [1] from equation [2] gives

$$3.5 = 10a$$

$$a = 0.35$$

Substituting for a in equation [1] gives

$$3.7 = 10 \times 0.35 + b$$

$$3.7 = 3.5 + b$$

$$3.7 - 3.5 = b$$

$$b = 0.2$$

The given equation therefore becomes:

$$E = 0.35W + 0.2$$

when

$$W = 12$$

$$E = 0.35 \times 12 + 0.2 = 4.2 + 0.2 = 4.4\text{ N}$$

Hence an effort of 4.4 N is needed to raise a load of 12 N.

b) A heating installation for one house consists of 5 radiators and 4 convector heaters and the cost of the installation is £540. In a second house 6 radiators and 7 convector heaters are used the cost of this installation being £804. In each house the installation costs are £100. Find the cost of a radiator and the cost of a convector heater.

For the first house the cost of the heaters is

$$£540 - £100 = £440$$

For the second house the cost of the heaters is

$$£804 - £100 = £704$$

Let £x be the cost of a radiator and £y be the cost of a convector heater.

For the first house, $5x + 4y = 440$ [1]

For the second house, $6x + 7y = 704$ [2]

Multiplying [1] by 6 gives

$$30x + 24y = 2640 \quad [3]$$

Multiplying [2] by 5 gives

$$30x + 35y = 3520 \quad [4]$$

Subtracting equation [3] from equation [4] gives

$$11y = 880$$

$$y = 80$$

Substituting for y in equation [1] gives

$$5x + 4 \times 80 = 440$$

$$5x = 120$$

$$x = 24$$

Therefore the cost of a radiator is £24 and the cost of a convector heater is £80.

Exercise 11.2

1) Two quantities M and N are connected by the formula $M = aN + b$ in which a and b are constants. When $N = 3, M = 6.5$ and when $N = 7$, $M = 12.5$. Find the values of a and b.

2) In an experiment to find the friction force F between two metallic surfaces when the load is W, the law connecting the two quantities was of the type $F = mW + b$. When $F = 2.5$, $W = 6$ and when $F = 3.1$, $W = 9$. Find the values of m and b. Hence find the value of F when $W = 12$.

3) A foreman and seven men together earn £1040 per week whilst two foremen and 17 men together earn £2464 per week. Find the earnings for a foreman and for a man.

4) For one installation 8 ceiling roses and 6 plugs are required the total cost of these items being £16. For a second installation 12 ceiling roses and 5 plugs are used the cost being £17.60. Find the cost of a ceiling rose and a plug.

5) Two quantities x and y are connected by the law $y = ax + b$. When $x = 2$, $y = 13$ and when $x = 5, y = 22$. Find the values of a and b.

6) In a certain lifting machine it is found that the effort, E, and the load, W are connected by the equation $E = aW + b$. An effort of 2.6 raises a load of 8, whilst an effort of 3.8 raises a load of 12. Find the values of the constants a and b and determine the effort required to raise a load of 15.

7) If 100 m of wire and 8 plugs cost £62 and 150 m of wire and 10 plugs cost £90 find the cost of 1 m of wire and the cost of a plug.

8) Find two numbers such that their sum is 27 and their difference is 3.

9) An alloy containing 8 cm^3 of copper and 7 cm^3 of tin has a mass of 121 g. A second alloy containing 9 cm^3 of copper and 11 cm^3 of tin has a mass of 158 g. Find the densities of copper and tin in g/cm^3.

10) A motorist travels x km at 40 km/h and y km at 50 km/h. The total time taken is $2\frac{1}{2}$ hours. If the time taken to travel $6x$ km at 30 km/h and $4y$ km at 50 km/h is 14 hours find x and y.

Self-Test 8

In the following questions state the letter (or letters) which correspond to the correct answer (or answers).

1) In the simultaneous equations:

$$\begin{aligned} 2x + 3y &= 17 \\ 3x + 4y &= 24 \end{aligned}$$

x is equal to 4. Hence the value of y is:

a $\frac{25}{3}$ b 75 c 3 d 4

2) In the simultaneous equations:

$$\begin{aligned} 2x - 3y &= -16 \\ 5y - 3x &= 25 \end{aligned}$$

x is equal to -5. Hence the value of y is:

a 2 b -2 c $\frac{26}{3}$ d 0

3) By eliminating x from the simultaneous equations:

$$\begin{aligned} 2x - 5y &= 8 \\ 2x - 3y &= -7 \end{aligned}$$

the equation below is obtained:

a $-8y = 1$ b $-2y = 15$ c $-8y = 15$ d $-2y = 1$

4) By eliminating y from the simultaneous equations:

$$\begin{aligned} 3x - 4y &= -10 \\ x + 4y &= 8 \end{aligned}$$

the equation below is obtained:

a $2x = -18$ b $4x = -18$ c $2x = -2$ d $4x = -2$

5) By eliminating x from the simultaneous equations:

$$\begin{aligned} 3x + 5y &= 2 \\ x + 3y &= 7 \end{aligned}$$

the equation below is obtained:

a $4y = -19$ b $8y = 9$ c $4y = 19$ d $4y = 5$

6) By eliminating y from the simultaneous equations:

$$2x - 4y = 3$$
$$3x + 8y = 7$$

the equation below is obtained:

a $7x = 13$ b $x = -1$ c $x = 1$ d $7x = 10$

7) The solutions to the simultaneous equations:

$$2x - 5y = 3$$
$$x - 3y = 1$$

are:

a $x = 4, y = 1$ b $y = 4, x = 1$
c $y = 4, x = 13$ d $x = 4, y = 3$

8) The solutions to the simultaneous equations:

$$3x - 2y = 5$$
$$4x - y = 10$$

are:

a $x = -3, y = 22$ b $x = -3, y = -22$
c $x = 3, y = 2$ d $x = 3, y = -2$

9) Two numbers, x and y are such that their sum is 18 and their difference is 12. The equations below will allow x and y to be found:

a $x + y = 18$, $y - x = 12$ b $x + y = 18$, $x - y = 12$
c $x - y = 18$, $x + y = 12$ $y - x = 18$, $x + y = 12$

10) A motorist travels x km at 50 km/h and y km at 60 km/h. The total time taken is 5 hours. If his average speed is 56 km/h then:

a $50x + 60y = 5$, $x + y = 280$ b $6x + 5y = 1500$, $x + y = 280$

c $\dfrac{x}{50} + \dfrac{y}{60} = 5$, $\dfrac{x+y}{5} = 56$ d $50x + 60y = 5$, $x + y = 280$

12. FORMULAE

fter reaching the end of this chapter you should be able to:

. Evaluate, by substitution of given data, simple formulae.

2. Transpose formulae.

EVALUATING FORMULAE

The statement $F = ma$ is described as a formula for F in terms of m and a. The value of F may be found by arithmetic after substituting the given values of m and a.

EXAMPLE 12.1

a) The formula $E = IR$ is used in electrical calculations. Find the value of E if $I = 6$ and $R = 4$

$$E = IR = 6 \times 4 = 24$$

b) The formula $P = \dfrac{Fs}{t}$ is used in mechanical engineering. Find the value of P when $F = 40$, $s = 15$ and $t = 3$

When $F = 40$, $s = 15$ and $t = 3$, then $P = \dfrac{40 \times 15}{3} = 200$

Exercise 12.1

1) The circumference of a circle is given by $C = \pi d$. Find C when $\pi = 3.14$ and $d = 7$

2) The current in an electrical circuit is found from the formula $I = \dfrac{V}{R}$. Find I when $V = 240$ and $R = 30$

3) The curved surface area of a cone is calculated from $A = \pi rl$. Find A when $\pi = 3.14$, $r = 2$ and $l = 9$

4) A formula used in automobile engineering is $E = \frac{(mv)^2}{2g}$. Find E when $m = 120$, $v = 25$ and $g = 10$

5) A formula used in geometry is $S = 90(2n - 4)$. Find S when $n = 5$

6) When electrical resistances are wired in parallel the formula from which total resistance is found is $\frac{1}{R} = \frac{1}{R_1} + \frac{1}{R_2}$. Find R when $R_1 = 5$ and $R_2 = 20$

7) The amount a metal bar expands is found by using the formula $l = l_0(1 + \alpha t)$. Find l when $l_0 = 30$ and $\alpha = 12 \times 10^{-6}$, t being 40

8) The formula $P = 2\pi NT$ is used in power calculations. Find P when $N = 200$, $T = 14$ and $\pi = \frac{22}{7}$

FORMULAE GIVING RISE TO SIMPLE EQUATIONS

In the formula $E = V + Inr$, E is called the subject of the formula. It may be that we are given values of E, V, n and r and we have to find the value of I. We can do this by forming a simple equation as shown in Example 12.2.

EXAMPLE 12.2

Find I from the formula $E = V + Inr$ when $E = 12$, $V = 10$, $r = 0.1$ and $n = 5$

Substituting the given values in the equation gives

$$
\begin{aligned}
12 &= 10 + I \times 5 \times 0.1 \\
12 &= 10 + 0.5I \\
0.5I &= 12 - 10 \\
0.5I &= 2 \\
I &= \frac{2}{0.5} \\
I &= 4
\end{aligned}
$$

EXAMPLE 12.3

If $C = \dfrac{nE}{R + nr}$, find the value of r when $C = 1.5$, $E = 3$, $n = 7$ and $R = 6.3$

Substituting the given values gives

$$1.5 = \frac{7 \times 3}{6.3 + 7r}$$

Multiplying both sides of the equation by $(6.3 + 7r)$ we have

$$1.5(6.3 + 7r) = 7 \times 3$$

$$\therefore \quad (1.5 \times 6.3) + (1.5 \times 7r) = 7 \times 3$$

$$\therefore \quad 9.45 + 10.5r = 21$$

Subtracting 9.45 from both sides we have

$$10.5r = 21 - 9.45$$

$$\therefore \quad 10.5r = 11.55$$

Dividing both sides by 10.5 gives

$$r = \frac{11.55}{10.5}$$

$$\therefore \quad r = 1.1$$

Exercise 12.2

1) Find F from the formula $P = Fv$ when $P = 300$ and $v = 40$

2) Find I from the formula $R = \dfrac{V}{I}$ when $R = 20$ and $V = 10$

3) Find h from the formula $A = \frac{1}{2}h(a + b)$ when $A = 20$, $a = 6$ and $b = 4$

4) Find h from the formula $v^2 = 2gh$ when $v = 30$ and $g = 10$

5) If $W = \dfrac{mv^2}{2g}$ find the value of m when $W = 100$, $v = 20$ and $g = 10$

6) The formula $v^2 = \dfrac{gar}{h}$ is used in connection with the speed of vehicles on a curved horizontal road. Calculate the value of r when $v = 40$, $a = 3$, $g = 10$ and $h = 1.5$

7) When n cells are connected in series the current is found from the formula $I = \dfrac{nE}{R+nr}$. Find E when $I = 1.5, n = 7, r = 1.1$ and $R = 0.3$

8) When two electrical resistances are wired in parallel the formula to find the total resistance is $\dfrac{1}{R_T} = \dfrac{1}{R_1} + \dfrac{1}{R_2}$. If $R_T = 2$ and $R_1 = 8$ find the value of R_2

TRANSPOSITION OF FORMULAE

Considering again the formula, $E = V + Inr$. We may be given several corresponding values of E, V, n and r and we want to find the corresponding values of I. We can, of course, find these values by the method shown in Example 12.2, but considerable time and effort will be spent in solving the resulting equations. Much of this time and effort would be saved if we could express the formula with I as the subject, because then we need only substitute the given values of E, V, n and r in the resulting formula.

The process of rearranging a formula so that one of the other symbols becomes the subject is called *transposing the formula.* The rules used in transposition are the same as those used in solving equations. The methods used are as follows:

Symbols Connected as a Product

EXAMPLE 12.4

a) Transpose the formula $F = ma$ to make a the subject.

Divide both sides by m. Then,

$$\frac{F}{m} = \frac{ma}{m}$$

or

$$\frac{F}{m} = a$$

or

$$a = \frac{F}{m}$$

b) Make h the subject of the formula $V = \pi r^2 h$.

Divide both sides by πr^2. Then

$$\frac{V}{\pi r^2} = \frac{\pi r^2 h}{\pi r^2}$$

or

$$\frac{V}{\pi r^2} = h$$

or

$$h = \frac{V}{\pi r^2}$$

Symbols Connected as a Quotient

EXAMPLE 12.5

a) Transpose $x = \dfrac{y}{b}$ for y.

Multiply both sides by b. Then

$$x \times b = \frac{y}{b} \times b$$

or

$$bx = y$$

or

$$y = bx$$

b) Transpose $M^3 = \dfrac{3x^2w}{p}$ for p.

Multiply both sides by p. Then

$$M^3 p = 3x^2 w$$

Divide both sides by M^3. Then

$$\frac{M^3 p}{M^3} = \frac{3x^2 w}{M^3}$$

or

$$p = \frac{3x^2 w}{M^3}$$

Symbols Connected by a Plus or Minus Sign

EXAMPLE 12.6

a) Transpose $x = 3y + 5$ for y.

Subtract 5 from both sides of the equation,

$$x - 5 = 3y$$

Divide both sides by 3:

$$\frac{x-5}{3} = y$$

or

$$y = \frac{x-5}{3}$$

b) Transpose $V = E + IR$ for I.

Subtract E from both sides:

$$V - E = IR$$

Divide both sides by R:

$$\frac{V-E}{R} = I$$

or

$$I = \frac{V-E}{R}$$

Formulae Containing Brackets

EXAMPLE 12.7

a) Transpose $y = a + \frac{x}{b}$ for x.

Subtract a from both sides:

$$y - a = \frac{x}{b}$$

Multiply both sides by b,

$$b(y-a) = x$$

or

$$x = b(y-a)$$

b) Transpose $l = a + (n-1)d$ for n.

Subtract a from both sides:

$$l - a = (n-1)d$$

Divide both sides by d:

$$\frac{l-a}{d} = n - 1$$

Add 1 to each side:

$$\frac{l-a}{d} + 1 = n$$

or

$$n = \frac{l-a}{d} + 1$$

Exercise 12.3

Transpose the following:

1) $C = \pi d$ for d

2) $S = \pi dn$ for d

3) $PV = c$ for V

4) $A = \pi rl$ for l

5) $v^2 = 2gh$ for h

6) $I = PRT$ for R

7) $x = \dfrac{a}{y}$ for y

8) $I = \dfrac{E}{R}$ for R

9) $x = \dfrac{u}{a}$ for u

10) $P = \dfrac{RT}{V}$ for T

11) $d = \dfrac{0.866}{N}$ for N

12) $S = \dfrac{ts}{T}$ for t

13) $\dfrac{M}{I} = \dfrac{E}{R}$ for R

14) $V = \dfrac{\pi d^2 h}{4}$ for h

15) $\dfrac{G\theta}{l} = \dfrac{T}{J}$ for J

16) $v = u + at$ for t

17) $n = p + cr$ for r

18) $y = ax + b$ for x

19) $y = \dfrac{x}{5} + 17$ for x

20) $L = a + (n-1)d$ for n

21) $a = b - cx$ for x

22) $D = B - 1.28d$ for d

23) $V = \dfrac{2R}{R-r}$ for r

24) $C = \dfrac{E}{R+r}$ for E

25) $A = \pi r(r+h)$ for h

26) $C = \dfrac{N-n}{2p}$ for N

27) $T = \dfrac{12(D-d)}{L}$ for d

28) $a = \dfrac{3}{4t+5}$ for t

Self-Test 9

In the following questions state the letter (or letters) corresponding to the correct answer (or answers).

1) If $v = u + at$, $u = 4$, $a = -2$ and $t = 3$, then the value of v is:
a 6 b 10 c 4 d -2 e -5

2) If $v = u + at$, $v = 12$, $u = 6$ and $t = 2$, then the value of a is:
a 3 b 1 c -3 d 4

3) Given the values $u = -2$, $v = 3$ and $s = 5$ and also that $s = \dfrac{(u+v)}{2}t$, then the value of t is:
a 2.5 b 20 c 7 d 10

4) The volume of a cone is given by $V = \dfrac{\pi r^2 h}{3}$. If $V = 25$ and $r = 2$, then the approximate value of h is:
a 6 b 2.2 c 100 d 2.5

5) If $v = u - at$ then t is equal to:
a $\dfrac{v-u}{a}$ b $\dfrac{u-v}{a}$ c $\dfrac{-v}{au}$ d $\dfrac{u}{av}$

6) If $A = 2\pi R(R+H)$ then H is equal to:
a $A - 2\pi R^2$ b $\dfrac{A}{2\pi R} - R$ c $R - \dfrac{A}{2\pi R}$ d $\dfrac{A - 2\pi R^2}{2\pi R}$

7) If $f = md(T-t)$ then t is equal to:
a $\dfrac{f}{md} - T$ b $\dfrac{f - mdT}{md}$ c $T - \dfrac{f}{md}$ d $\dfrac{mdT - f}{md}$

8) If $S = 90(2n - 4)$ then n is equal to:

a $\frac{S}{180} + 4$ b $\frac{S}{180} - 2$ c $\frac{S}{180} + 2$ d $\frac{S + 360}{180}$

9) If $F = \frac{W(v - u)}{gt}$ then u is equal to:

a $v - \frac{WF}{gt}$ b $\frac{Fgt}{W} - v$ c $v - \frac{Fgt}{W}$ d $Fgt - Wv$

10) If $x = 2y - \frac{w}{v}$ then v is equal to:

a $\frac{w}{2y - x}$ b $\frac{2y - x}{w}$ c $\frac{-w}{x - 2y}$ d $\frac{-w}{2y - x}$

13. GRAPHS

After reaching the end of this chapter you should be able to:

1. *Define a right-handed pair of axes.*
2. *Label axes.*
3. *Determine scales from given data and mark axes accordingly.*
4. *Plot a point accurately given its coordinates.*
5. *Use a given straight line graph through the origin to find corresponding pairs of values on the two axes.*
6. *Convert given data into another system of related units which are directly proportional, e.g. imperial to metric.*
7. *Read values of y for given values of x from linear and non-linear graphs.*
8. *Identify dependent and independent variables.*
9. *State the relationship between two variables which are (i) directly proportional, (ii) inversely proportional.*
10. *Calculate the coefficient of proportionality from given data.*
11. *State that for inverse proportionality t product of the variables is constant.*
12. *Solve problems involving for examp (i) Hooke's Law, (ii) Boyle's Law, (iii) Charl Law, (iv) Ohm's Law, and determine t coefficients of proportionality.*
13. *Plot three points from coordinates det mined from an equation of the fo $y = mx + c$, where m and c are given nume cal values and draw a straight line through t points.*
14. *Determine the intercept on the y-axis a relate it to the c-value.*
15. *Determine the gradient of the straight li and relate it to the m-value.*
16. *Distinguish between lines having positi negative and zero gradients.*

INTRODUCTION

In newspapers, business and technical reports, and government publications, use is made of pictorial illustrations to compare quantities of the same kind. These diagrams help the reader to understand what deductions can be drawn from the quantities represented in the diagrams. The most common form of diagram is the graph.

AXES OF REFERENCE

To plot a graph we take two lines at right angles to each other (Fig. 13.1). These lines are called the axes of reference. Their intersection, the point O is called the origin.

In plotting a graph we may have to include values which are positive and negative. To represent these on a graph we make use of the number scales used in directed numbers. When the axes are numbered in the way shown in Fig. 13.1, we have a right-handed pair of axes. The horizontal axis is often called the x-axis and the vertical axis the y-axis.

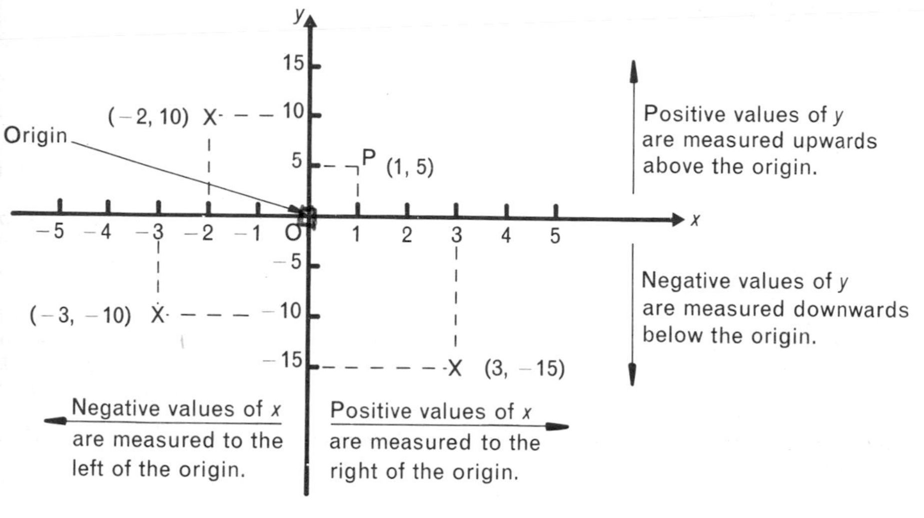

Fig. 13.1

COORDINATES

Coordinates are used to mark the points on a graph. In Fig. 13.1 the point P has been plotted so that $x = 1$ and $y = 5$. The values 1 and 5 are said to be the rectangular coordinates of P. For brevity we say that P is the point $(1, 5)$.

In plotting graphs we may have to include coordinates which are positive and negative. To represent these on a graph we make use of the number scales used in directed numbers. As well as the point $(1, 5)$ the points $(3, -15)$, $(-2, 10)$ and $(-3, -10)$ are plotted in Fig. 13.1.

AXES AND SCALES

The location of the axes and the scales along each axis should be chosen so that all the points may be plotted with the greatest possible accuracy. The scales should be as large as possible but they must be chosen so that they are easy to read. The most useful scales are 1, 2 and 5 units to 1 large square on the graph paper.

Some multiples of these such as 10, 20, 100 units etc. per large square are also suitable. Note that the scales chosen need not be the same on both axes.

EXAMPLE 13.1

If the rate of exchange between the £ and the franc is £1 = 8 francs construct a graph to show the value of the franc in £'s and find from the graph the value of 60 francs in £'s, and £5 in francs.

The first step is to draw up a table of corresponding values as follows:

£	0	2	4	6	8	10	12
Francs	0	16	32	48	64	80	96

The graph is shown plotted in Fig. 13.2 and it will be seen to be a straight line. If one quantity is directly proportional to another (as they are in this case) a straight line always results passing through the origin.

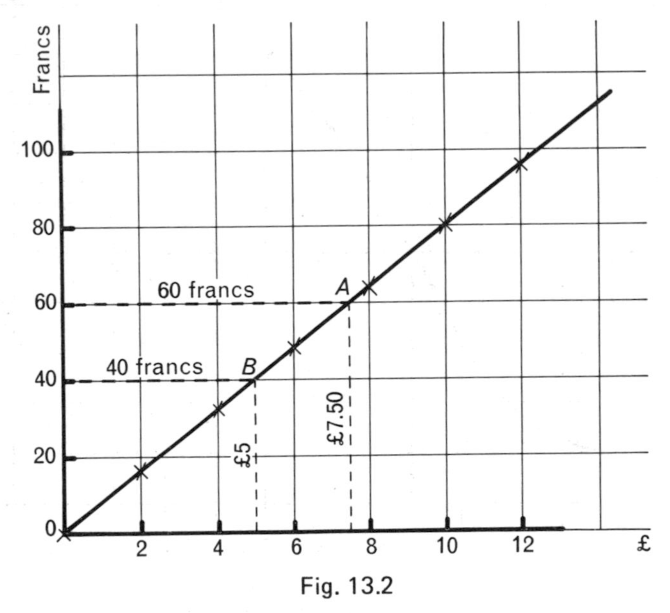

Fig. 13.2

EXAMPLE 13.2

The table below gives corresponding values of x and y. Plot this information and from the graph find:

a) the value of y when $x = -3$,

b) the value of x when $y = 2$

x	-4	-2	0	2	4	6
y	-2.0	-1.6	0	1.4	2.5	3.0

The graph is shown plotted in Fig. 13.3 and it is a smooth curve. This means that there is a definite law (or equation) connecting x and y. We can therefore use the graph to find corresponding values of x and y between those given in the original table of values. By using the constructions shown in Fig. 13.3:

a) the value of y is -1.9 when $x = -3$,

b) the value of x is 3 when $y = 2$

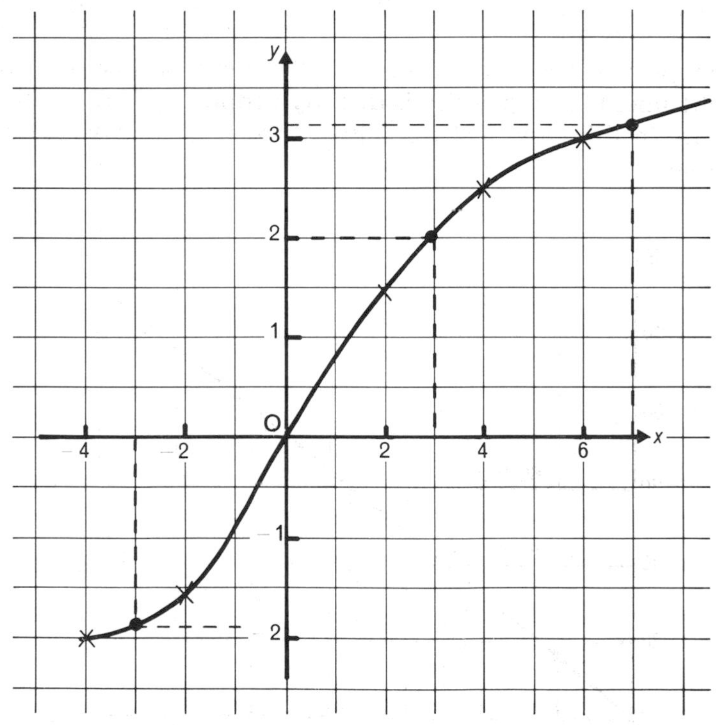

Fig. 13.3

Using a graph in this way to find values of x and y not given in the original table of values is called *interpolation*. If we extend the curve so that it follows the general trend we can estimate corresponding values of x and y which lie *just beyond* the range of the given values. Thus in Fig. 13.3 by extending the curve we can find the probable value of y when $x = 7$. This is found to be 3.2.

Finding a probable value in this way is called *extrapolation*. An extrapolated value can usually be relied upon but in some cases it may contain a substantial amount of error. Extrapolated values must therefore be used with care.

It must be clearly understood that interpolation and extrapolation can only be used if the graph is a straight line or a smooth curve.

EXAMPLE 13.3

Corresponding values of x and y are shown in the table below.

x	0	10	20	30	40	50
y	20.0	22.0	23.5	24.4	25.0	25.4

Illustrate this relationship on a graph.

Looking at the range of values for y, we see that they range from 20.0 to 25.4. We can therefore make 20.0 the starting point on the vertical axis as shown in Fig. 13.4. By doing this a larger scale may be used on the y-axis thus resulting in a more accurate graph. The graph is again a smooth curve and hence there is a definite equation connecting x and y.

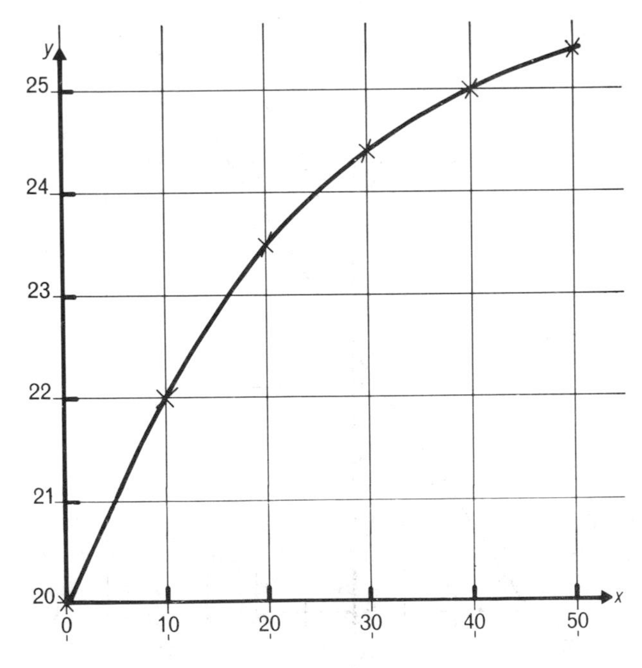

Fig. 13.4

GRAPHS OF SIMPLE EQUATIONS

Consider the equation:

$$y = 2x + 5$$

We can give x any value we please and so calculate a corresponding value for y. Thus,

when $x = 0$	$y = 2 \times 0 + 5 = 5$	
when $x = 1$	$y = 2 \times 1 + 5 = 7$	
when $x = 2$	$y = 2 \times 2 + 5 = 9$	and so on.

The value of y therefore depends on the value allocated to x. We therefore call y the *dependent variable.* Since we can give x any value we please, we call x the *independent variable.* It is usual to mark the values of the independent variable along the horizontal x-axis and the values of the dependent variable are then marked off along the vertical y-axis.

EXAMPLE 13.4

Draw the graph of $y = 2x - 5$ for values of x between -3 and 4

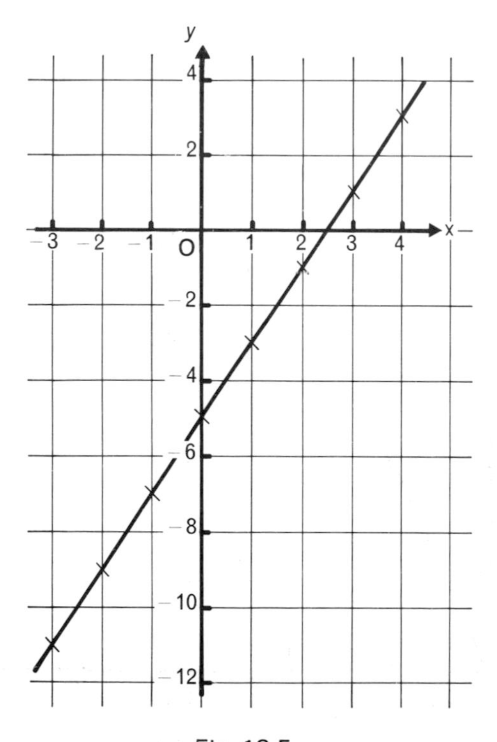

Fig. 13.5

Having decided on some values for x we calculate the corresponding values for y by substituting in the given equation. Thus,

$$\text{when } x = -3, \; y = 2\times(-3)-5 = -6-5 = -11$$

For convenience the calculations are tabulated as shown below.

x	-3	-2	-1	0	1	2	3	4
$2x$	-6	-4	-2	0	2	4	6	8
-5	-5	-5	-5	-5	-5	-5	-5	-5
$y = 2x-5$	-11	-9	-7	-5	-3	-1	1	3

A graph may now be plotted using these values of x and y (Fig. 13.5). The graph is a straight line. Equations of the type $y = 2x-5$, where the highest powers of the variables, x and y, is the first are called equations of the *first degree*. All equations of this type give graphs which are straight lines and hence they are often called *linear equations.* In order to draw graphs of linear equations we need only take two points. It is safer, however, to take three points, the third point acting as a check on the other two.

EXAMPLE 13.5

By means of a graph show the relationship between x and y in the equation $y = 5x+3$. Plot the graph between $x = -3$ and $x = 3$

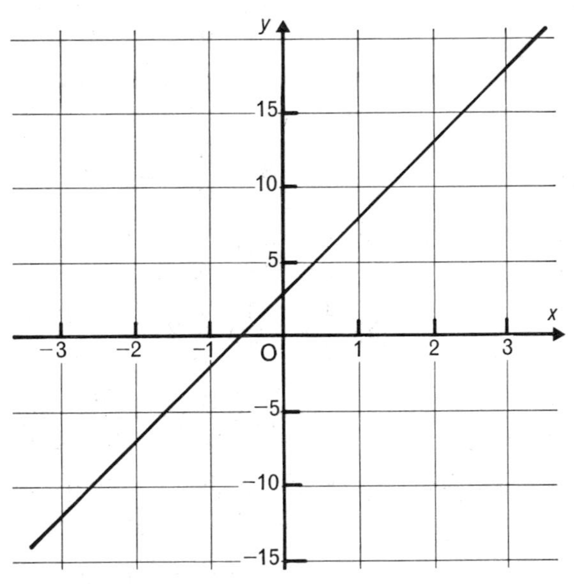

Fig. 13.6

Since this is a linear equation we need only take three points.

x	-3	0	$+3$
$y = 5x + 3$	-12	3	$+18$

The graph is shown in Fig. 13.6.

THE LAW OF A STRAIGHT LINE

In Fig. 13.7, the point B is any point on the line shown and has coordinates x and y. Point A is where the line cuts the y-axis and has coordinates $x = 0$ and $y = c$.

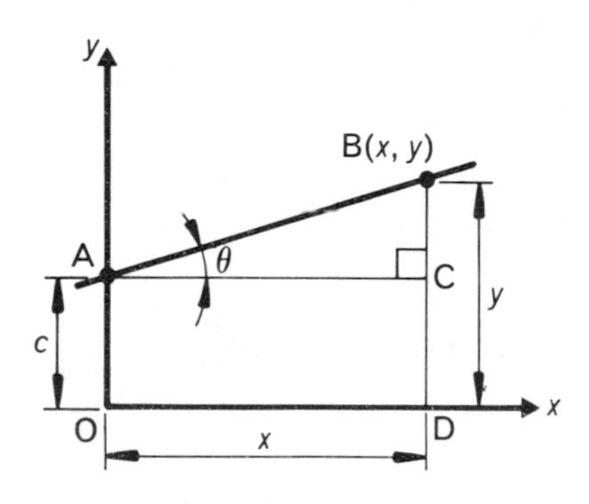

Fig. 13.7

In ΔABC $\qquad \dfrac{BC}{AC} = \tan\theta$

$\therefore \qquad BC = (\tan\theta).AC$

but also $\qquad y = BC + CD = (\tan\theta).AC + CD$

$\therefore$

$$\boxed{y = mx + c}$$

This equation is called the law of a straight line

m is called the *gradient of the line.*

c is called the *intercept on the y-axis.* Care must be taken as this only applies if the origin (i.e. the point (0, 0)) is at the intersection of the axes.

In mathematics the gradient of a line is defined as the tangent of the angle that the line makes with the horizontal, and is denoted by the letter m.

Hence in Fig. 13.7 the gradient $= m = \tan\theta = \dfrac{\text{BC}}{\text{AC}}$

(Care should be taken not to confuse this with the gradient given on maps, railways, etc. which is the sine of the angle (not the tangent) — e.g. a railway slope of 1 in 100 is one unit vertically for every 100 units measured along the slope.)

Fig. 13.8 shows the difference between positive and negative gradients.

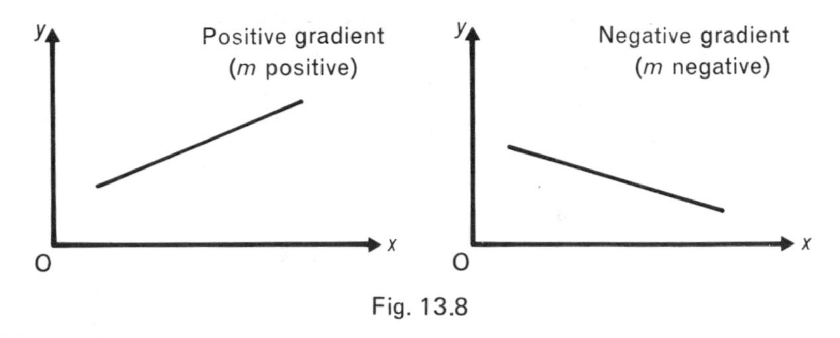

Fig. 13.8

Summarising:

The standard equation, or law, of a straight line is $y = mx + c$

where m is the gradient

and c is the intercept on the y-axis,

providing the origin is at the intersection of the axes.

OBTAINING THE STRAIGHT LINE LAW OF A GRAPH

When it is convenient to arrange the origin, i.e. the point $(0, 0)$, at the intersection of the axes the values of gradient m and intercept c may be found directly from the graph as shown in Example 13.6.

EXAMPLE 13.6

Find the law of the straight line shown in Fig. 13.9.

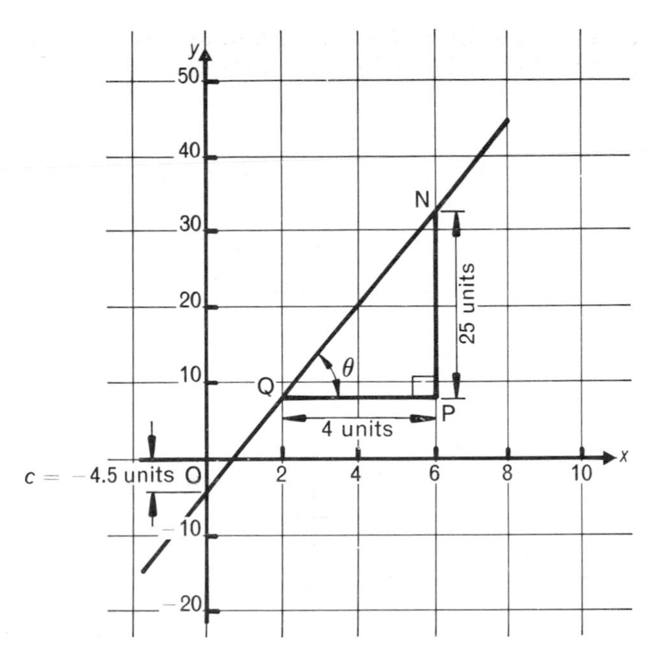

Fig. 13.9

To find gradient *m*. Take any two points Q and N on the line and construct the right-angled triangle QPN. This triangle should be of reasonable size, since a small triangle will probably give an inaccurate result. Note that if we can measure to an accuracy of 1 mm using an ordinary rule, then this error in a length of 20 mm is much more significant than the same error in a length of 50 mm.

The lengths of NP and QP are then found using the scales of the x and y axes. Direct lengths of these lines as would be obtained using an ordinary rule, e.g. both in millimetres, must *not* be used —the scales of the axes must be taken into account.

$$\therefore \qquad \text{Gradient } m = \tan\theta = \frac{\text{NP}}{\text{QP}} = \frac{25}{4} = 6.25$$

To find intercept *c*. This is measured again using the scale of the y-axis.

$\therefore$ intercept $\qquad c = -4.5$

The law of the straight line.

The standard equation is $\qquad y = mx + c$

$\therefore$ the required equation is $\quad y = 6.25x + (-4.5)$

i.e. $\qquad y = 6.25x - 4.5$

When the origin is not at the intersection of the axes the 'two point' method shown in Examples 13.7 and 13.8 may be used.

EXAMPLE 13.7

Find the law of the straight line shown in Fig. 13.10.

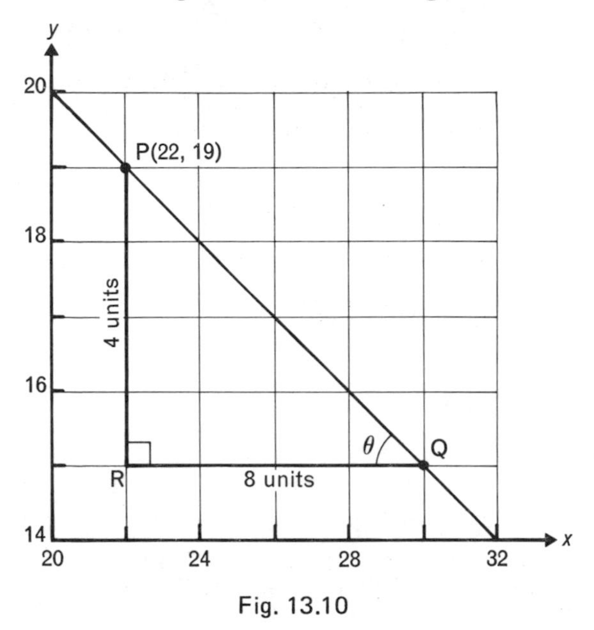

Fig. 13.10

Two things should be noted about the graph in Fig. 13.10:

(a) the gradient is negative because the straight line slopes downward to the right,

(b) the origin is not at the intersection of the x- and y-axes.

To find the gradient m we proceed in a similar way to Example 13.6. We take two points P and Q which lie on the line and draw the right-angled triangle PQR. The lengths of PR and QR are then found using the scales on the x- and y-axes. From the graph, PR is found to be 4 units and QR 8 units. Hence

$$\text{Gradient } m = -\tan\theta = -\frac{\text{PR}}{\text{QR}} = -\frac{4}{8} = -0.5$$

Using the standard equation for a straight line shows that

$$y = -0.5x + c \qquad [1]$$

To find the value of c we choose a point such as P which lies on the line and find its coordinates. From the graph we find the coordinates of P to be (22, 19). Substituting the values $x = 22$ and $y = 19$ into equation [1] gives

$$19 = -0.5 \times 22 + c$$

or $$19 = -11 + c$$

$\therefore$ $$c = 30$$

Hence the law of the straight line is $y = -0.5x + 30$.

We may now use this law to determine corresponding values of the two variables.

Thus when $x = 26$, $y = -0.5 \times 26 + 30 = -13 + 30 = 17$

and when $y = 18$, $18 = -0.5x + 30$

or $$0.5x = 12$$

$\therefore$ $$x = 24$$

GRAPHS OF EXPERIMENTAL DATA

Readings which are obtained as a result of an experiment will usually contain errors owing to inaccurate measurement and other experimental errors. If the points, when plotted, show a trend towards a straight line or a smooth curve this is usually accepted and the best straight line or curve drawn. In this case the line will not pass through some of the points and an attempt must be made to ensure an even spread of these points above and below the line or the curve.

One of the most important applications of the straight line law is the determination of a law connecting two quantities when values have been obtained from an experiment as Example 13.8 illustrates.

EXAMPLE 13.8

During a test to find how the power of a lathe varied with the depth of cut results were obtained as shown in the table. The speed and feed of the lathe were kept constant during the test.

Depth of cut, d (mm)	0.51	1.02	1.52	2.03	2.54	3.0
Power, P (W)	0.89	1.04	1.14	1.32	1.43	1.55

Show that the law connecting d and P is of the form $P = ad + b$ and find the law. Hence find the value of d when P is 1.2 watts.

The standard equation of a straight line is $y = mx + c$. It often happens that the variables are *not* x and y. In this example d is used instead of x and is plotted on the horizontal axis, and P is used instead of y and is plotted on the vertical axis.

Similarly the gradient is a instead of m, and b is used instead of c.

On plotting the points (Fig. 13.11) it will be noticed that they deviate slightly from a straight line. Since the data are experimental we must expect errors in observation and measurement and hence a slight deviation from a straight line must be expected.

The points, therefore, approximately follow a straight line and we can say that the equation connecting P and d is of the form $P = ad + b$.

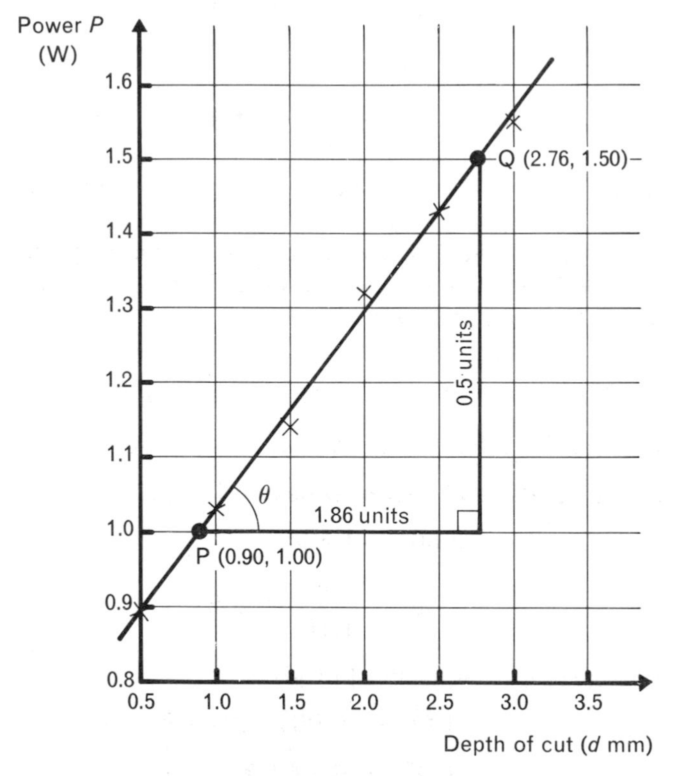

Fig. 13.11

To find the gradient a we draw the right-angled triangle PQR and find the lengths PR and QR. From the graph we find PR to be 1.86 units and QR to be 0.5 units. Hence

$$a = \tan\theta = \frac{QR}{PR} = \frac{0.5}{1.86} = 0.27$$

The equation $P = ad + b$ now becomes

$$P = 0.27d + b$$

Choosing the point Q(2.76, 1.50) which lies on the line, we have

$$1.50 = 0.27 \times 2.76 + b$$

or $$1.50 = 0.7 + b$$

∴ $$b = 0.76$$

Hence the required law is $P = 0.27d + 0.76$

When $P = 1.2$, then

$$1.2 = 0.27d + 0.76$$

∴ $$0.27d = 0.45$$

∴ $$d = \frac{0.45}{0.27} = 1.67$$

Hence when P is 1.2 watts the depth of cut, d, is 1.67 mm.

Alternatively, we can find the values of a and b by taking two points which lie on the line and finding their coordinates. Thus in Fig. 13.11, the points P(0.90, 1.00) and Q(2.76, 1.50) have been chosen.

The point P(0.90, 1.00) lies on the line hence $1.00 = 0.90a + b$ [1]

The point Q(2.76, 1.50) lies on the line hence $1.50 = 2.76a + b$ [2]

Subtracting equation [1] from equation [2] gives

$$0.50 = 1.86a$$

∴ $$a = \frac{0.50}{1.86} = 0.27$$

Substituting for a in equation [1] gives

$$1.00 = 0.90 \times 0.27 + b$$

or $$1.00 = 0.24 + b$$

∴ $$b = 0.76$$

Therefore, as before, the equation of the straight line is

$$P = 0.27d + 0.76$$

Exercise 13.1

1) The rate of exchange between the mark and the £ is £1 = 5 marks. Construct a graph to show the value of the mark in £s up to £50. From the graph find the value of £22 in marks and 180 marks in £s.

2) The values of the resistance (R ohm) of a length of copper wire at different temperatures ($t\,^\circ$C) are shown in the table below.

t	20	30	40	50	60	70
R	1.086	1.129	1.172	1.215	1.258	1.301

Illustrate the relationship on a graph and find the resistance of the wire when the temperature is 55°C. (Plot R vertically.)

3) The forces required to pull blocks of various mass along a horizontal surface are shown in the table below.

Mass of block (kg)	1	2	3	4	5	6
Force (newtons)	3.2	6.2	9.2	12.2	15.2	18.2

Plot a graph of this information with the force vertical. From the graph find the force needed to move a block having a mass of 4.4 kg.

4) The table below gives the load (W kg) and the corresponding values of the effort (E kg) for a lifting machine.

W	2	4	8	10	12	14	20
E	4	5	7	8	9	10	13

Plot a graph from this table with W plotted along the horizontal axis and hence find the effort required to lift a load of 6 kg. What load will an effort of 11 kg lift?

5) It was found that a metal filament lamp gave the following values of the resistance (R ohms) at various voltages (V volts).

V	60	80	90	120	150
R	92	116	128	164	200

Plot a graph of this information with R plotted along the vertical axis. Hence find the resistance when the voltage is 100 volts and the voltage when the resistance is 104 ohms.

Draw graphs of the following simple equations:

6) $y = x + 2$ taking values of x between -3 and 2

7) $y = 2x + 5$ taking values of x between -4 and 4

8) $y = 3x - 4$ taking values of x between -4 and 3

9) $y = 5 - 4x$ taking values of x between -2 and 4

The following equations represent straight lines. State in each case the gradient of the line and the intercept on the y-axis.

10) $y = x + 3$

11) $y = -5x - 2$

12) $y = -3x + 4$

13) $y = 4x - 3$

14) Find the values of m and c if the straight line $y = mx + c$ passes through the point $(-2, 5)$ and has a gradient of 4.

15) Find the values of m and c if the straight line $y = mx + c$ passes through the point $(3, 4)$ and the intercept on the y-axis is -2.

16) The following table gives values of x and y which are connected by an equation of the type $y = mx + c$. Plot the graph and from it find the values of m and c.

x	2	4	6	8	10	12
y	10	16	22	28	34	40

The tables below give corresponding values of x and y which are connected by equations of the type $y = mx + c$. In each case plot a graph and from it determine the equation of the line.

17)

x	2	4	5	8	12
y	1	7	10	19	31

18)

x	10	12	15	19	24
y	20	16	10	2	-8

19)

x	-2	-1	1	4	6
y	2	-1	-7	-16	-22

20)

x	8	10	13	15	18
y	24	25	26.5	27.5	29

21) The table below gives corresponding values of x and y. Plot the graph and from it find its gradient and intercept on the y-axis.

x	4	7	9	11	15
y	11	17	21	25	33

22) The following are observed values of P and Q.

P	5.0	7.0	8.8	11.6	15.0	19.2	24.0	32.0
Q	13.6	17.6	22.2	28.0	35.5	47.4	56.1	74.6

Plot a graph of these readings with P plotted vertically. Show that the law connecting P and Q is of the type $P = aQ + b$ and find values for a and b.

23) In an experiment carried out with a lifting machine the effort E and the load W were found to have the values given in the table.

W	10	30	50	60	80	100
E	8.9	19.1	29	33	45	54

Plot a graph of this data with E vertically and hence show that $E = mW + c$. Find values for m and c.

24) The following figures show how the resistance of a conductor varies as the temperature increases:

Temperature (t °C)	25	50	75	100	150
Resistance (R ohms)	20.7	21.5	22.3	23.0	24.5

Plot these results with R vertical and hence show that R and t are connected by an equation of the type $R = at + b$. Hence find values for a and b. Find the value of R when $t = 60°$.

25) Two quantities W and P are connected as shown in the following table of values:

W	28	50	59	67	74	79	84
P	2	5.4	6.9	8.0	9.1	9.9	10.7

Draw a graph of W against P, plotting W vertically. Hence show that W and P are connected by a law of the type $W = mP + c$. Find suitable values for m and c and find the value of P when W is 71.

DIRECT VARIATION

The statement that y is proportional to x (often written $y \propto x$) means that the graph of y against x is a straight line passing through the origin (Fig. 13.12). We may write

$$y = kx$$

where k is called the constant of proportionality. (Note that k gives the gradient of the line.)

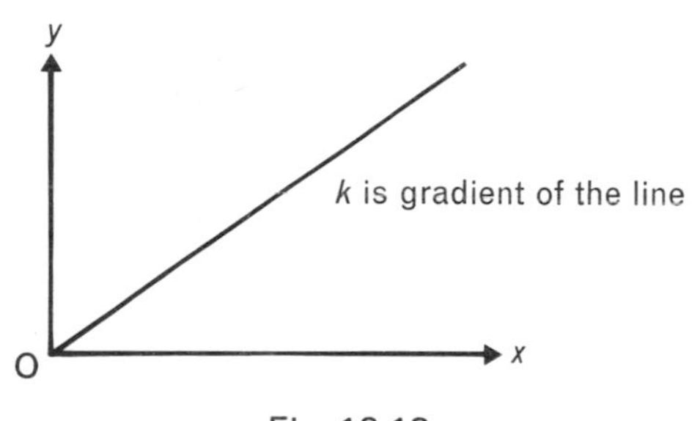

Fig. 13.12

The ratio of y to x is equal to k, and y is said to vary directly as x. Hence direct variation means that if x is doubled, then y is also doubled. If x is halved then y is halved, and so on.

Some examples of direct variation are:

(a) The circumference of a circle is directly proportional to its diameter. ($c = \pi d$, π being the constant of proportionality.)

(b) The cost of metal is directly proportional to the quantity bought. ($C = kQ$, k being the constant of proportionality. Note that in this case k is the price per unit quantity.)

EXAMPLE 13.9

If y is directly proportional to x, and $y = 2$ when $x = 5$, find:

a) the value of the constant of proportionality,

b) the value of y when $x = 6$,

c) the value of x when $y = 8$.

a) Since $y \propto x$, then $y = kx$.

We are given that $y = 2$ when $x = 5$

Hence $$2 = 5k$$

$$\therefore \quad k = \frac{2}{5} = 0.4$$

b) Since $k = 0.4$, then $$y = 0.4x$$

Thus when $x = 6$, $$y = 0.4 \times 6 = 2.4$$

c) When $y = 8$, $$8 = 0.4x$$

$$\therefore \quad x = \frac{8}{0.4} = 20$$

EXAMPLE 13.10

Hooke's Law states that up to the limit of proportionality, stress is directly proportional to strain. If the strain is 0.005 when the stress is 20 N/mm^2, find the strain when the stress is 30 N/mm^2.

Since $$\text{stress} \propto \text{strain}$$

we may write $$\text{stress} = k \times (\text{strain})$$

When stress = 20, strain = 0.005,

then $$20 = 0.005k$$

$$\therefore \quad k = \frac{20}{0.005} = 4000$$

Thus $$\text{stress} = 4000 \times (\text{strain})$$

When stress = 30,

then $$30 = 4000 \times (\text{strain})$$

$$\therefore \quad \text{strain} = \frac{30}{4000} = 0.0075$$

INVERSE VARIATION

If y is inversely proportional to x, then

$$y = \frac{k}{x}$$

If we plot a graph of y against $\dfrac{1}{x}$, we obtain a straight line passing through the origin (Fig. 13.13).

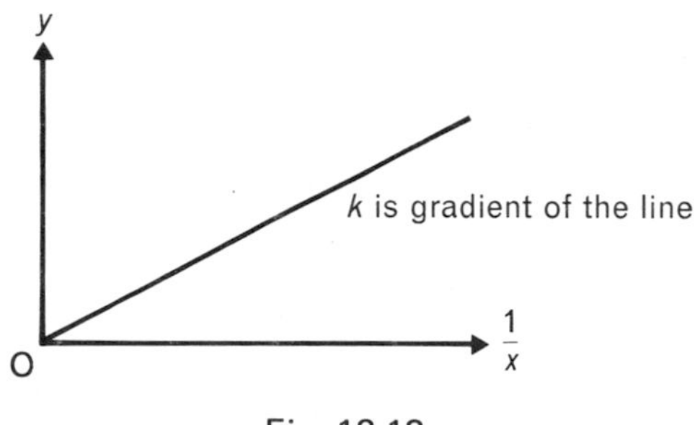

Fig. 13.13

EXAMPLE 13.11

If S varies inversely as P and $S = 4$ when $P = 3$, find the value of P when $S = 16$

We are given that

$$S = \frac{k}{P}$$

When $S = 4$, $P = 3$,

then $$4 = \frac{k}{3}$$

$\therefore$ $$k = 12$$

Thus $$S = \frac{12}{P}$$

When $S = 16$,

then $$16 = \frac{12}{P}$$

$\therefore$ $$P = \frac{12}{16} = 0.75$$

EXAMPLE 13.12

Boyle's Law states that the volume, V, of a fixed mass of gas is inversely proportional to the pressure, p, provided the temperature remains constant. If $V = 125\,\text{cm}^3$ when $p = 755\,\text{mm}$ of mercury, find V when $p = 760\,\text{mm}$ of mercury.

We are given that

$$V = \frac{k}{p}$$

When $V = 125$ and $p = 755$,

then $$125 = \frac{k}{755}$$

$\therefore$ $$k = 125 \times 755$$

Thus $$V = \frac{125 \times 755}{p}$$

When $p = 760$,

then $$V = \frac{125 \times 755}{760} = 124$$

Hence when the pressure is 760 mm of mercury the volume of gas is 124 cm^3.

In Example 13.12 we may transpose the equation

$$V = \frac{k}{p}$$

to give $$pV = k$$

that is, the product of p and V is constant. This leads to an alternative method of solving the problem.

Let $p_1 = 755$, $V_1 = 125, p_2 = 760$ and $V_2 =$ the required volume.

Then $$p_1V_1 = p_2V_2$$

$\therefore$ $$V_2 = \frac{p_1V_1}{p_2} = \frac{755 \times 125}{760} = 124$$

EXAMPLE 13.13

If a quantity M varies inversely as a quantity R and $M = 8$ when $R = 6$, find the value of R when $M = 12$.

Let $M_1 = 8$, $R_1 = 6$, $M_2 = 12$ and $R_2 =$ the required value of R,

then $$MR = \text{a constant}$$

or $$M_1R_1 = M_2R_2$$

giving $$R_2 = \frac{M_1R_1}{M_2} = \frac{8 \times 6}{12} = 4$$

Exercise 13.2

1) M varies directly as Q. If $M = 8$ when $Q = 4$ find:

(a) the value of the constant of proportionality

(b) the value of M when $Q = 5$

(c) the value of Q when $M = 16$

2) Ohm's Law states that the current, I, passing through a wire is directly proportional to the potential difference, V, between its ends. If $V = 12$ when $I = 2$, find:

(a) the constant of proportionality

(b) I when $V = 10$

(c) V when $I = 5$

3) P varies inversely as Q. If $P = 4$ when $Q = 3$ find:

(a) the constant of proportionality

(b) the value of P when $Q = 6$

(c) the value of Q when $P = 6$

4) The electrical resistance, R ohms, of a wire of given length is inversely proportional to the area of the wire, $A\ \text{mm}^2$. If R is 4.25 ohms when A is $3\ \text{mm}^2$, find the value of R when $A = 7\ \text{mm}^2$.

5) The volume, $V\ \text{m}^3$, of a gas is directly proportional to the temperature θ K, if the pressure is kept constant. If $V = 20\ \text{m}^3$ when $\theta = 290$ K, find V when $\theta = 350$ K.

6) If the temperature is kept constant, the pressure of a gas, p mm of mercury, is inversely proportional to the volume of gas, $V\ \text{m}^3$. If $p = 77$ mm of mercury when $V = 1.21\ \text{m}^3$, find p when $V = 1.5\ \text{m}^3$.

7) The volume $V\ \text{m}^3$ of a gas is directly proportional to the temperature, θ K, at constant pressure. To what temperature must 2 litres of air at 290 K be heated to increase its volume to 3 litres?

8) If the current, I amperes, in a conductor is inversely proportional to the resistance, R ohms, and $R = 16$ ohms when $I = 15$ amperes, find:

(a) I when $R = 20$ ohms,

(b) R when $I = 10$ amperes.

14. ANGLES AND STRAIGHT LINES

After reaching the end of this chapter you should be able to:

1. *Define degree as 1/360th of a revolution and minute as 1/60th of a degree.*
2. *Recognise acute, obtuse, reflex and right angles.*
3. *Identify vertically opposite pairs of angles.*
4. *Identify for given parallel lines and a transverse (i) alternate angles, (ii) correspondi angles, (iii) interior angles, (iv) supplementa angles.*
5. *Identify parallel lines through relationship given pairs of angles.*

ANGLES

When two lines meet at a point they form an angle. The size of the angle depends only upon the amount of opening between the lines. It does not depend upon the lengths of the lines forming the angle. In Fig. 14.1 the angle A is larger than the angle B despite the fact that the lengths of the arms are shorter.

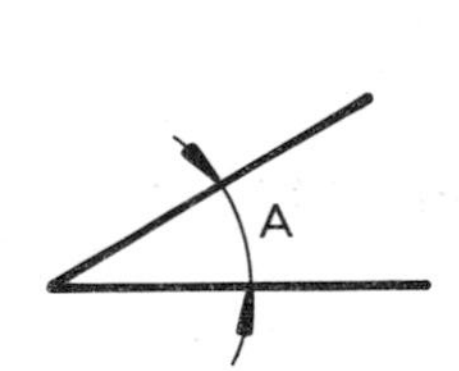

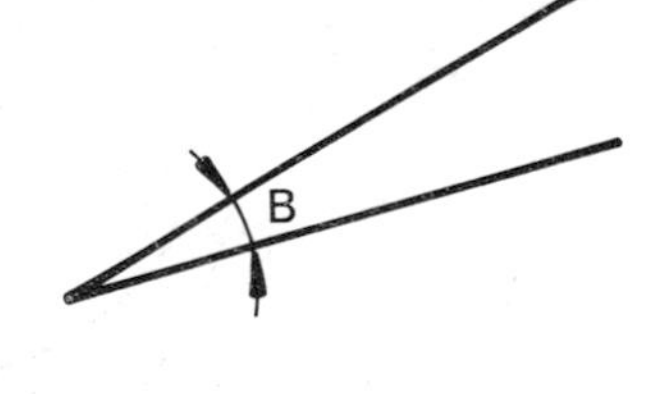

Fig. 14.1

ANGULAR MEASUREMENT

An angle may be looked upon as the amount of rotation or turning. In Fig. 14.2 the line OA has been turned about O until it takes up the position OB. The angle through which the line has turned is the amount of opening between the lines OA and OB.

If the line OA is rotated until it returns to its original position it will have described one revolution. Hence we can measure an angle as a fraction of a revolution. Fig. 14.3 shows a circle divided up into 36 equal parts. The first division is split up into 10 equal parts so that each small division is $\frac{1}{360}$ of a complete revolution. We call this division a *degree*.

Thus $\qquad 1 \text{ degree} = \frac{1}{360} \text{ of a revolution}$

or $\qquad 360 \text{ degrees} = 1 \text{ revolution}$

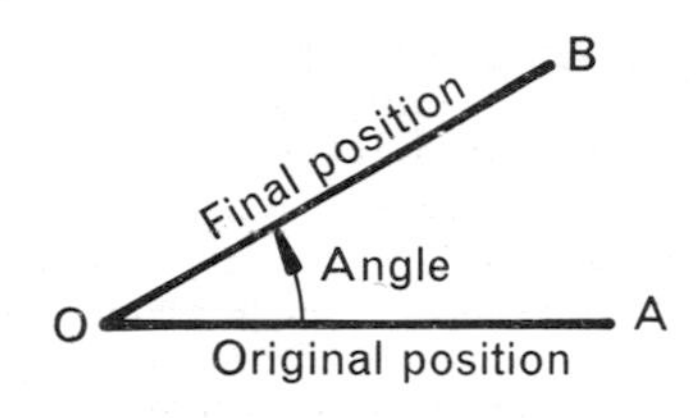

Fig. 14.2

When writing angles we write seventy degrees as 70°. The small ° at the right-hand corner of the figure replaces the word degrees. Thus 87° reads 87 degrees.

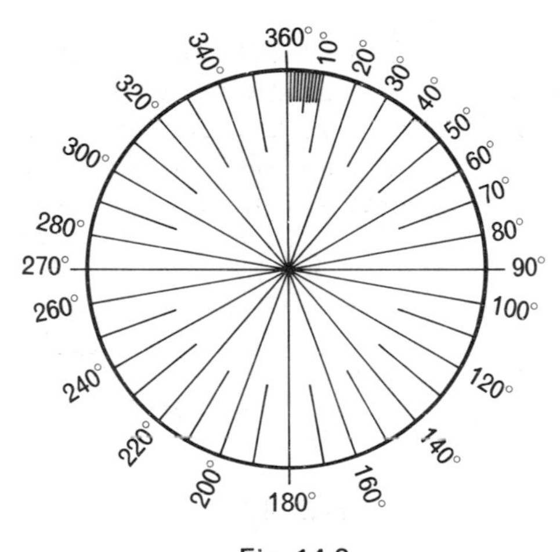

Fig. 14.3

The right angle is one quarter of a revolution and hence it contains $\frac{1}{4}$ of $360° = 90°$. Two right angles contain 180° and three right angles contain 270°.

EXAMPLE 14.1

Find the angle in degrees corresponding to $\frac{1}{8}$ of a revolution.

$$1 \text{ revolution} = 360°$$

$$\therefore \quad \tfrac{1}{8} \text{ revolution} = \tfrac{1}{8} \times 360° = 45°$$

EXAMPLE 14.2

Find the angle in degrees coresponding to 0.6 of a revolution.

$$1 \text{ revolution} = 360^\circ$$

$$0.6 \text{ revolution} = 0.6 \times 360^\circ = 216^\circ$$

For some purposes the degree is too large a unit and it is subdivided into minutes and seconds so that:

$$60 \text{ seconds} = 1 \text{ minute}$$

$$60 \text{ minutes} = 1 \text{ degree}$$

$$360 \text{ degrees} = 1 \text{ revolution}$$

An angle of 25 degrees 7 minutes 30 seconds is written $25^\circ 7' 30''$.

EXAMPLE 14.3

a) Add together $22^\circ 35'$ and $49^\circ 42'$.

$$\begin{array}{r} 22^\circ 35' \\ \underline{49^\circ 42'} \\ 72^\circ 17' \end{array}$$

The minutes 35 and 42 add up to 77 minutes which is $1^\circ 17'$. The 17 is written in the minutes column and 1° carried over to the degrees column. The degrees 22, 49 and 1 add up to 72 degrees.

b) Subtract $17^\circ 49'$ from $39^\circ 27'$.

$$\begin{array}{r} 39^\circ 27' \\ \underline{17^\circ 49'} \\ 21^\circ 38' \end{array}$$

We cannot subtract $49'$ from $27'$ so we borrow 1 from the 39° making it 38°. The $27'$ now becomes $27' + 60' = 87'$. Subtracting $49'$ from $87'$ gives $38'$ which is written in the minutes column. The degree column is now $38^\circ - 17^\circ = 21^\circ$.

c) Convert $49^\circ 38'$ to degrees and decimals of a degree.

Since $\qquad 1 \text{ degree} = 60 \text{ minutes}$

then $\qquad 38 \text{ minutes} = \dfrac{38}{60} \text{ degree} = 0.633 \text{ degree}$

$\therefore \qquad 49^\circ 38' = 49.633^\circ$

d) Convert 54.638° into degrees, minutes and seconds.

Since $\quad 1 \text{ degree} = 3600 \text{ seconds}$

then $\quad 0.638 \text{ degree} = 0.638 \times 3600 = 2297 \text{ seconds}$

Since $\quad 60 \text{ seconds} = 1 \text{ minute}$

then $\quad 2297 \text{ seconds} = \dfrac{2297}{60} = 38\tfrac{17}{60} = 38'17''$

$\therefore \quad 54.638^\circ = 54^\circ 38' 17''$

Exercise 14.1

1) How many degrees are there in $1\frac{1}{2}$ right angles?

2) How many degrees are there in $\frac{3}{5}$ of a right angle?

3) How many degrees are there in $\frac{2}{3}$ of a right angle?

4) How many degrees are there in 0.7 of a right angle?

Find the angle in degrees corresponding to the following:

5) $\frac{1}{20}$ revolution

6) $\frac{3}{8}$ revolution

7) $\frac{4}{5}$ revolution

8) 0.8 revolution

9) 0.3 revolution

10) 0.25 revolution

Add together the following angles:

11) $11^\circ 8'$ and $17^\circ 29'$

12) $25^\circ 38'$ and $43^\circ 45'$

13) $8^\circ 38' 49''$ and $5^\circ 43' 45''$

14) $27^\circ 4' 52''$ and $35^\circ 43' 19''$

15) $72^\circ 15' 4''$, $89^\circ 27' 38''$ and $17^\circ 28' 43''$

Subtract the following angles:

16) $8^\circ 2'$ from $29^\circ 5'$

17) $17^\circ 28'$ from $40^\circ 16'$

18) $12^\circ 34' 16''$ from $20^\circ 18' 12''$

19) $0^\circ 7' 15''$ from $6^\circ 2' 5''$

20) $48^\circ 19' 21''$ from $85^\circ 17' 32''$

Convert the following to degrees and decimals of a degree:

21) $17^\circ 26'$

22) $28^\circ 7' 38''$

23) $83^\circ 0' 12''$

24) $87^\circ 58' 29''$

(where necessary state the answer correct to 3 decimal places).

Convert the following to degrees, minutes and seconds:

25) 28.391° 26) 47.46°

27) 58.939° 28) 5.016°

(where necessary state the answer correct to the nearest second.)

TYPES OF ANGLES

An *acute angle* (Fig. 14.4) is less than 90°.

An *obtuse angle* (Fig. 14.5) lies between 90° and 180°.

A *reflex angle* (Fig. 14.6) is greater than 180°.

A *right angle* (Fig. 14.7) is equal to 90°.

Complementary angles are angles whose sum is 90°.

Supplementary angles are angles whose sum is 180°.

An acute angle

Fig. 14.4

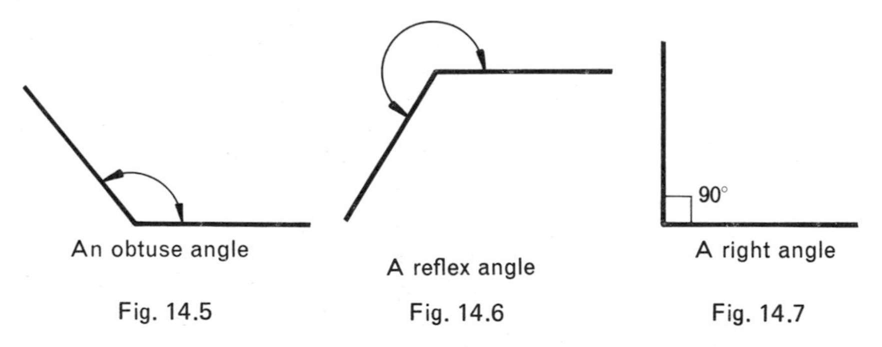

An obtuse angle — Fig. 14.5

A reflex angle — Fig. 14.6

A right angle — Fig. 14.7

PROPERTIES OF ANGLES AND STRAIGHT LINES

(1) *The total angle on a straight line is* 180° (Fig. 14.8). The angles A and B are called adjacent angles. They are also supplementary.

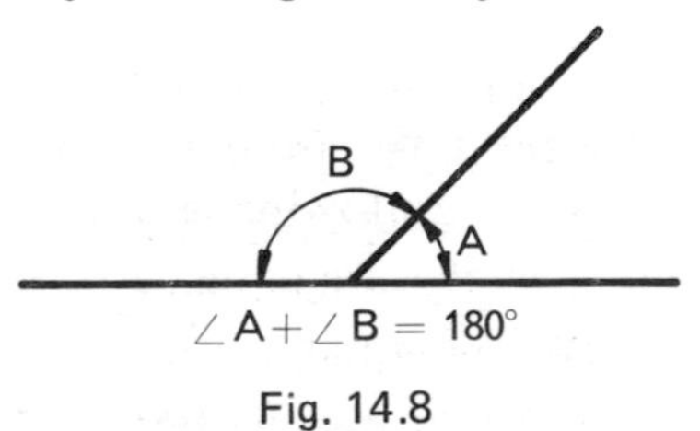

Fig. 14.8

(2) *When two straight lines intersect the opposite angles are equal* (Fig. 14.9). The angles A and C are called vertically opposite angles. Similarly the angles B and D are also vertically opposite angles.

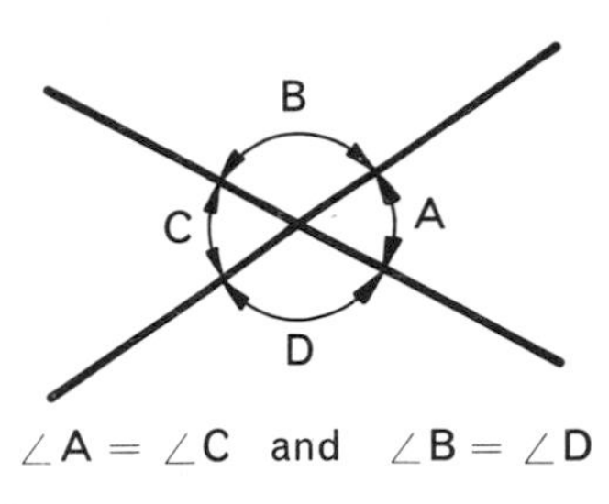

$\angle A = \angle C$ and $\angle B = \angle D$

Fig. 14.9

(3) *When two parallel lines are cut by a transversal* (Fig. 14.10).

(a) *The corresponding angles are equal*

$$A = L;\ B = M;\ C = P;\ D = Q$$

(b) *The alternate angles are equal* $D = M;\ C = L.$

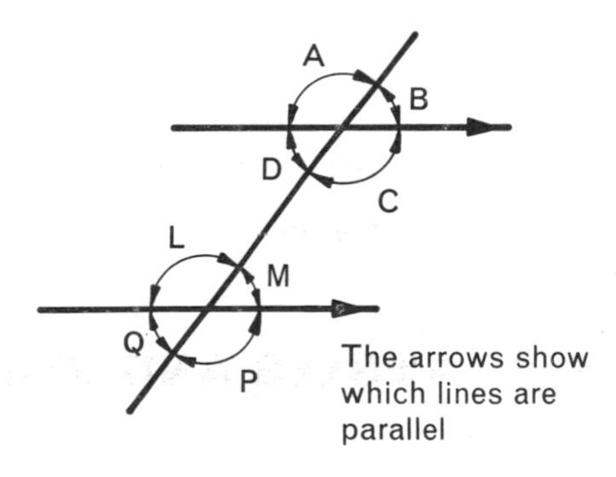

Fig. 14.10

(c) Interior angles are supplementary $D + L = 180°; C + M = 180°$.
Conversely if two straight lines are cut by a transversal the lines are parallel if any *one* of the following is true:

(i) Two corresponding angles are equal.
(ii) Two alternate angles are equal.
(iii) Two interior angles are supplementary.

EXAMPLE 14.4

a) Find the angle A shown in Fig. 14.11.

$$\angle B = 180° - 138° = 42°$$

$$\angle B = \angle A \text{ (corresponding angles)}$$

$$\angle A = 42°$$

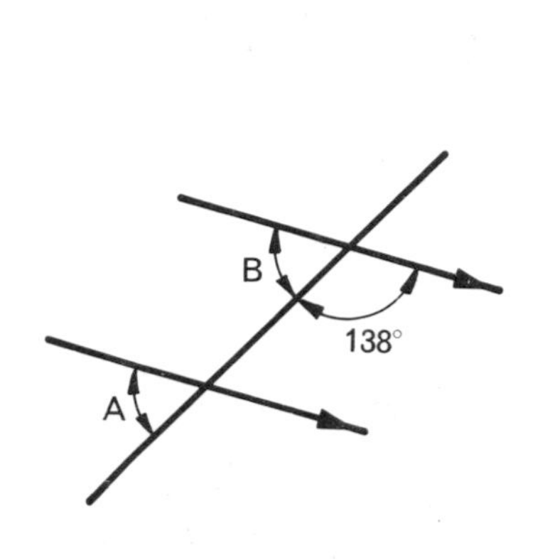

Fig. 14.11

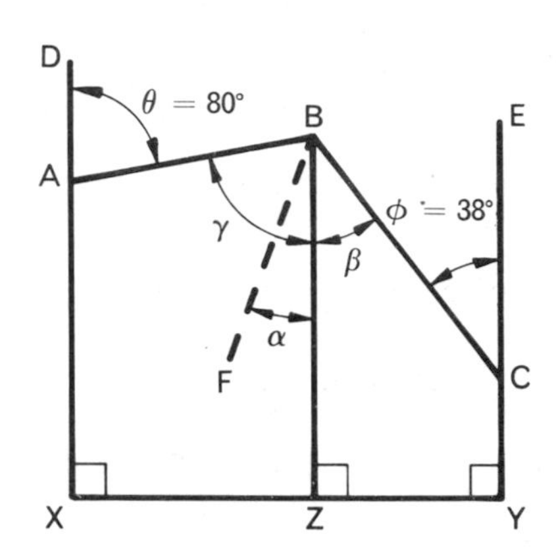

Fig. 14.12

b) In Fig. 14.12 the line BF bisects $\angle ABC$. Find the value of the angle α.

The lines AX, BZ and EY are all parallel because they lie at right angles to the line XY.

Now $\phi = \beta$ (alternate angles: BZ||EY)

$\therefore$ $\beta = 38°$ (since $\phi = 38°$)

Also $\theta = \gamma$ (alternate angles: XD||BZ)

$\therefore$ $\gamma = 80°$ (since $\theta = 80°$)

Also $\angle ABC = \gamma + \beta = 80° + 38° = 118°$

But $\angle FBC = 118° \div 2 = 59°$ (since BF bisects $\angle ABC$)

Thus $\beta + \alpha = 59°$

$\therefore$ $38° + \alpha = 59°$

Hence $\alpha = 59° - 38° = 21°$

Exercise 14.2

1) Find α in Fig. 14.13.

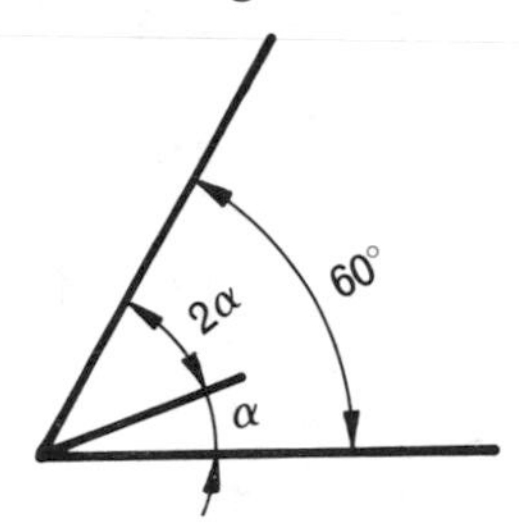

Fig. 14.13

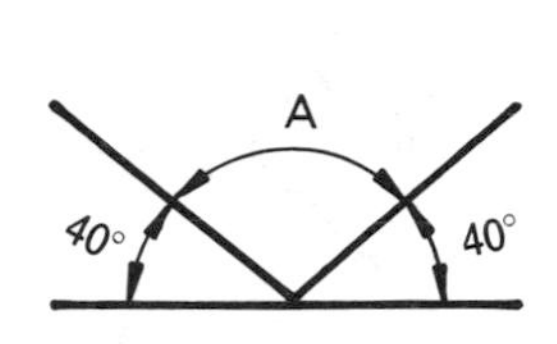

Fig. 14.14

2) Find A in Fig. 14.14.

3) Find β in Fig. 14.15.

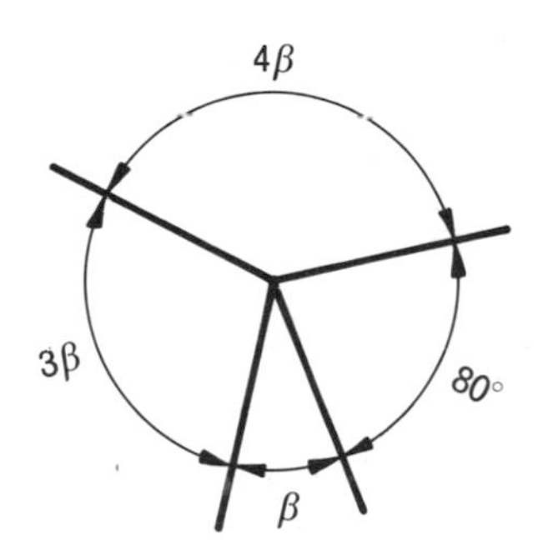

Fig. 14.15

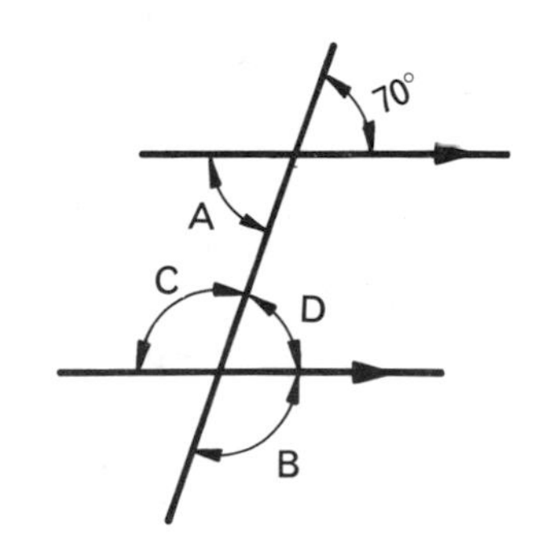

Fig. 14.16

4) In Fig. 14.16 find A, B, C and D.

5) In Fig. 14.17 find A.

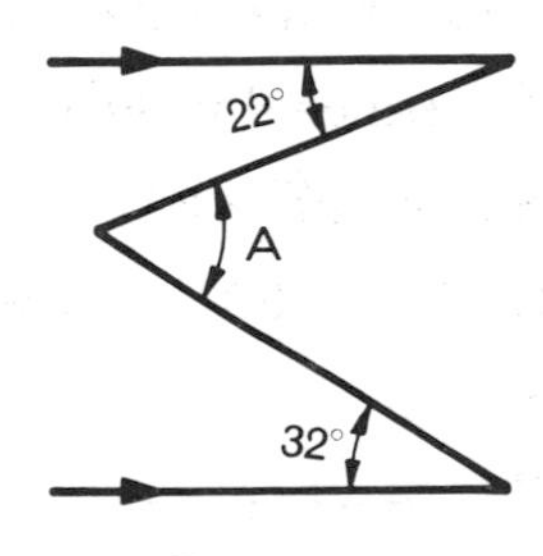

Fig. 14.17

6) In Fig. 14.18 prove that AB is parallel to ED.

7) Find A in Fig. 14.19.

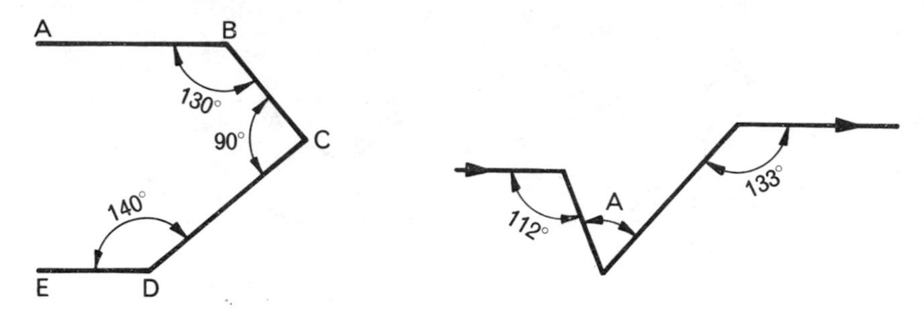

Fig. 14.18

Fig. 14.19

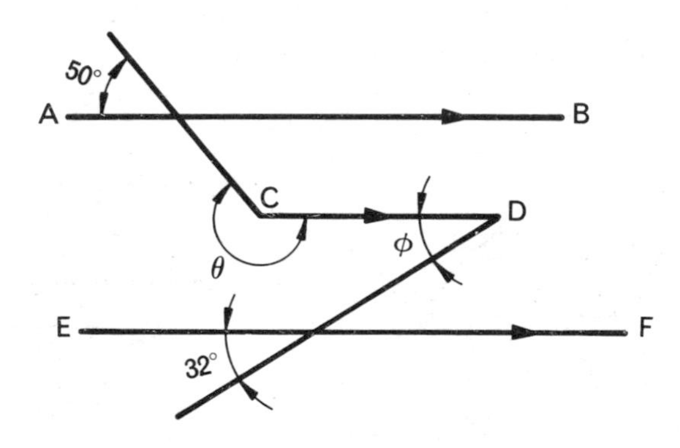

Fig. 14.20

8) In Fig. 14.20 the lines AB, CD and EF are parallel. Find the values of θ and ϕ.

9) Find the angle γ in Fig. 14.21.

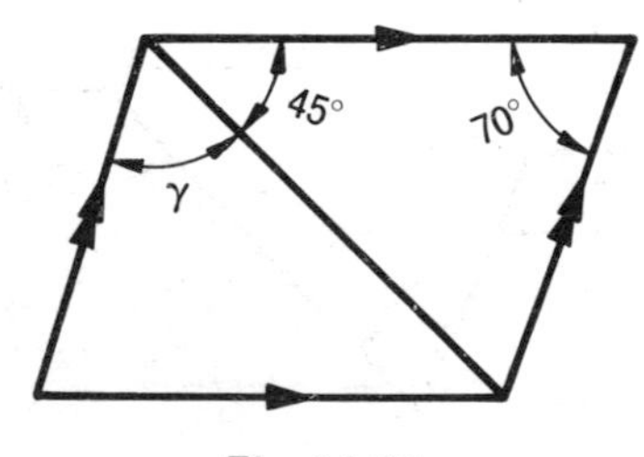

Fig. 14.21

10) Find α in Fig. 14.22.

11) Find the angle A of the dovetail shown in Fig. 14.23.

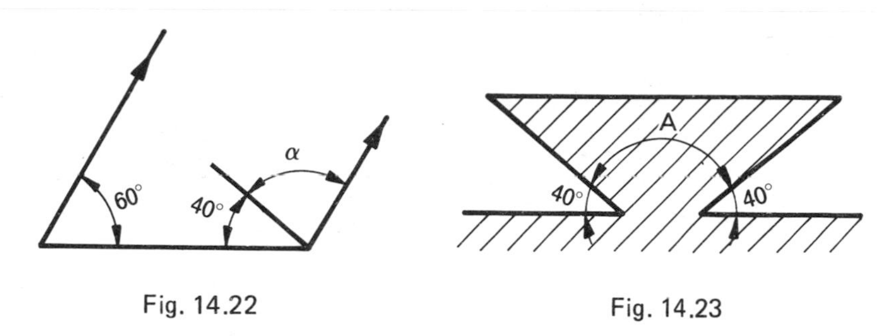

Fig. 14.22

Fig. 14.23

12) In the circular plate shown in Fig. 14.24 a sector of 54° is cut out and the remainder divided to give 8 equally spaced holes. Find the angle θ between the holes.

13) Find angle θ in Fig. 14.25.

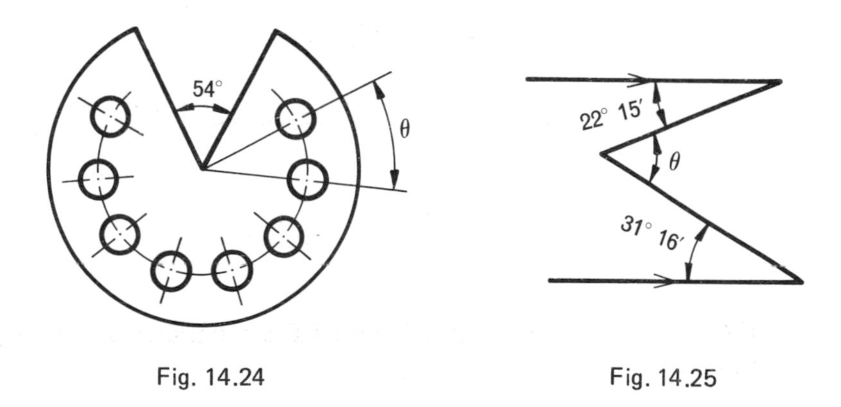

Fig. 14.24

Fig. 14.25

14) Find angle A for the part of a template shown in Fig. 14.26.

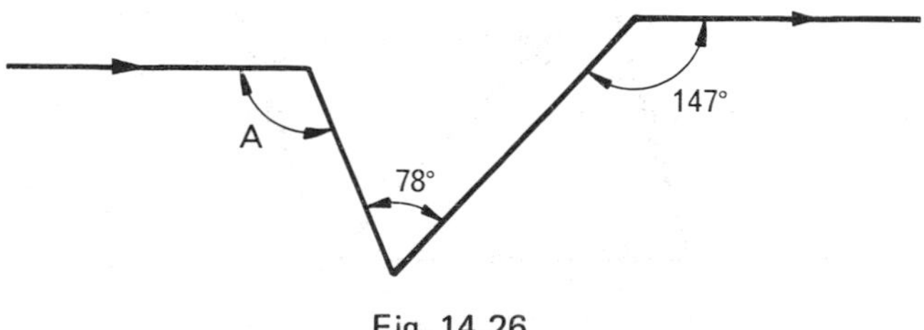

Fig. 14.26

15) The point angle of a chisel is 55° and it is inclined to give a clearance angle of 10° (see Fig. 14.27). Find

(a) the angle of inclination,
(b) the rake angle.

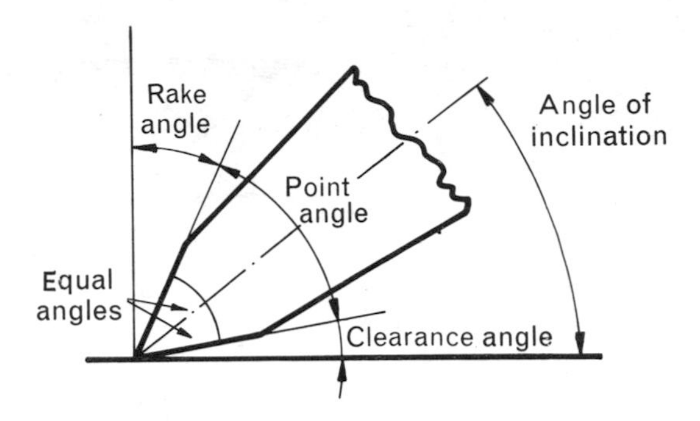

Fig. 14.27

SUMMARY

a) Angles are measured in degrees, minutes and seconds so that:

$$60 \text{ seconds} = 1 \text{ minute}$$
$$60 \text{ minutes} = 1 \text{ degree}$$
$$360 \text{ degrees} = 1 \text{ revolution}$$

b) An acute angle is less than 90°. An obtuse angle lies between 90° and 180°. A reflex angle is greater than 180°. A right angle equals 90°.

c) Complementary angles are angles whose sum is 90°.

d) Supplementary angles are angles whose sum is 180°.

e) The total angle on a straight line is 180°.

f) When two straight lines intersect the vertically opposite angles are equal.

g) When two parallel lines are cut by a transversal the corresponding and alternate angles are equal. Also, the interior angles are supplementary.

Self-Test 10

In the following state the letter (or letters) corresponding to the correct answer (or answers).

1) The angle shown in Fig. 14.28 is:

   a acute **b** right **c** reflex **d** obtuse

2) The angle α shown in Fig. 14.29 is equal to:

   a 120° **b** 60° **c** neither of these

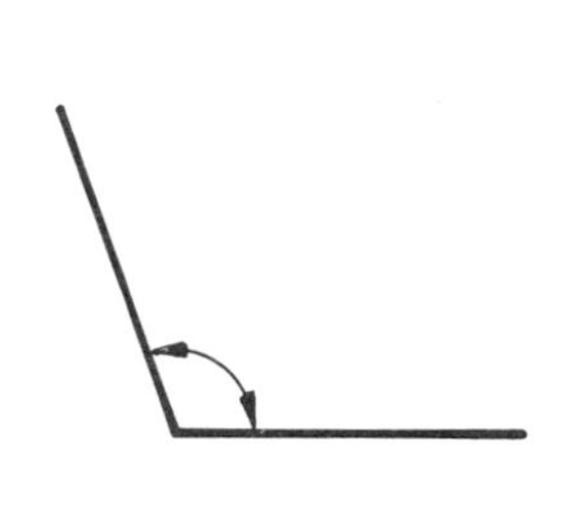

Fig. 14.28

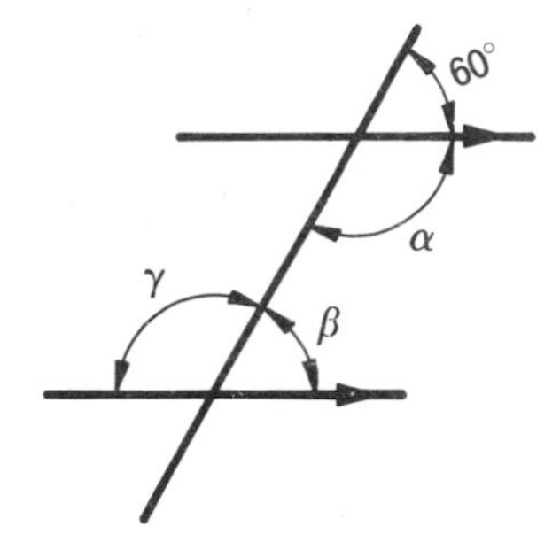

Fig. 14.29

3) The angle β shown in Fig. 14.29 is equal to:

   a 120° **b** 60° **c** neither of these

4) The angle γ shown in Fig. 14.29 is equal to:

   a 120° **b** 60° **c** neither of these

5) In Fig. 14.30:

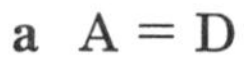

   a A = D **b** A = E **c** E = B **d** A = C

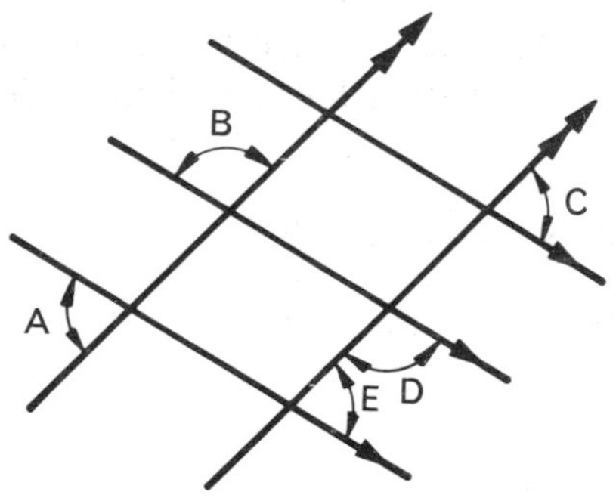

Fig. 14.30

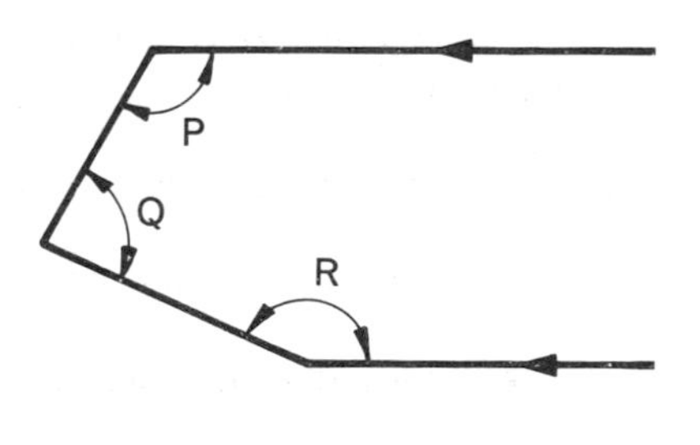

Fig. 14.31

6) In Fig. 14.31:

   a Q = P + R **b** P + Q + R = 360°
   c Q = R − P **d** Q = 360° − P − R

7) In Fig. 14.32:

a $\alpha = \beta$ b $\alpha = 180^\circ - \beta$
c $\alpha = \beta - 180^\circ$ d $\alpha + \beta = 180^\circ$

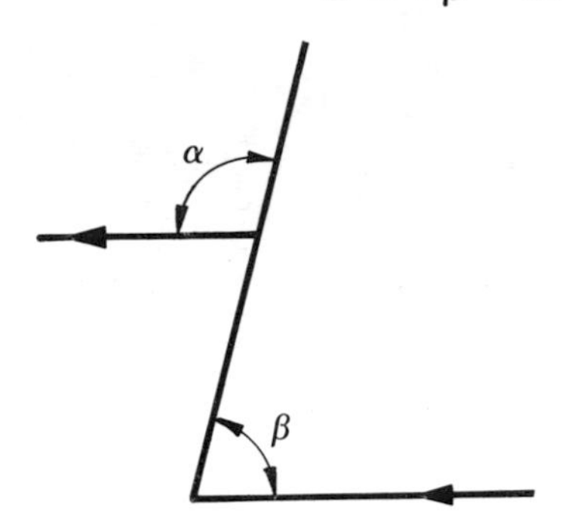

Fig. 14.32

8) A reflex angle is:

a less than 90° b greater than 90°
c greater than 180° d equal to 180°

9) Angles whose sum is 180° are called:

a complementary angles b alternate angles
c supplementary angles d corresponding angles

15. TRIANGLES

After reaching the end of this chapter you should be able to:

1. *Recall the angle sum of a triangle.*
2. *Identify types of triangles as acute-angled, right-angled obtuse-angled, isosceles and equilateral.*
3. *Identify complementary angles.*
4. *Calculate any third side of a right-angled triangle using Pythagoras' theorem.*
5. *Construct a right angle using the sides of 3:4:5 triangle.*
6. *Compare two triangles for similarity or congruency.*
7. *Determine an unknown side or an angle of a second triangle using 6.*
8. *Construct a triangle given (i) three sides, (ii) two sides and the included angle, (iii) one side and two angles, (iv) one side, hypotenuse and right angle.*

TYPES OF TRIANGLES

(1) An *acute-angled* triangle has all its angles less than 90° (Fig. 15.1).

(2) A *right-angled* triangle has one of its angles equal to 90°. The side opposite to the right angle is the longest side and it is called the hypotenuse (Fig. 15.2).

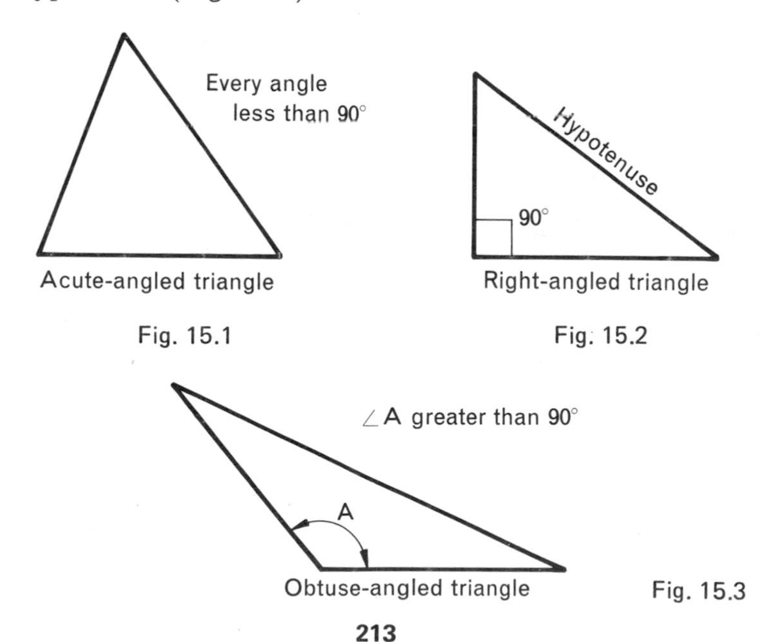

Acute-angled triangle

Fig. 15.1

Right-angled triangle

Fig. 15.2

Obtuse-angled triangle

Fig. 15.3

(3) An *obtuse-angled* triangle has one angle greater than 90° (Fig. 15.3).

(4) A *scalene* triangle has all three sides of different length.

(5) An *isosceles* triangle has two sides and two angles equal. The equal angles lie opposite to the equal sides (Fig. 15.4).

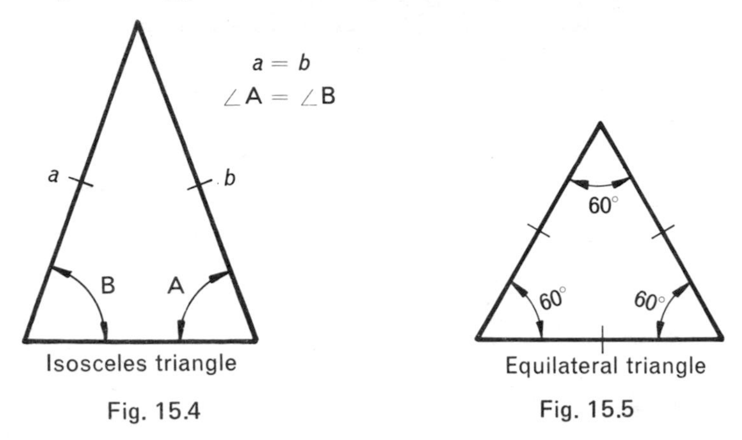

Isosceles triangle

Fig. 15.4

Equilateral triangle

Fig. 15.5

(6) An *equilateral* triangle has all its sides and angles equal. Each angle of the triangle is 60° (Fig. 15.5).

ANGLE PROPERTY OF TRIANGLES

The sum of the angles of a triangle are equal to 180° (Fig. 15.6).

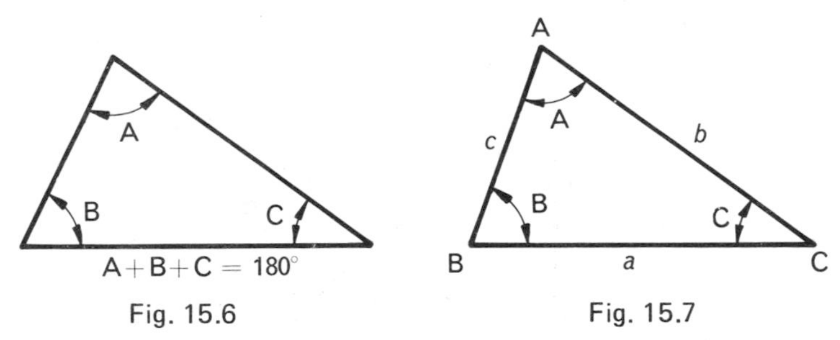

Fig. 15.6

Fig. 15.7

STANDARD NOTATION FOR A TRIANGLE

Fig. 15.7 shows the *standard notation for a triangle*. The three vertices are marked A, B and C. The angles are called by the same letter as the vertices (see diagram). The side a lies opposite the angle A, b lies opposite the angle B and c lies opposite the angle C.

PYTHAGORAS' THEOREM

In any right-angled triangle the square on the hypotenuse is equal to the sum of the squares on the other two sides. In the diagram (Fig. 15.8)

$$AC^2 = AB^2 + BC^2$$

or

$$b^2 = a^2 + c^2$$

The hypotenuse is the longest side and it always lies opposite to the right angle. Thus in Fig. 15.8 the side b is the hypotenuse since it lies opposite to the right angle at B. It is worth remembering that triangles with sides of 3, 4, 5; 5, 12, 13; 7, 24, 25 are right-angled triangles.

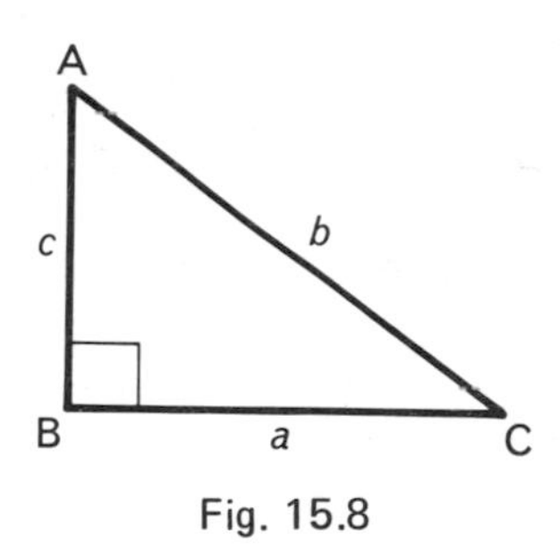

Fig. 15.8

EXAMPLE 15.1

a) In $\triangle ABC, \angle B = 90°, a = 4.2$ m and $c = 3.7$ m (Fig. 15.9). Find b.

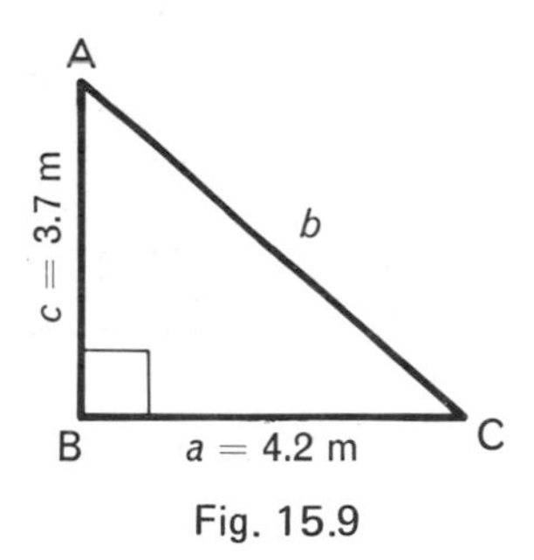

Fig. 15.9

By Pythagoras' Theorem,

$$b^2 = a^2 + c^2$$

$$\therefore \quad b^2 = 4.2^2 + 3.7^2 = 17.64 + 13.69 = 31.33$$

Thus $$b = \sqrt{31.33} = 5.60 \text{ m}$$

b) In ΔABC, $\angle A = 90°$, $a = 64$ mm and $b = 52$ mm. Find c (Fig. 15.10).

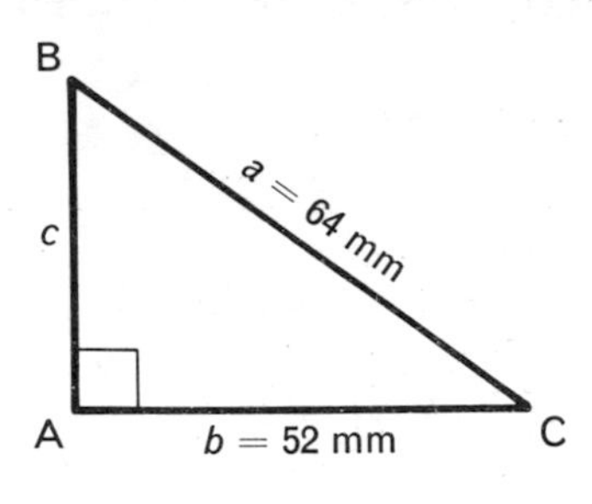

Fig. 15.10

Now $\quad a^2 = b^2 + c^2$

or $\quad c^2 = a^2 - b^2 = 64^2 - 52^2 = 4096 - 2704 = 1392$

$\therefore \quad c = \sqrt{1392} = 37.3$ mm

PROPERTIES OF THE ISOSCELES TRIANGLE

The most important properties of an isosceles triangle is that the perpendicular dropped from the apex to the unequal side:

(a) Bisects the unequal side. Thus in Fig. 15.11, BD = CD.

(b) Bisects the apex angle. Thus in Fig. 15.11, $\angle BAD = \angle CAD$.

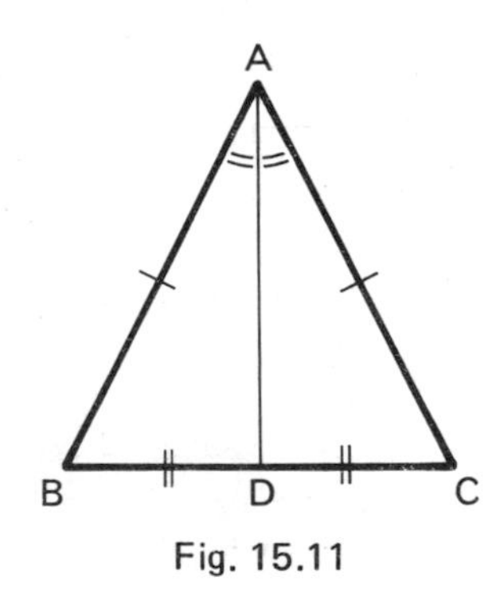

Fig. 15.11

EXAMPLE 15.2

Four holes are bored in a plate as shown in Fig. 15.12. If D is midway between B and C find the distance between A and D.

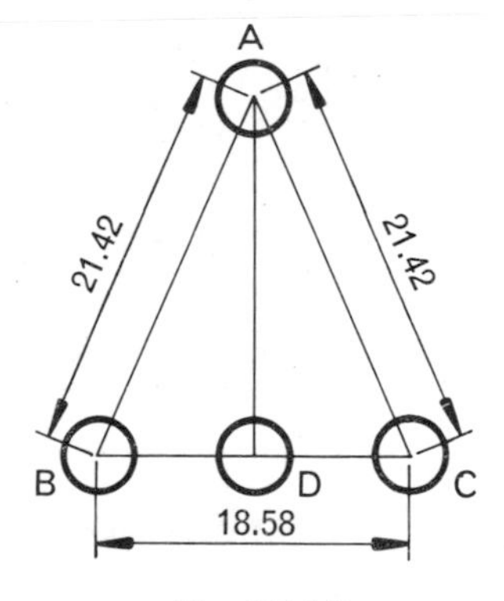

Fig. 15.12

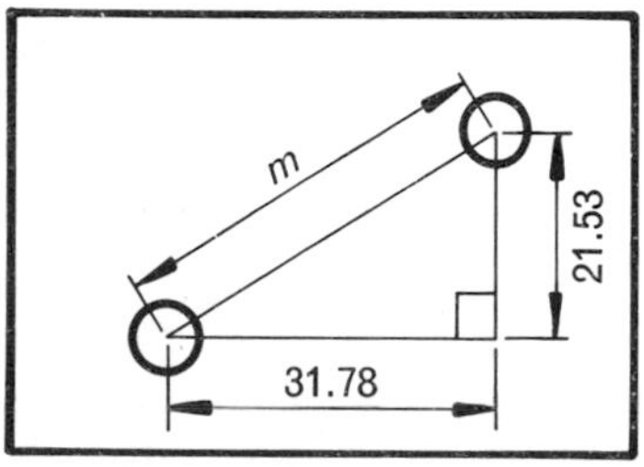

Fig. 15.13

ΔABC is isosceles since it has two equal sides. The line AD which bisects the base BC is therefore perpendicular to BC.

Therefore ΔACD is right-angled and thus:

$$AC^2 = CD^2 + AD^2$$

or

$$AD^2 = AC^2 - CD^2 = 21.42^2 - 9.29^2 = 458.8 - 86.3$$
$$= 372.5$$

$$\therefore \quad AD = \sqrt{372.5} = 19.30 \text{ mm}$$

Exercise 15.1

1) (a) In ΔABC, $\angle B = 90°$, $a = 6$ m and $b = 7$ m. Find c.
   (b) In ΔABC, $\angle C = 90°$, $a = 6.1$m and $b = 3.4$ m. Find c.
   (c) In ΔABC, $\angle A = 90°$, $a = 53$ mm and $b = 48$ mm. Find c.

2) Two holes are bored in a plate to the dimensions shown in Fig. 15.13. To check the holes dimension m is required. What is this dimension?

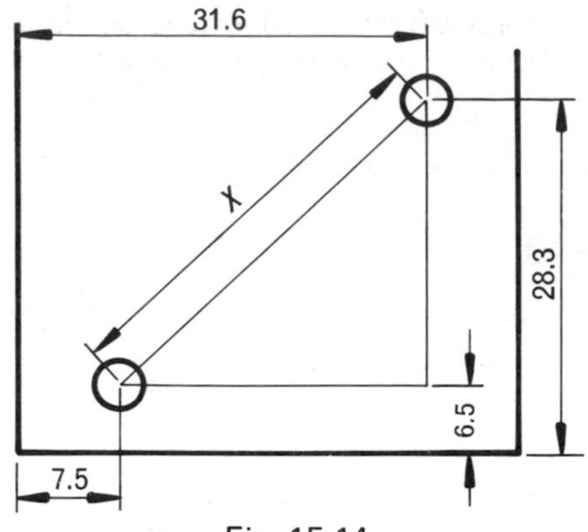

Fig. 15.14

3) Fig. 15.14 shows part of a drawing. If the holes are drilled correctly, what should be dimension x?

4) Fig. 15.15 shows a round bar of 30 mm diameter which has a flat milled on it. Find the width of the flat.

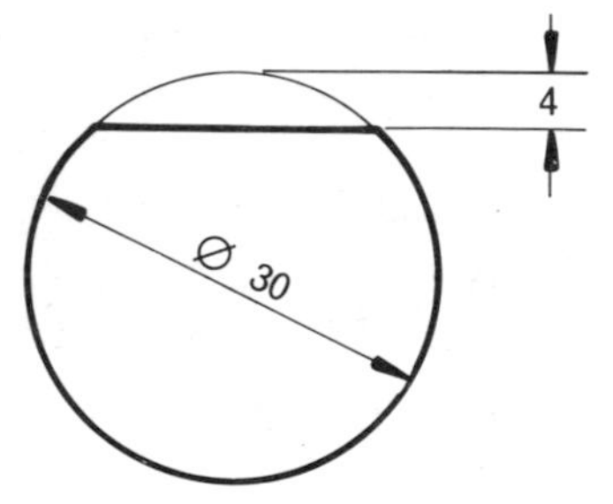

Fig. 15.15

5) Fig. 15.16 shows a bar which has two opposite flats milled on it. Find the distance d between the flats.

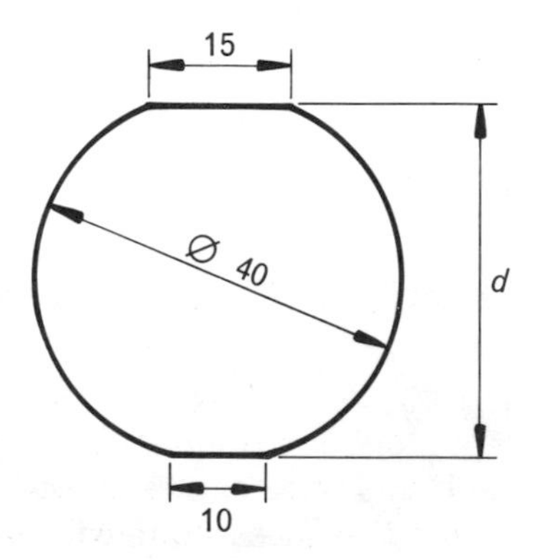

Fig. 15.16

6) A rectangular beam 50 mm × 40 mm is to be cut from a round bar (Fig. 15.17). What is the smallest diameter of bar that can be used?

7) Find the dimensions x and y shown in Fig. 15.18.

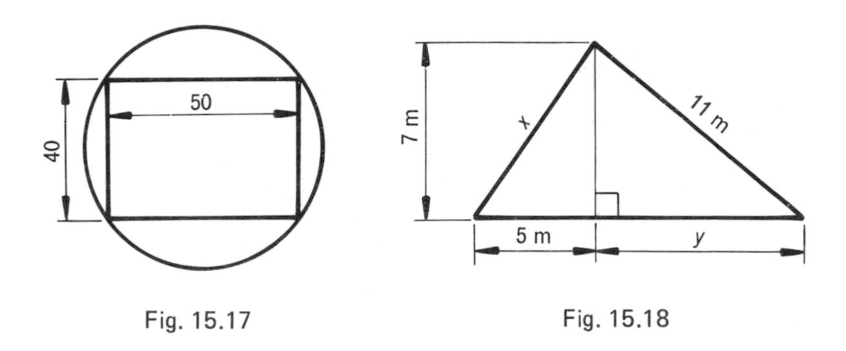

Fig. 15.17

Fig. 15.18

8) Fig. 15.19 shows the conditions that apply when a milling cutter approaches the work. Calculate the approach distance d.

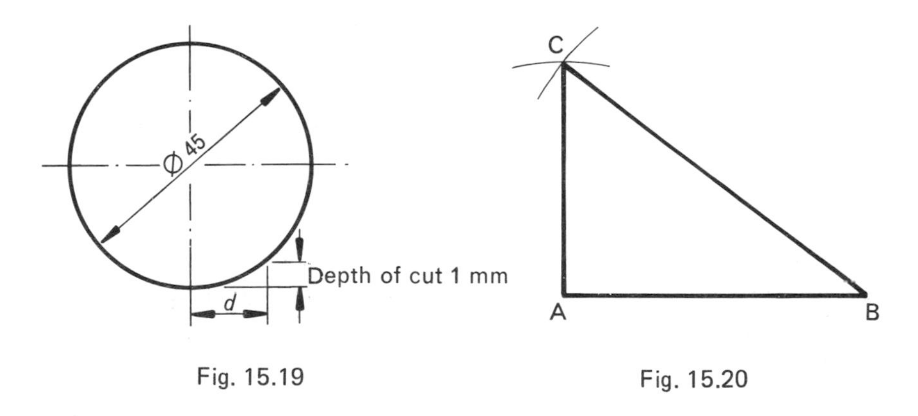

Fig. 15.19

Fig. 15.20

CONSTRUCTING A RIGHT ANGLE

A 3, 4, 5, triangle may be used to construct a right angle as shown in Fig. 15.20. AB is drawn 40 mm long. With centre B and radius 50 mm draw an arc and with centre A and radius 30 mm draw a second arc to cut the first at C. Join AC and BC. ΔABC is then a right-angled triangle with the right angle at A.

CONGRUENT TRIANGLES

Two triangles are said to be congruent if they are equal in every respect. Thus in Fig. 15.21 the triangles ABC and XYZ are congruent because:

$$\left.\begin{aligned} AC &= XZ \\ AB &= XY \\ BC &= ZY \end{aligned}\right\} \quad \text{and} \quad \left\{\begin{aligned} \angle B &= \angle Y \\ \angle C &= \angle Z \\ \angle A &= \angle X \end{aligned}\right.$$

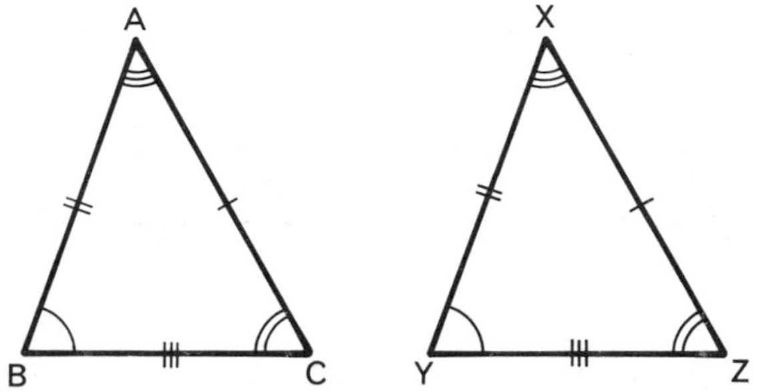

Fig. 15.21

Note that the angles which are equal lie opposite to the corresponding sides.

If two triangles are congruent they will also be equal in area. The notation used to express the fact that $\triangle ABC$ is congruent to $\triangle XYZ$ is $\triangle ABC \equiv \triangle XYZ$.

For two triangles to be congruent the six elements of one triangle (three sides and three angles) must be equal to the six elements of the second triangle. However to prove that two triangles are congruent it is not necessary to prove all six equalities. Any of the following are sufficient to prove that two triangles are congruent.

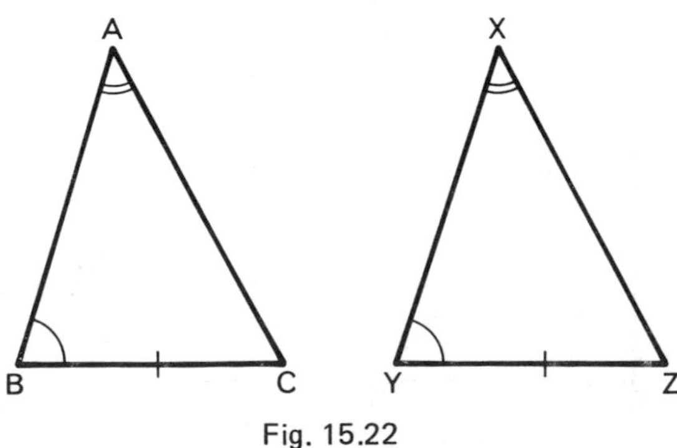

Fig. 15.22

(1) *One side and two angles in one triangle equal to one side and two similarly located angles in the second triangle* (Fig. 15.22).

(2) *Two sides and the angle between them in one triangle equal to two sides and the angle between them in the second triangle* (Fig. 15.23).

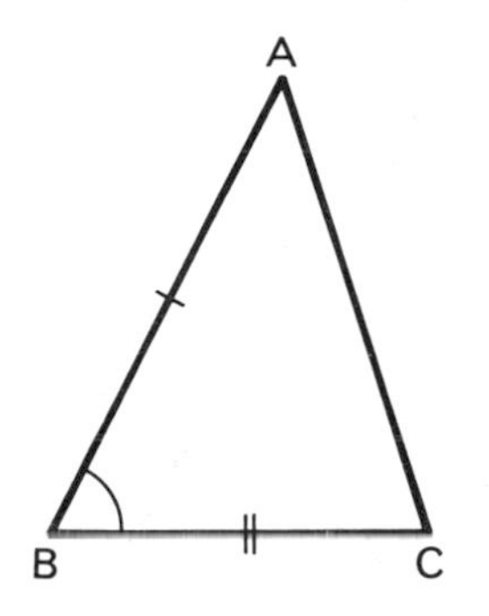

Fig. 15.23

(3) *Three sides of one triangle equal to three sides of the other triangle* (Fig. 15.24).

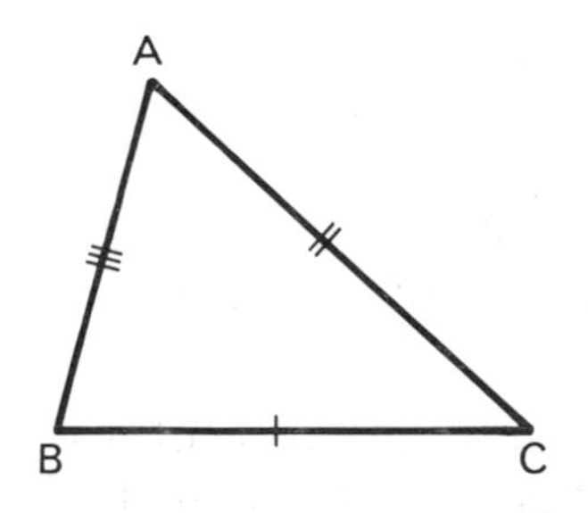

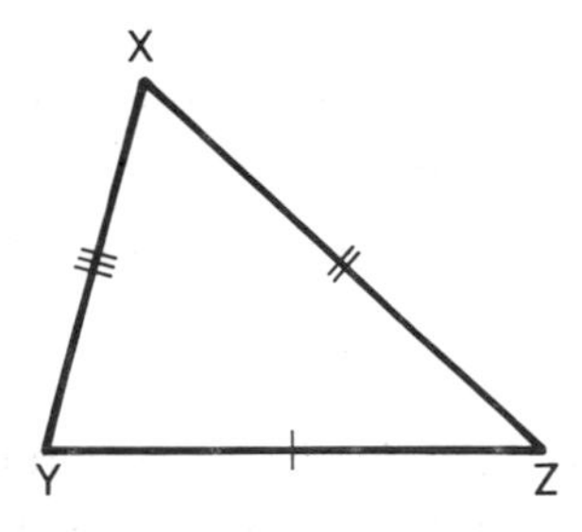

Fig. 15.24

(4) *In right-angled triangles if the hypotenuses are equal and one other side in each triangle are also equal* (Fig. 15.25).

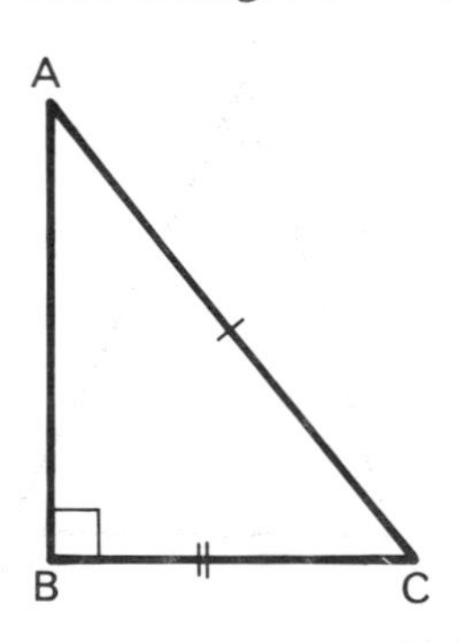

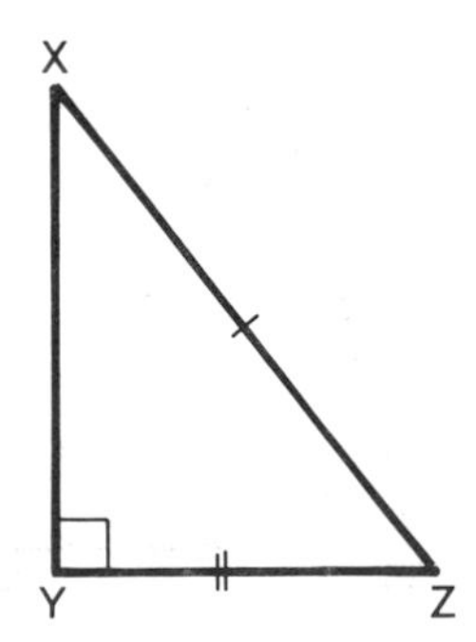

Fig. 15.25

Note that three equal angles are not sufficient to prove congruency and neither are two sides and a non-included angle. An included angle is an angle between the two equal sides of the triangles (e.g. $\angle ABC$ and $\angle XYZ$ in Fig. 15.24).

EXAMPLE 15.3

a) The mid-points of the sides MP and ST of ΔLMP and ΔRST are X and Y respectively. If $LM = RS$, $MP = ST$ and $LX = RY$ prove that $\Delta LMP \equiv \Delta RST$.

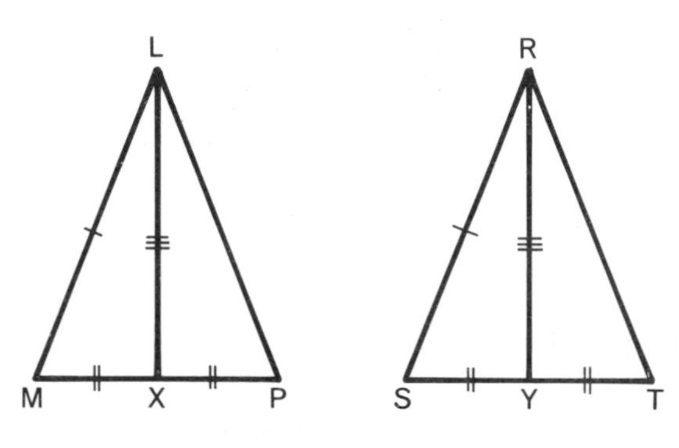

Fig. 15.26

Referring to Fig. 15.26.

$$\Delta LMX \equiv \Delta RSY \quad \text{(corresponding sides equal)}$$

$$\therefore \quad \angle M = \angle S$$

In Δs LMP and RST:

$$LM = RS; \; MP = ST; \; \angle M = \angle S.$$

That is, two sides and the included angle in ΔLMP equal the two sides and the included angle in ΔRST. Hence $\Delta LMP \equiv \Delta RST$.

b) The diagonals of the quadrilateral XYZW intersect at O. Given that $OX = OW$ and $OY = OZ$ prove that $XY = WZ$.

Referring to Fig. 15.27.

In Δs XOY and WOZ

$$OX = OW \quad \text{and} \quad OY = OZ \quad \text{(given)}$$

$$\alpha = \beta \quad \text{(vertically opposite angles)}$$

Hence the two sides and the included angle in ΔXOY equal two sides and the included angle in ΔWOZ. Hence $\Delta XOY \equiv \Delta WOZ$.

$$\therefore \qquad XY = WZ$$

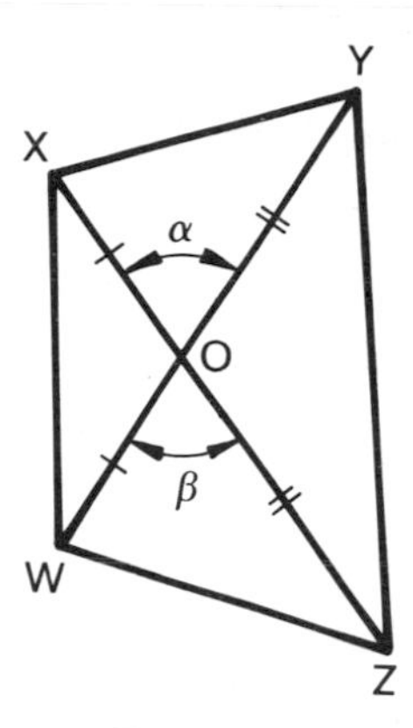

Fig. 15.27

Exercise 15.2

1) Two straight lines PQ and RS cut at X. If $PX = RX$ and $\angle SPX = \angle QRX$ prove that $\Delta SPX \equiv \Delta QRX$.

2) If two straight lines PQ and XY bisect each other, prove that $PX = YQ$.

3) State which of the following must be congruent triangles and which need not necessarily be congruent triangles:

(a) ΔABC and ΔDEF in which $AB = DE$, $BC = EF$ and $\angle C = \angle F = 90°$.

(b) ΔKLM and ΔPQR in which $\angle K = \angle P$, $\angle L = \angle Q$ and $\angle M = \angle R$.

(c) ΔSTU and ΔWXZ in which $\angle S = \angle W$, $SU = XZ$ and $\angle U = \angle Z$.

4) D is a point on the base BC of an isosceles triangle ABC in which $AB = AC$. The triangle ADE is drawn so that $AD = AE$, $\angle DAE = \angle BAC$ and D and E are on opposite sides of AC. Prove that:

(a) $\angle BAD = \angle CAE$;

(b) Δs BAD and CAE are congruent.

5) PQRST is a pentagon (a five sided figure) in which PQ = PT, QR = TS and ∠PQR = ∠PTS. Prove that:

(a) PR = PS;

(b) ∠QRS = ∠TSR.

6) In the equilateral triangle ABC a line through A meets BC at R and a line through B meets CA at S such that BR = CS. Prove △ABR = △BCS. If AR and BS intersect at T, prove that ∠RTB = 60°.

7) D is a point on the hypotenuse BC of a right-angled triangle ABC such that DB = DA. Prove that D is the mid-point of BC.

8) In Fig. 15.28, AE = BE and ∠C = ∠D = ∠AEB = 90°. Prove △ADE ≡ △ECB.

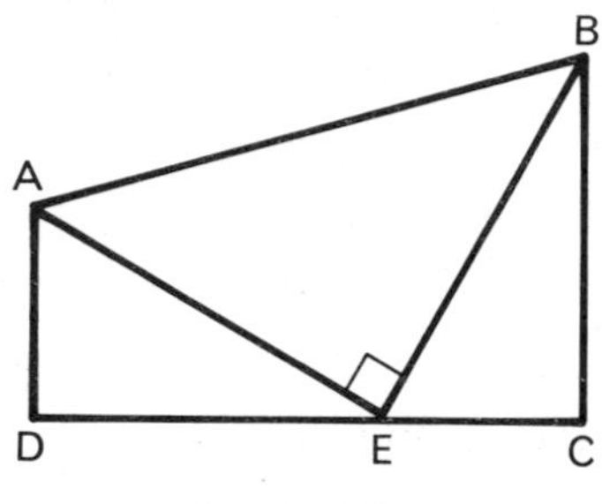

Fig. 15.28

SIMILAR TRIANGLES

Triangles which are equi-angular are said to be similar triangles. Thus in Fig. 15.29, if

$$\angle A = \angle X, \quad \angle B = \angle Y \quad \text{and} \quad \angle C = \angle Z$$

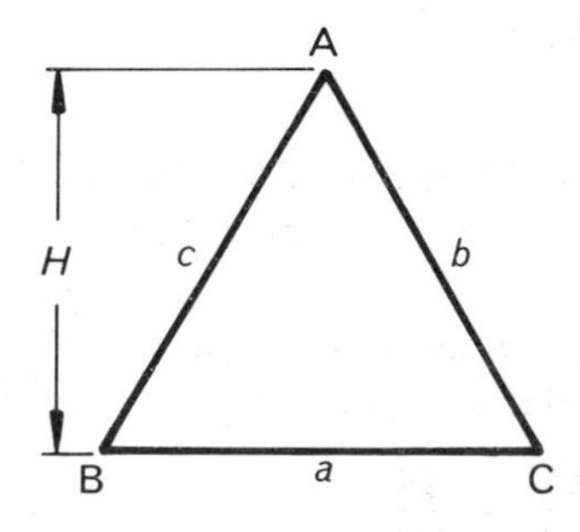

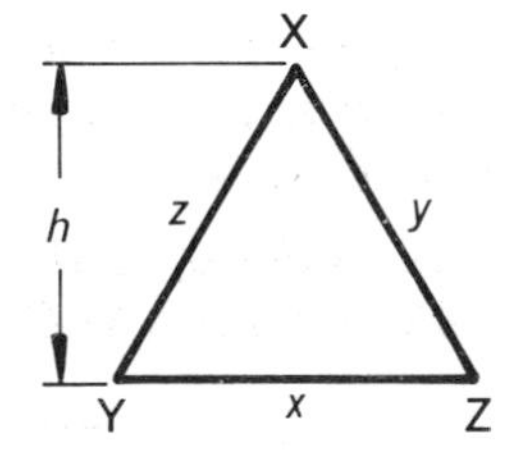

Fig. 15.29

the triangles ABC and ZYX are similar. In similar triangles the ratios of corresponding sides are equal. Thus for the triangles shown in Fig. 15.29.

$$\frac{a}{x} = \frac{b}{y} = \frac{c}{z} = \frac{H}{h}$$

Note that by corresponding sides we mean the sides opposite to the equal angles. It helps in solving problems on similar triangles if we write the two triangles with the equal angles under each other. Thus in Δs ABC and XYZ if $\angle A = \angle X, \angle B = \angle Y$ and $\angle C = \angle Z$ we write $\dfrac{ABC}{XYZ}$.

The equations connecting the sides of the triangles are then easily obtained by writing any two letters in the first triangle over any two corresponding letters in the second triangle. Thus,

$$\frac{AB}{XY} = \frac{AC}{XZ} = \frac{BC}{YZ}$$

In Fig. 15.30 to prove ΔABC is similar to ΔXYZ it is sufficient to prove any one of the following:

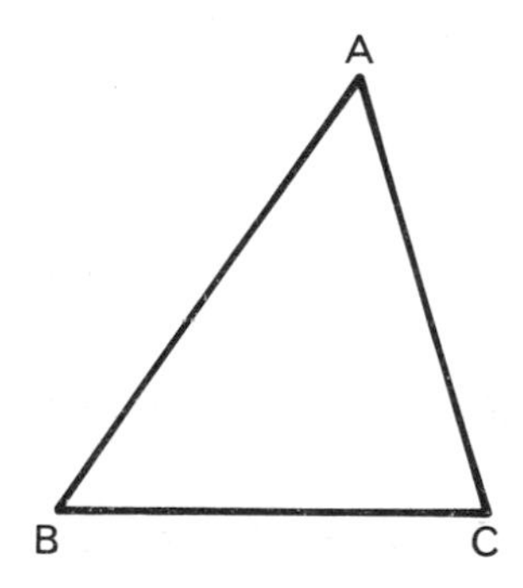

Fig. 15.30

(1) *Two angles in* ΔABC *equal to two angles in* ΔXYZ. For instance, the triangles are similar if $\angle A = \angle X$ and $\angle B = \angle Y$, since it follows that $\angle C = \angle Z$.

(2) *The three sides of* ΔABC *are proportional to the corresponding sides of* ΔXYZ. Thus ΔABC is similar to ΔXYZ if

$$\frac{AB}{XY} = \frac{AC}{XZ} = \frac{BC}{YZ}$$

(3) *Two sides in* ΔABC *are proportional to two sides in* ΔXYZ *and the angles included between these sides in each triangle are equal.* Thus ΔABC is similar to ΔXYZ if

$$\frac{AB}{XY} = \frac{AC}{XZ} \quad \text{and} \quad \angle A = \angle X$$

EXAMPLE 15.4

a) In Fig. 15.31 find the dimension marked x.

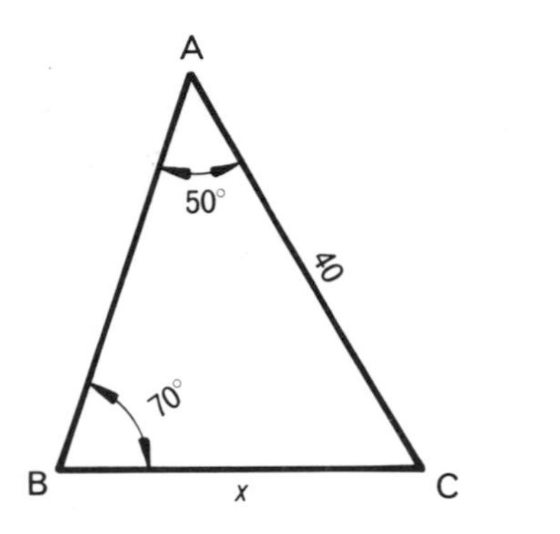

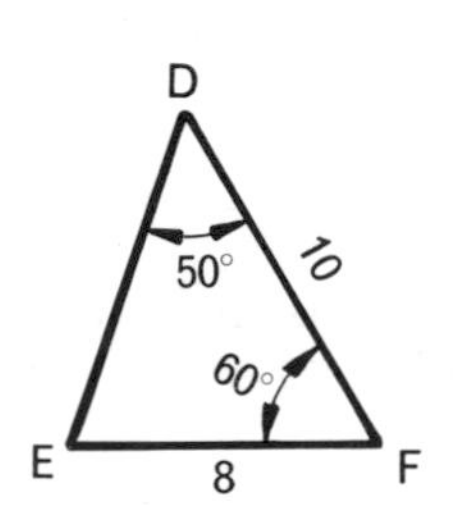

Fig. 15.31

In ΔABC, angle C $= 180° - 50° - 70° = 60°$
In ΔDEF, angle E $= 180° - 50° - 60° = 70°$

Therefore ΔABC and ΔDEF are similar.

Thus $$\frac{40}{10} = \frac{x}{8} \quad \text{or} \quad 320 = 10x$$

$\therefore$ $$x = \frac{320}{10} = 32\,\text{mm}$$

b) In Fig. 15.32 prove that Δs PTS and PQR are similar and calculate the length of TS.

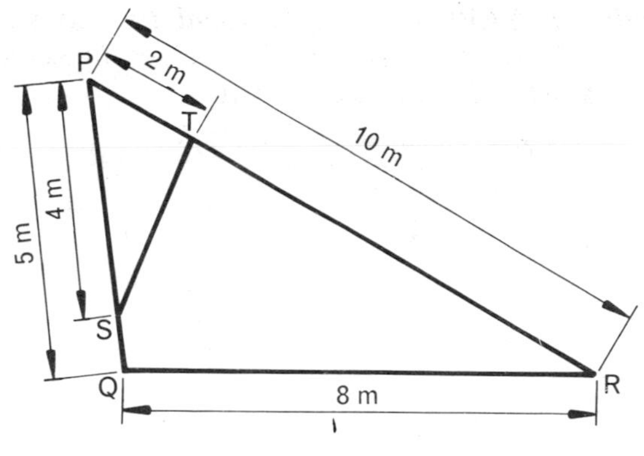

Fig. 15.32

In ∆s PTS and PQR:

$$\frac{PS}{PR} = \frac{4}{10} = 0.4$$

and
$$\frac{PT}{PQ} = \frac{2}{5} = 0.4$$

∴
$$\frac{PS}{PR} = \frac{PT}{PQ}$$

Also ∠P is common to both triangles and it is the included angle between PS and PT in ∆PTS and PR and PQ in ∆PQR. Hence ∆s PTS and PQR are similar. Writing $\frac{\Delta PTS}{\Delta PQR}$ we see that

$$\frac{TS}{QR} = \frac{PT}{PQ}$$

or
$$\frac{TS}{8} = \frac{2}{5}$$

∴
$$TS = \frac{2 \times 8}{5} = 3.2\,\text{m}$$

Exercise 15.3

1) In Fig. 15.33 find BC.

2) In ∆s ABC and PQR, ∠A = ∠R and ∠C = ∠P. State the ratios between the sides of the two triangles.

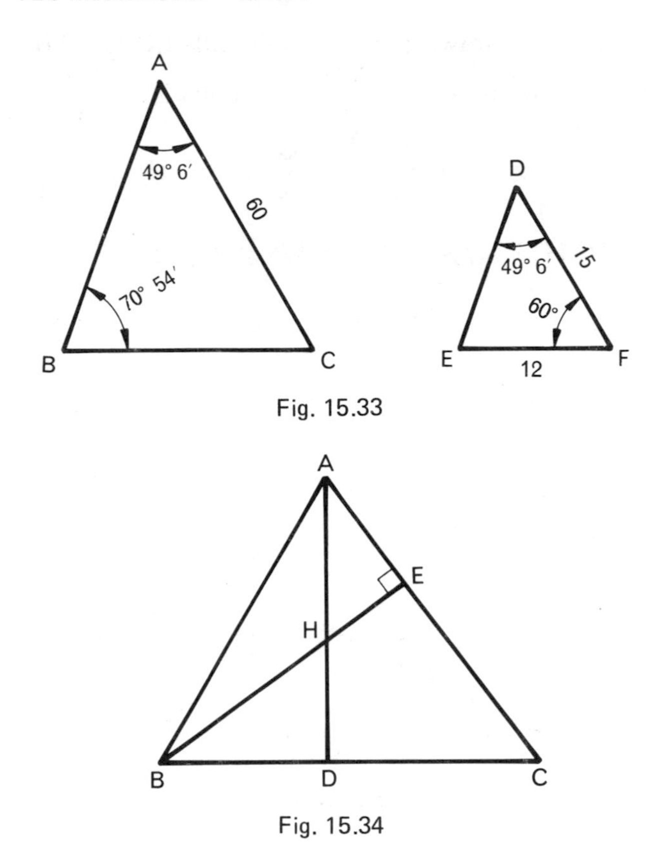

Fig. 15.33

Fig. 15.34

3) In Fig. 15.34 calculate AE and EH, if BH = 5 m, DH = 3 m, BD = 4 m and AH = 4 m.

4) A triangle ABC has sides AB = 50 mm and BC = 20 mm. Calculate the length of the side corresponding to BC in a similar triangle if that corresponding to AB = 40 mm.

5) In the parallelogram ABCD, the point P is taken on the side AB so that AP = 2PB. The lines BD and PC intersect at O. Prove that Δs PBO and CDO are similar and hence calculate $\frac{OP}{OC}$.

6) WXYZ is a parallelogram. A line through W meets ZY at T and XY produced at U. Prove that Δs WZT and UYT are similar. If ZT = 30 mm, TY = 20 mm and UY = 10 mm, find WZ.

7) ABCD is a trapezium in which AB is parallel to DC. AB = 3 m, DC = 6 m and the diagonal BD = 7.8 m. If BD and AC meet at K, calculate KB. If X is the mid-point of BD and the line parallel to DC through X meets AC at Y, calculate XY.

8) The line MN is drawn parallel to the side BC of ΔABC to meet AB at M and AC at N so that $\frac{MN}{BC} = \frac{1}{4}$. Calculate $\frac{AM}{MB}$.

CONSTRUCTION OF TRIANGLES

(1) To construct a triangle given the lengths of each of the three sides.

Construction: Suppose $a = 60$ mm, $b = 30$ mm and $c = 40$ mm. Draw BC = 60 mm. With centre B and radius 40 mm draw a circular arc. With centre C and radius 30 mm draw a circular arc to cut the first arc at A. Join AB and AC. Then ABC is the required triangle (Fig. 15.35).

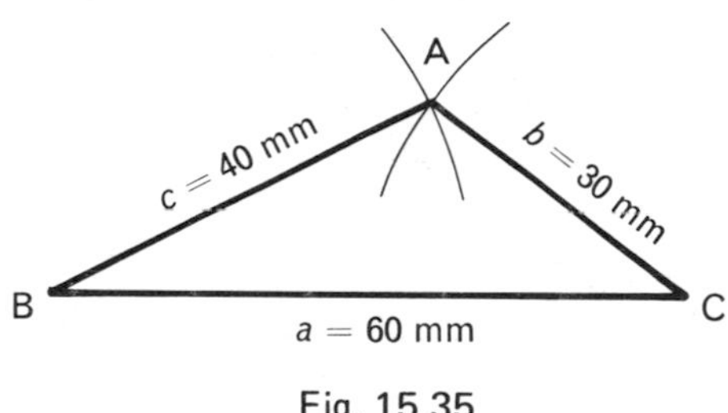

Fig. 15.35

(2) To construct a triangle given two sides and the included angle between the two sides.

Construction: Suppose $b = 50$ mm, $c = 60$ mm and $\angle A = 60°$. Draw AB = 60 mm and draw AX such that $\angle$BAX = 60°. Along AX mark off AC = 50 cm. Then ABC is the required triangle (Fig. 15.36).

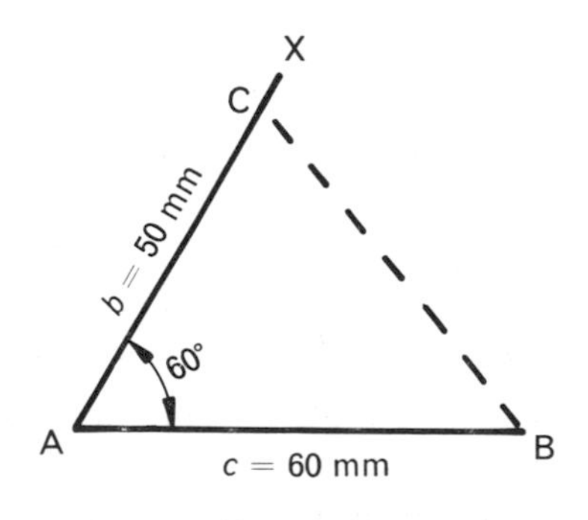

Fig. 15.36

(3) To construct a triangle given one side and two angles.

Suppose $c = 40$ mm, $A = 30°$ and $B = 50°$. Draw AB = 40 mm and draw AX such that ∠BAX = 30°. Draw BY such that ∠ABY = 50°. The intersection of AX and BY gives the point C and ABC is the required triangle (Fig. 15.37).

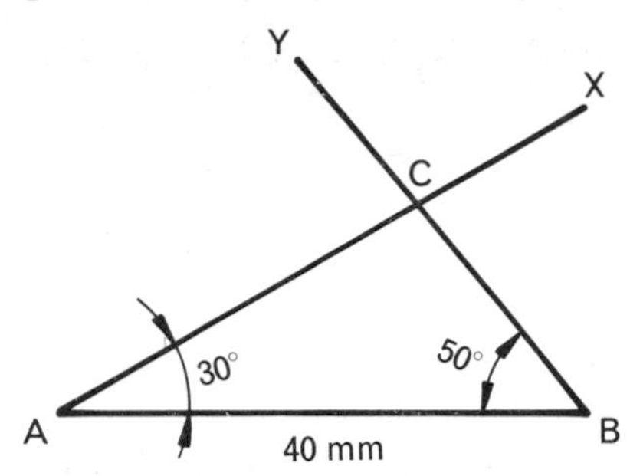

Fig. 15.37

(4) To construct a triangle given one side, the hypotenuse and the right angle.

Suppose $A = 90°$, $a = 60$ mm and $c = 50$ mm. Draw AB = 50 mm. With centre O and radius OA draw a circle to cut AB at Y. Draw the diameter YW. Join AW and produce it to X. ∠BAX is then a right angle. With centre B and radius 60 mm draw an arc to cut AX at C and join BC. ABC is then the required triangle (Fig. 15.38).

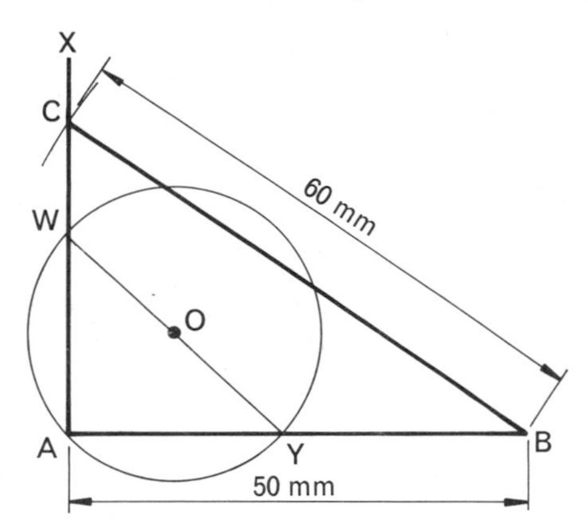

Fig. 15.38

Exercise 15.4

1) Draw △ABC given $a = 50$ mm, $b = 60$ mm and $c = 80$ mm. State A.

2) Draw △ABC given $a = 20$ mm, $b = 60$ mm and $c = 70$ mm. State B.

3) Draw ΔABC given A = 50°, b = 80 mm and c = 60 mm. State a.

4) Draw ΔABC given C = 70°, a = 50 mm and c = 70 mm. State A.

5) Draw ΔABC given A = 40°, B = 70° and c = 50 mm. State b.

6) Draw ΔABC given A = 60°, C = 50° and a = 40 mm. State c.

7) Draw ΔABC given A = 90°, c = 40 mm and a = 50 mm. State B.

8) Draw ΔABC given A = 90°, b = 50 mm and a = 80 mm. State c.

SUMMARY

a) An acute-angled triangle has all its angles less than 90°.

b) A right-angled triangle has one of its angles equal to 90°. The longest side lies opposite to the right angle and is called the hypotenuse.

c) An obtuse-angled triangle has one angle greater than 90°.

d) A scalene triangle has three sides of different lengths.

e) An isosceles triangle has two sides and two angles equal. The equal angles lie opposite to the equal sides.

f) An equilateral triangle has all its sides and angles equal.

g) The sum of the angles of a triangle equal 180°.

h) Pythagoras' Theorem states 'In any right-angled triangle the square on the hypotenuse is equal to the sum of the squares on the other two sides'.

i) Two triangles are congruent if they are equal in every respect. Any of the following are sufficient to prove that two triangles are congruent:

   (i) One side and two angles in one triangle equal to one side and two similarly located angles in the second triangle.

   (ii) Two sides and the angle between them in one triangle equal to two sides and the angle between them in the second triangle.

   (iii) Three sides of one triangle equal in length to three sides of the second triangle.

   (iv) The hypotenuse and one side in a right-angled triangle equal to the hypotenuse and one side in a second right-angled triangle.

j) Triangles which are equi-angular are called similar triangles. In Δs ABC and XYZ, if A = X, B = Y and C = Z,

$$\frac{AB}{XY} = \frac{AC}{XZ} = \frac{BC}{YZ}$$

To prove ΔABC is similar to ΔXYZ it is sufficient to prove any one of the following:

(i) Two angles in ΔABC equal to two angles in ΔXYZ.

(ii) Three sides of ΔABC proportional to the three corresponding sides of ΔXYZ.

(iii) Two sides of ΔABC proportional to two sides of ΔXYZ and the angles between these sides in each triangle equal.

Self-Test 11

State the letter (or letters) corresponding to the correct answer (or answers).

1) The triangle shown in Fig. 15.39 is:

a acute-angled b obtuse-angled c scalene d isosceles

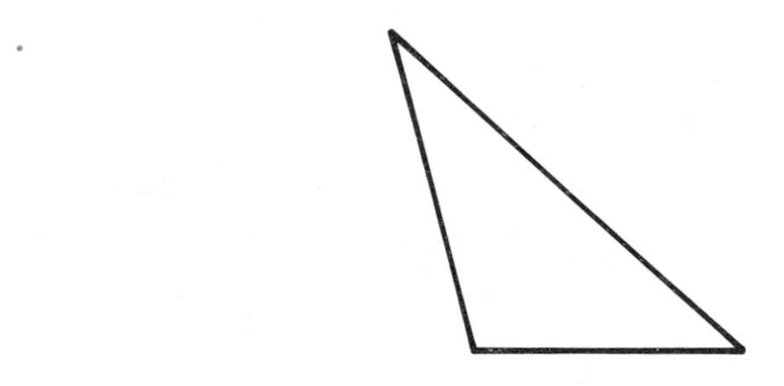

Fig. 15.39

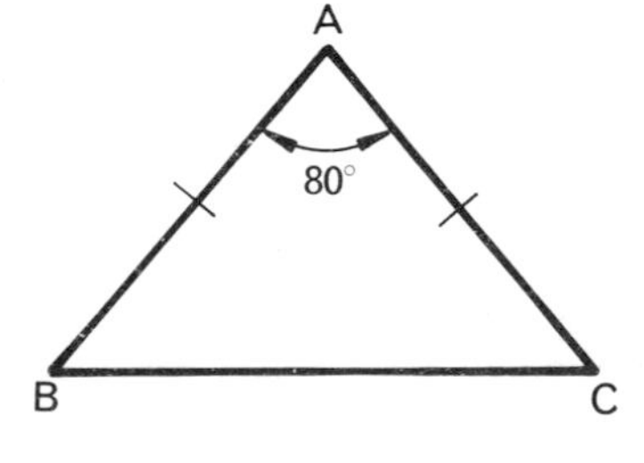

Fig. 15.40

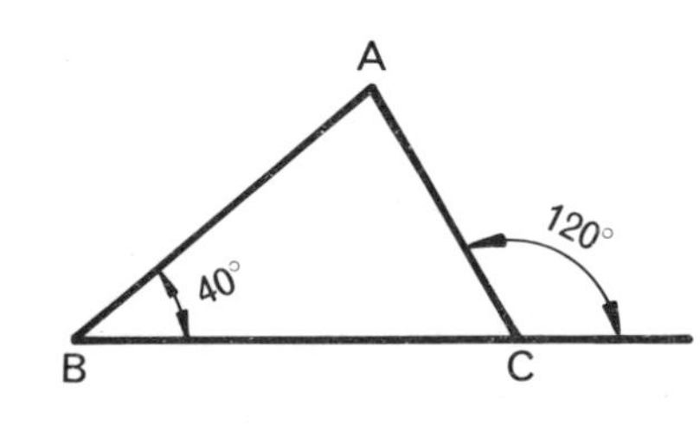

Fig. 15.41

2) In Fig. 15.40, ∠B is equal to:

a 80° b 40° c 50° d 90°

3) In Fig. 15.41, ∠A is equal to:

a 20° b 40° c 60° d 80°

4) In Fig. 15.42, θ is equal to:

a 140° b 70° c 80° d 60°

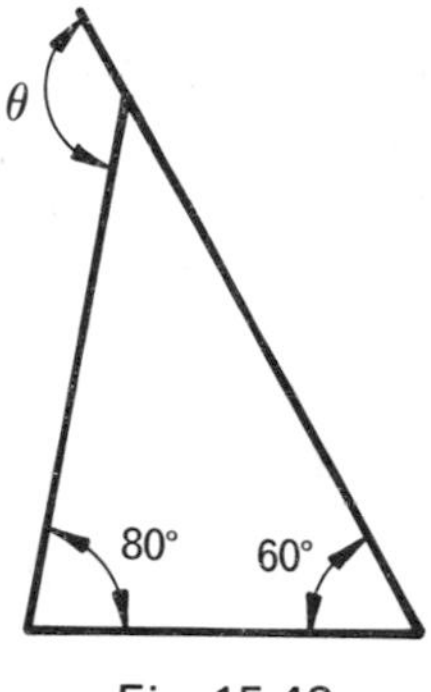

Fig. 15.42

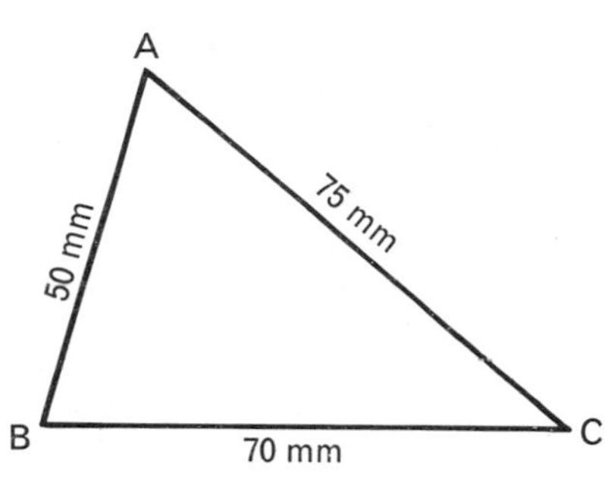

Fig. 15.43

5) In Fig. 15.43, the largest angle of the triangle is:

a ∠A b ∠B c ∠C

6) A triangle is stated to have sides whose lengths are 50 mm, 80 mm and 140 mm.

a It is possible to draw the triangle
b It is impossible to draw the triangle

7) In Fig. 15.44, one of the following is not correct. Which one?

a ∠BAD = ∠DAC b ∠ABD = ∠ACD
c BD = DC AD = BD

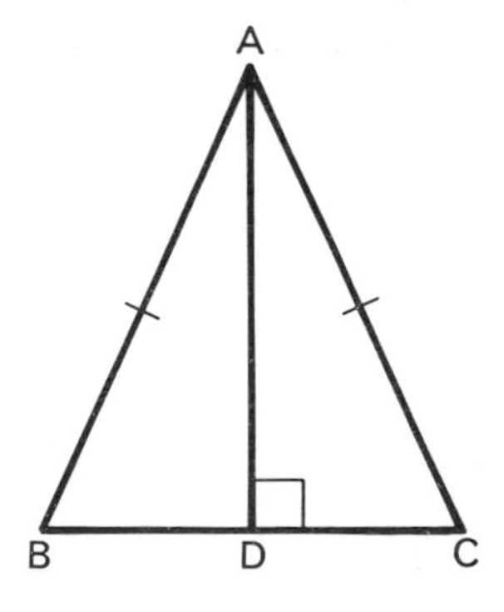

Fig. 15.44

8) In Fig. 15.45, α is equal to:

a 17° b 48° c $8^\circ 30'$ d 25°

9) In Fig. 15.46, ϕ is equal to:

a 80° b 60° c 40° d 100°

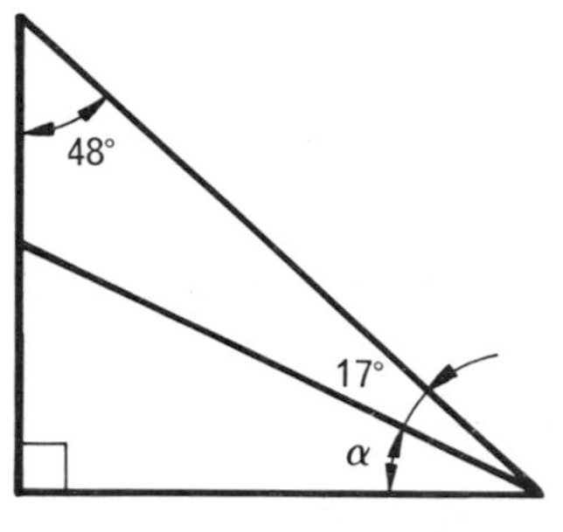

Fig. 15.45

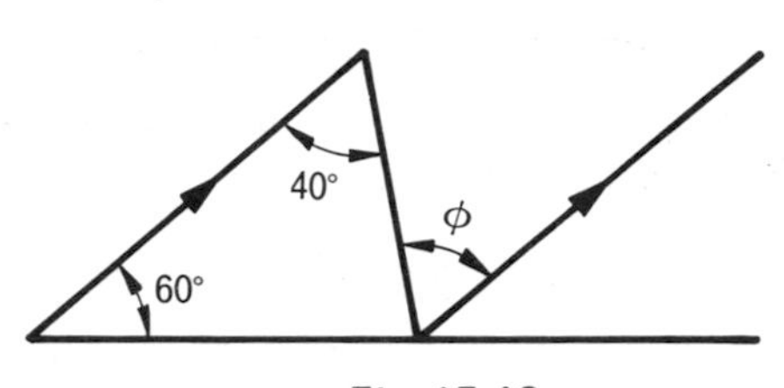

Fig. 15.46

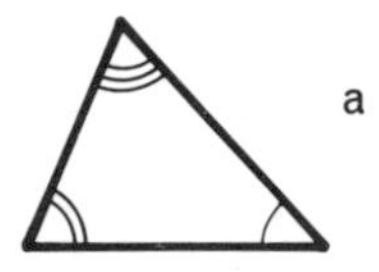
a
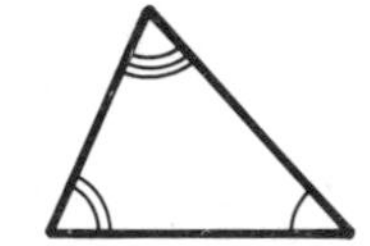

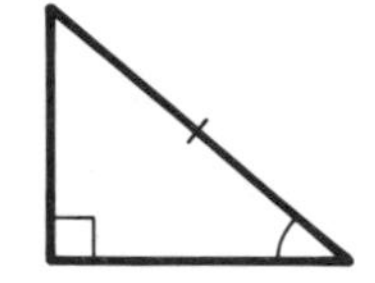
c
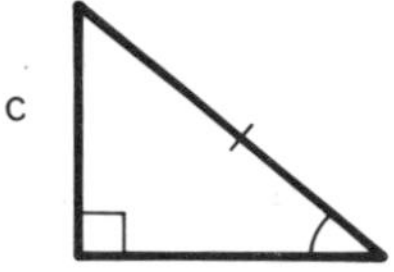

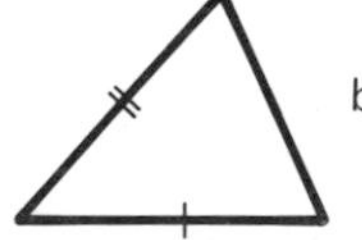
b
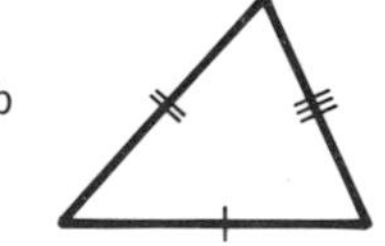

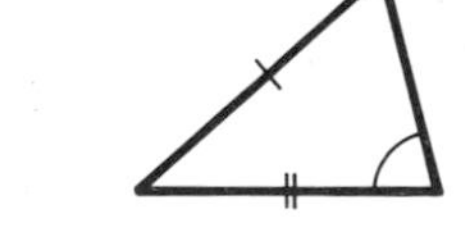
d
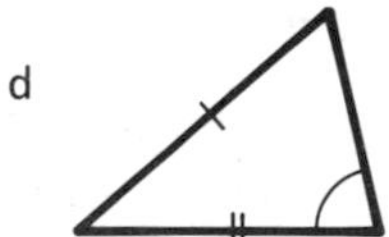

Fig. 15.47

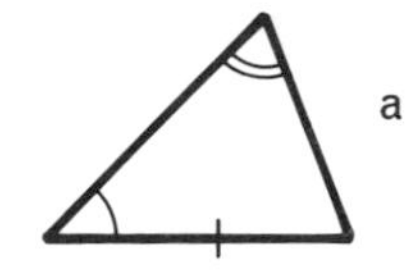
a
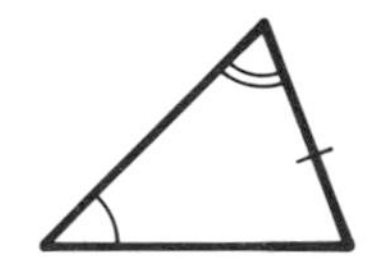

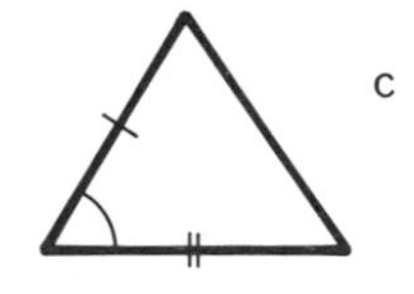
c
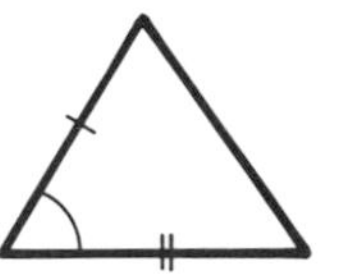

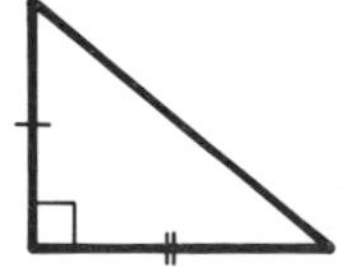
b
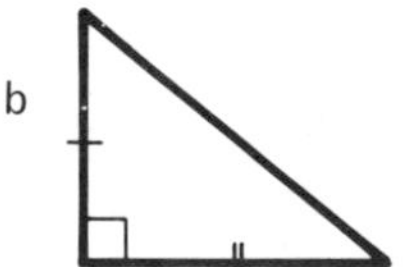

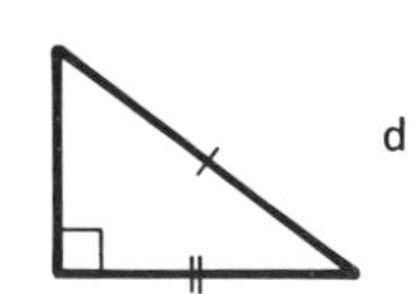
d
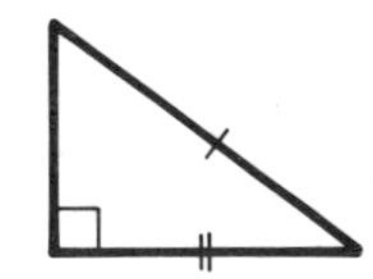

Fig. 15.48

10) Two angles of a triangle are $(2\theta - 40)^\circ$ and $(3\theta + 10)^\circ$. The third angle is therefore:

a $(210 - 5\theta)^\circ$ **b** $(230 + \theta)^\circ$ **c** $(220 - 5\theta)^\circ$

11) The three angles of a triangle are $(2\alpha + 20)^\circ$, $(3\alpha + 20)^\circ$ and $(\alpha + 20)^\circ$. The value of α is:

a 60° **b** 40° **c** 20° **d** 10°

12) In Fig. 15.47 state the letter which corresponds to those triangles which are congruent.

13) In Fig. 15.48 state the letter which corresponds to those triangles which are congruent.

14) In Fig. 15.49, AD = BC and AC = DB. Hence:

a $\angle DAC = \angle DBC$ **b** $\angle ADB = \angle DBC$
c $\angle ADC = \angle BCD$ **d** $\angle ADC = \angle BDC$

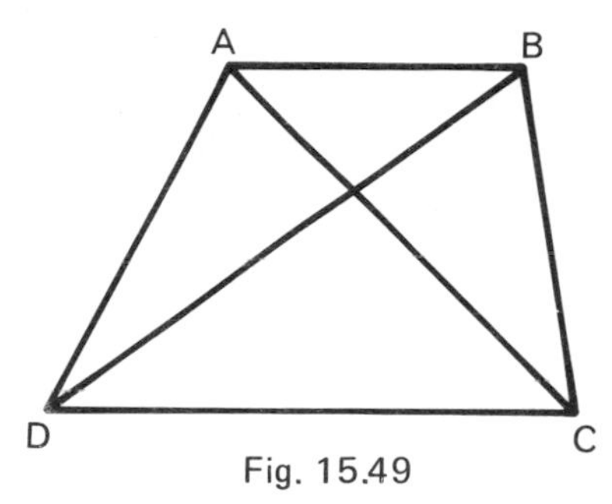

Fig. 15.49

15) In Fig. 15.50 ΔDEC is equilateral and ABCD is a square. $\angle DEA$ is therefore:

a 30° **b** 15° **c** 45° **d** 20°

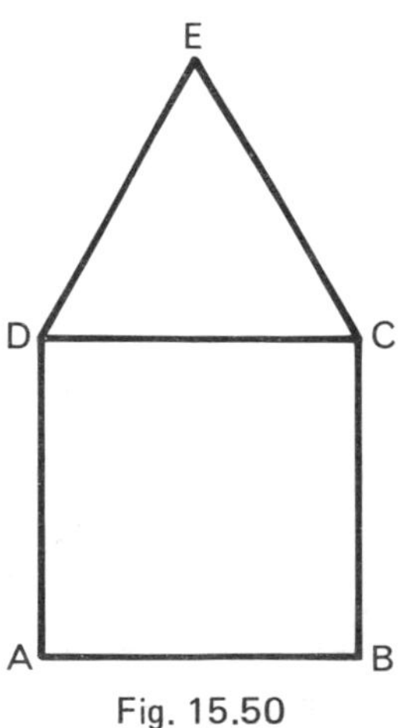

Fig. 15.50

16) In Fig. 15.51 two straight lines bisect each other at X. Therefore:

a $\Delta AXC \equiv \Delta DXB$ b $\Delta ABD \equiv \Delta ABC$
c $\Delta ADX \equiv \Delta CXB$ d $\angle CAX = \angle XDB$

17) The triangles shown in Fig. 15.52 are:

a congruent b similar c neither of these

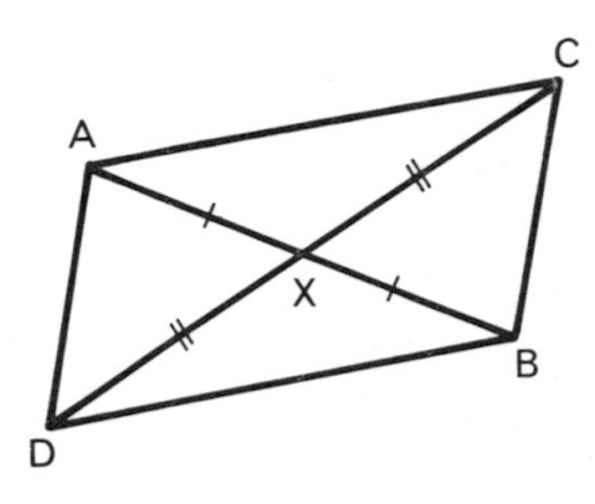

Fig. 15.51

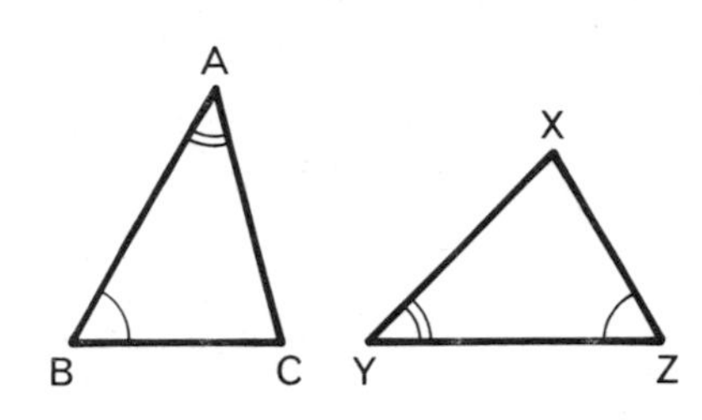

Fig. 15.52

18) If the triangles ABC and XYZ shown in Fig. 15.52 are similar, then:

a $\frac{AC}{XY} = \frac{XZ}{BC}$ b $\frac{AC}{XY} = \frac{BC}{XZ}$ c $\frac{BC}{AB} = \frac{YZ}{XZ}$ d $\frac{BC}{AB} = \frac{XZ}{YZ}$

19) In Fig. 15.53 if $\frac{AB}{XY} = \frac{AC}{XZ}$ and $\angle B = \angle Y$ then:

a $\frac{AB}{XY} = \frac{BC}{YZ}$ b $\angle A = \angle X$ c $\angle C = \angle Z$

d none of the foregoing are necessarily true

20) In Fig. 15.54, $\angle A = \angle X$ and $\angle B = \angle Y$. Hence:

a $XY = 6\frac{7}{8}$ m b $XY = 17\frac{3}{5}$ m
c $YZ = 19\frac{1}{5}$ m d $YZ = 7\frac{1}{2}$ m

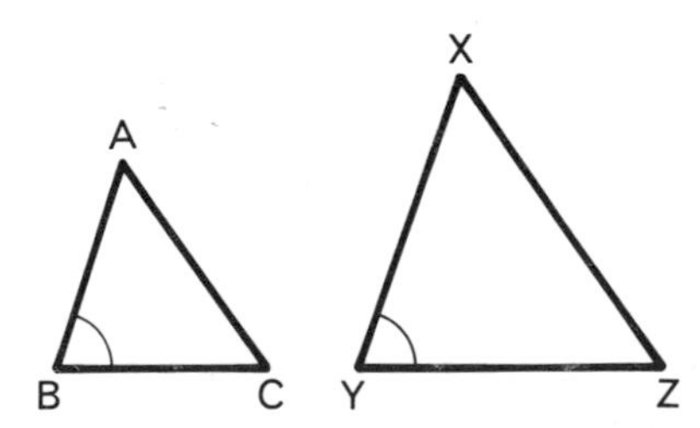

Fig. 15.53

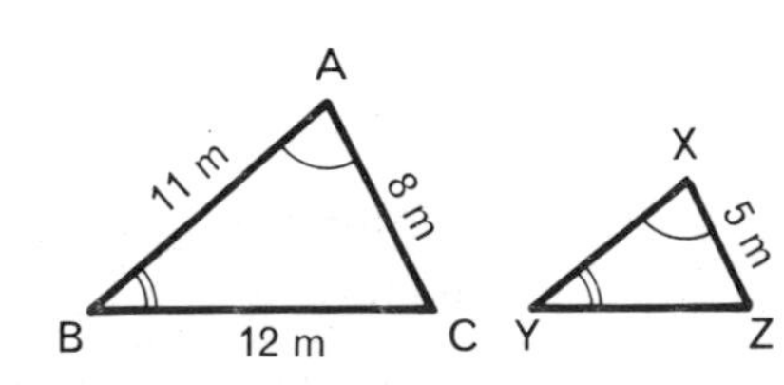

Fig. 15.54

16. THE CIRCLE

fter reaching the end of this chapter you should be able to:

1. Identify radius, diameter, circumference, chord, tangent, sector, segment and arc of a circle.
2. Solve simple problems relating circumference, radius and diameter of circles.
3. State the relationship between the angle at the circumference and the angle at the centre of a circle.
4. Recognise that the angle between a tangent and the radius of a circle is a right angle.

Fig. 16.1 shows the main components of a circle.

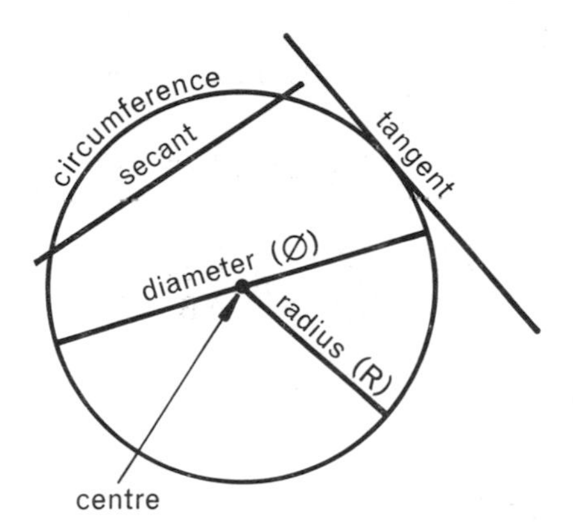

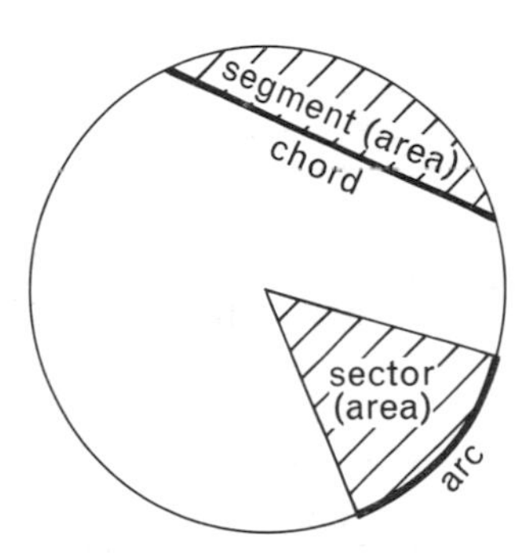

Fig. 16.1

RELATION BETWEEN DIAMETER, RADIUS AND CIRCUMFERENCE

Whatever size of circle is chosen the ratio $\frac{\text{circumference}}{\text{diameter}}$ is always constant and has a value 3.141 592 653 . . .

The value of this ratio is so important that it has been given a special symbol π (the Greek letter 'pi').

Hence $\frac{\text{circumference}}{\text{diameter}} = \pi$

$\therefore$ circumference $= \pi \times$ diameter $= 2 \times \pi \times$ radius

or $C = \pi d = 2\pi r$

THE NUMERICAL VALUE OF π

π cannot be given an exact value and care has to be taken in choosing the correct number of decimal places to obtain the required accuracy in any calculation.

Remember that any answer is only as accurate as the least accurate of the figures used. For instance, it is useless to use $\pi = 3.14$ for accurate workshop calculations. The following may act as a guide:

$\pi = 3.14$ (correct to 3 significant figures). This is the value used for slide rule calculations and $\frac{22}{7}$ is sometimes used instead.

$\pi = 3.142$ (correct to 4 significant figures). This is the value used for calculations with 4-figure log. tables.

$\pi = 3.141\,593$ (correct to 7 significant figures). This is the value which MUST be used for accurate workshop calculations (including angles correct to the nearest 'second'). Electronic calculators or 7-figure log. tables will generally be used for such calculations.

On a scientific calculator use of the $\boxed{\pi}$ key will give a value to a number of significant figures, depending on the number of digits on the display. The answer should then be rounded off to the number of significant figures consistent with the least accurate figure of the given data.

EXAMPLE 16.1

The diameter of a circle is 280 mm. What is its circumference?

Now $C = \pi d$

Thus $C = \pi \times 280 \text{ mm}$

$= 880 \text{ mm}$ correct to three significant figures

EXAMPLE 16.2

Find the radius of a circle which has a circumference of 15 m.

Now $C = 2\pi r$

$$\therefore \qquad r = \frac{C}{2\pi}$$

$$= \frac{15}{2 \times \pi}\text{ m}$$

$$= 2.39\text{ m} \quad \text{correct to three significant figures}$$

Strictly the answer should only be stated correct to two significant figures, since that was the accuracy of the given data, namely 15, but it is usual to assume that the measurement was 15.0 which is three-significant-figure accuracy.

EXAMPLE 16.3

A wheel 715 mm diameter makes 30 revolutions about a fixed centre. How far does a point on the rim travel?

Now Distance travelled in 1 revolution $= \pi \times \text{diameter}$

$= 3.142 \times 715\text{ mm}$

$= 2247\text{ mm}$

$\therefore$ Distance travelled in 30 revolutions $= 30 \times 2247\text{ mm}$

$= 67\,410\text{ mm}$

$= 67.4\text{ m}$

Exercise 16.1

Find the circumference of the circles in questions 1 to 8

1) Radius 21 m
2) Radius 350 mm
3) Radius 43 m
4) Radius 3.16 m
5) Diameter 280 mm
6) Diameter 85 mm
7) Diameter 8.423 m
8) Diameter 1400 mm
9) A wheel has a diameter of 560 mm. How far, in metres, will a point on the rim travel in 50 revolutions?
10) A circular flower bed has a circumference of 64 m. What is its radius?
11) Find the diameter of a circle whose circumference is 110 km.
12) Find the radius of a circle whose circumference is 956 mm.

ANGLES IN CIRCLES

The angle which an arc of a circle subtends at the centre is twice the angle which the arc subtends at the circumference.

Thus in Fig. 16.2, $\angle AOB = 2 \times \angle APB$.

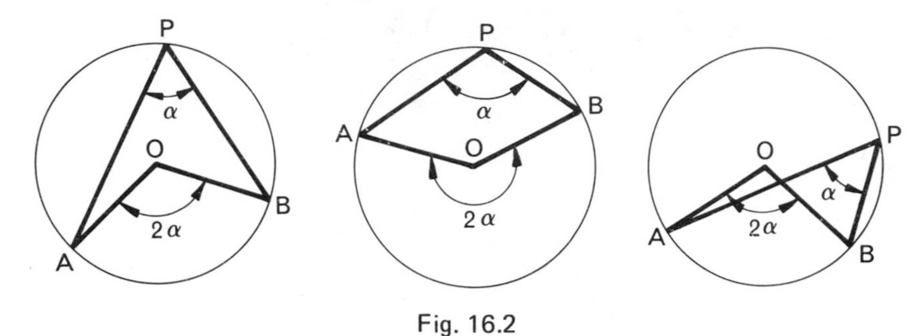

Fig. 16.2

If a triangle is inscribed in a semi-circle, the angle opposite the diameter is a right angle (Fig. 16.3).

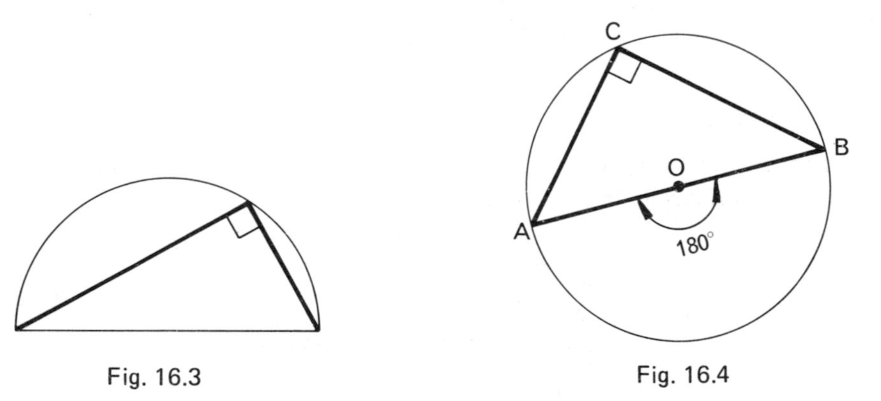

Fig. 16.3

Fig. 16.4

This theorem follows from the fact that *the angle subtended by an arc at the centre is twice the angle subtended by the arc at the circumference.*

Thus in Fig. 16.4:

angle subtended at the centre by the arc AB = 180°,
hence the angle subtended at the circumference = 90°.

EXAMPLE 16.4

In Fig. 16.5, O is the centre of the circle. If $\angle AOB = 60°$, find $\angle ACB$.

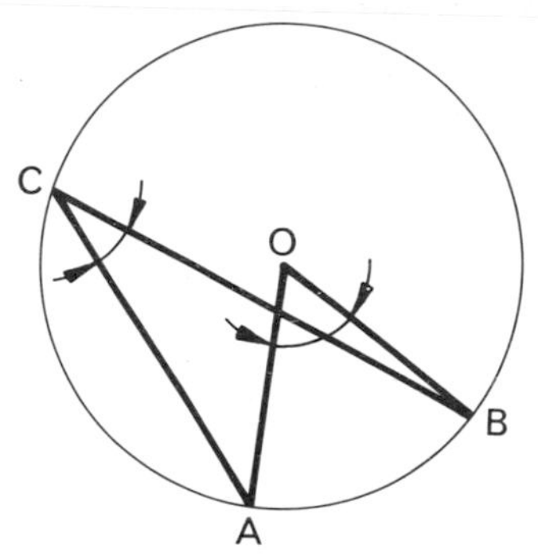

Fig. 16.5

Since $\angle AOB$ is the angle subtended by the arc AB at the centre O and $\angle ACB$ is the angle subtended by AB at the circumference:

Now $\qquad \angle AOB = 2 \times \angle ACB$

but $\qquad \angle AOB = 60°$

$\therefore \qquad \angle ACB = 30°$

EXAMPLE 16.5

The three holes A, B and C are to be marked off. Find the diameter of the pitch circle (Fig. 16.6).

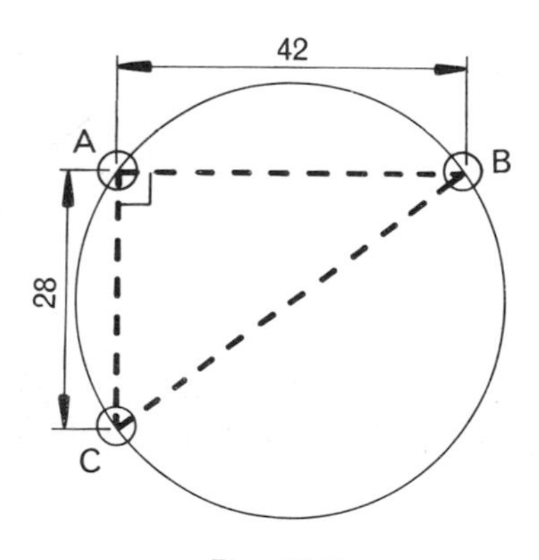

Fig. 16.6

Since $\angle CAB$ is a right angle, this is the angle in a semi-circle and hence BC is a diameter.

In ΔABC, by Pythagoras,

$$BC^2 = AB^2 + AC^2 = 42^2 + 28^2$$

$$= 1764 + 784 = 2548$$

$$\therefore \quad BC = \sqrt{2548} = 50.5\,\text{mm}$$

The diameter of the pitch circle is therefore 50.5 mm.

A tangent to a circle is at right angles to a radius drawn to the point of tangency (Fig. 16.7).

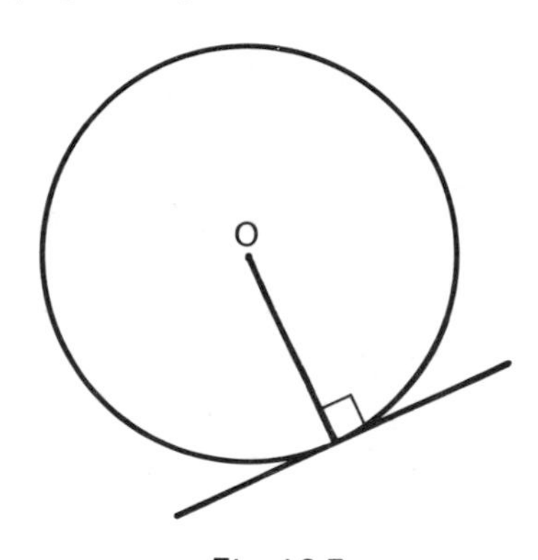

Fig. 16.7

EXAMPLE 16.6

a) Fig. 16.8 shows a belt passing over two pulleys. Calculate the length of a straight portion of the belt.

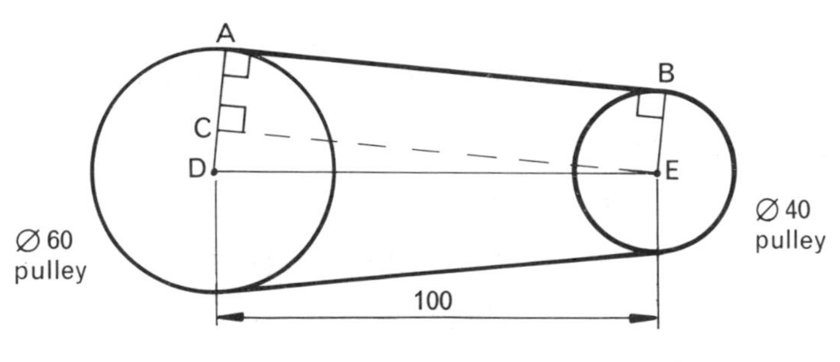

Fig. 16.8

Draw CE parallel to AB.

then $\angle CAB = 90^\circ$ (angle between a radius and a tangent)

but $\angle CAB = \angle DCE$ (since AB is || to CE)

$\therefore$ $\angle DCE = 90^\circ$

In the right-angled triangle DCE,

$$DE = 100\,\text{mm}, \quad CD = 30-20 = 10\,\text{mm}$$

and $DE^2 = CD^2 + CE^2$

or $CE^2 = DE^2 - CD^2 = 100^2 - 10^2$

$\therefore$ $CE = \sqrt{9900} = 99.5\,\text{mm}$

Since CE = AB this is the length of a straight portion.

b) Fig. 16.9 shows a round component supported in a vee block. Find the distance x.

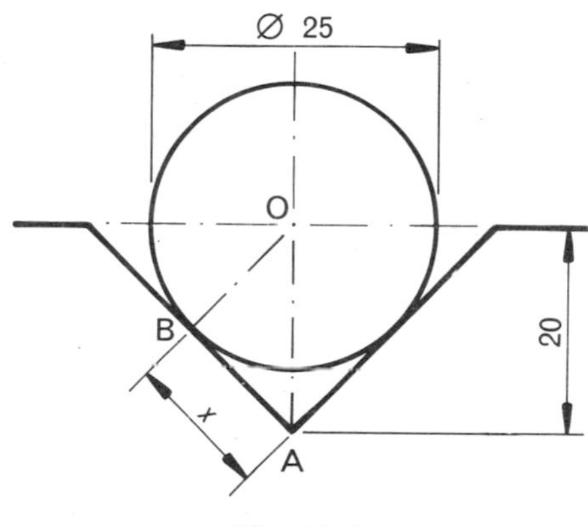

Fig. 16.9

In $\triangle AOB$, $\angle ABO = 90^\circ$ (angle between a radius and a tangent).

By Pythagoras' Theorem,

$$AB^2 = OA^2 - OB^2$$

or $x^2 = 20^2 - 12.5^2 = 243.75$

$\therefore$ $x = \sqrt{243.75} = 15.6\,\text{mm}$

Therefore the distance x is 15.6 mm.

If two circles touch internally or externally, then the line which passes through their centres also passes through the point of tangency (Fig. 16.10).

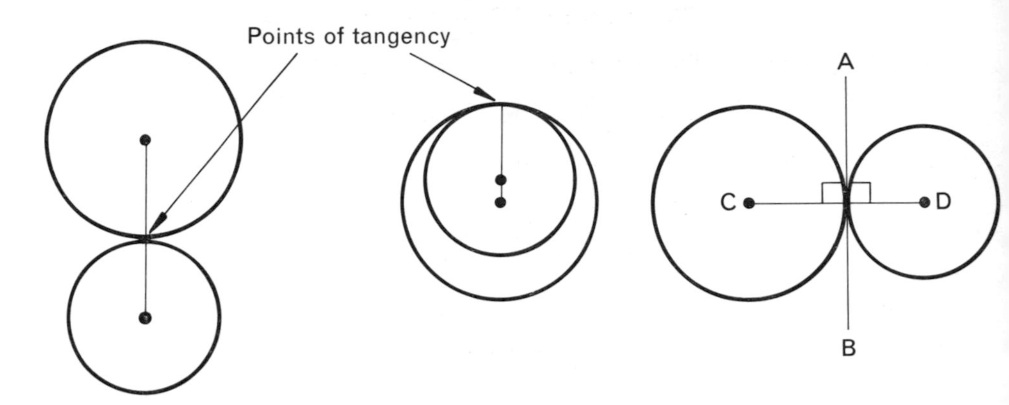

Fig. 16.10

Fig. 16.11

This theorem follows from the fact that the angle between a tangent and a radius drawn to the point of tangency is a right angle. Thus in Fig. 16.11, since AB is tangential to both circles then the line CD joining the centres of the circles must be a straight line.

EXAMPLE 16.7

Two plugs are placed together and a third one is placed on top of the other two (see Fig. 16.12). Find the distance h.

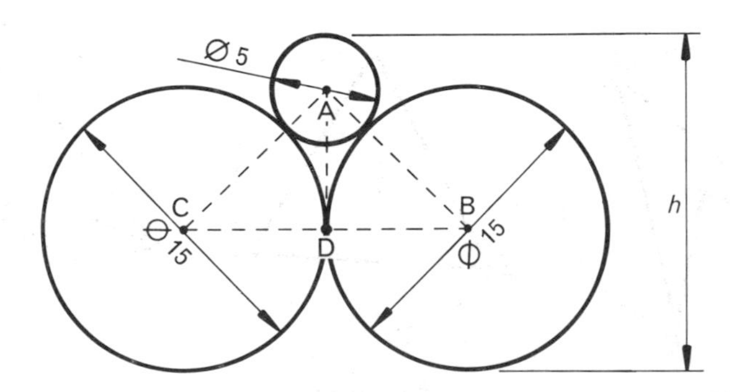

Fig. 16.12

Because the three circles are tangential to each other,

$$AC = 7.5 + 2.5 = 10\,\text{mm}$$

$$AB = 7.5 + 2.5 = 10\,\text{mm}$$

$$BC = 7.5 + 7.5 = 15\,\text{mm}$$

ΔABC is isosceles. The perpendicular AD bisects the base CB.

$$\therefore \quad CD = \tfrac{1}{2} \times 15 = 7.5\,\text{mm}$$

In ΔACD, by Pythagoras' Theorem,

$$AD^2 = AC^2 - CD^2 = 10^2 - 7.5^2$$

$$\therefore \quad AD = \sqrt{43.75} = 6.62\,\text{mm}$$

Thus

$$h = AD + 2.5 + 7.5$$

$$= 6.62 + 10 = 16.6\,\text{mm}$$

Exercise 16.2

1) In Fig. 16.13 if $\angle AOB = 76^\circ$, find $\angle ACB$.

2) ABC is an equilateral triangle inscribed in a circle whose centre is O. Find $\angle BOC$ (Fig. 16.14).

3) ABC is an isosceles triangle inscribed in a circle whose centre is O. If $\angle AOB = 116^\circ$, find $\angle ABC$ (Fig. 16.15).

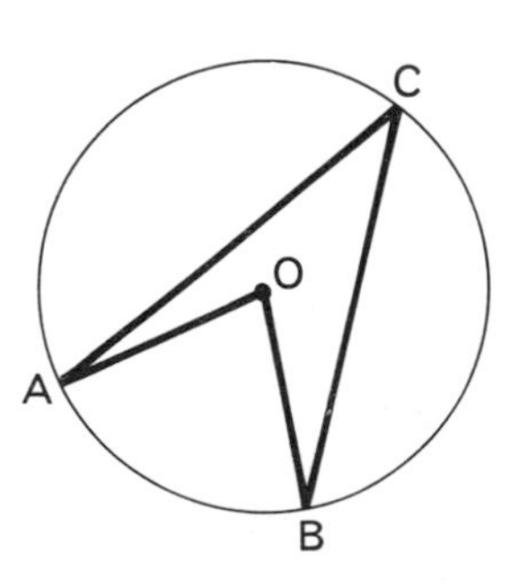

Fig. 16.13

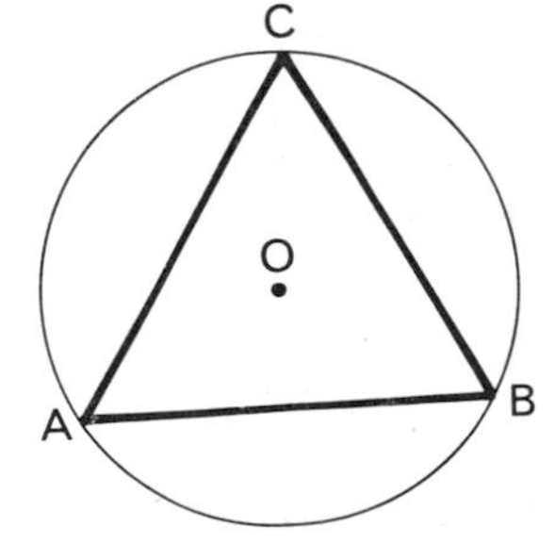

Fig. 16.14

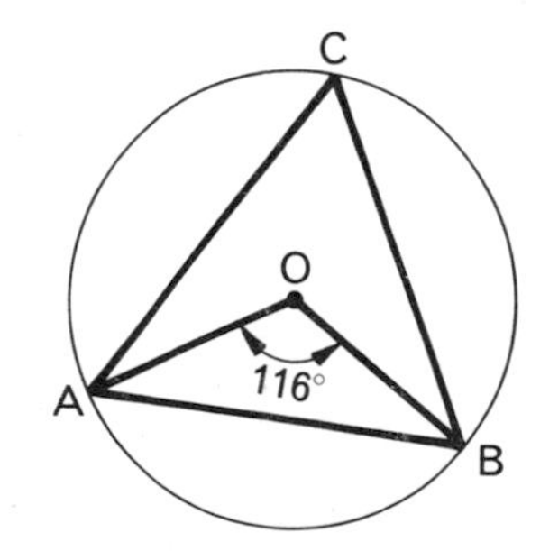

Fig. 16.15

4) Two plugs are placed on a surface table touching each other as shown in Fig. 16.16. Calculate the distance x.

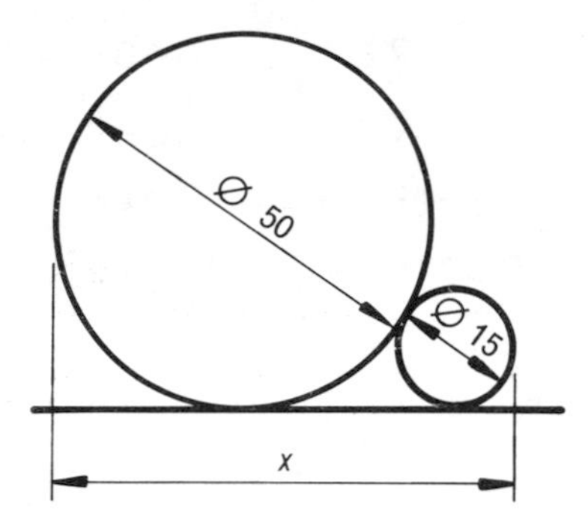

Fig. 16.16

5) Calculate the distance h shown in Fig. 16.17.

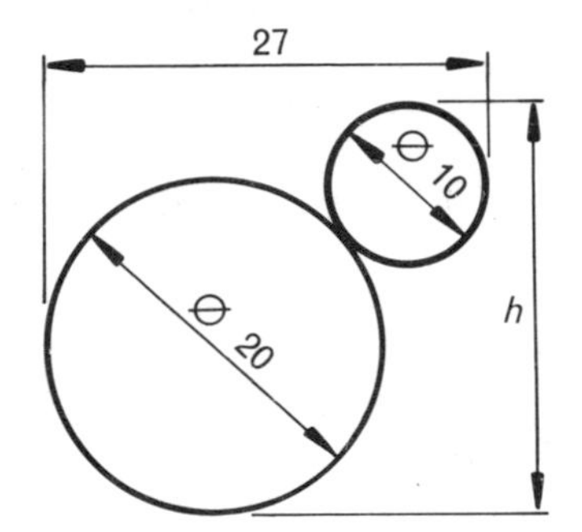

Fig. 16.17

6) Three plugs are arranged as shown in Fig. 16.18. Calculate h.

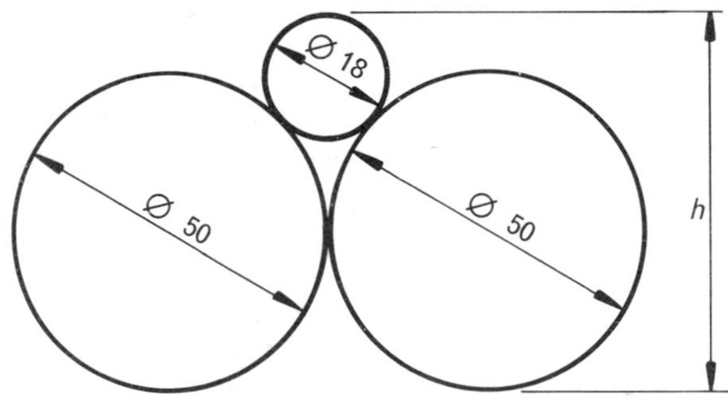

Fig. 16.18

7) The set-up for finding the radius of a circular component is shown in Fig. 16.19. If the radius of the component $R = 108.6$ mm, find the dimension d.

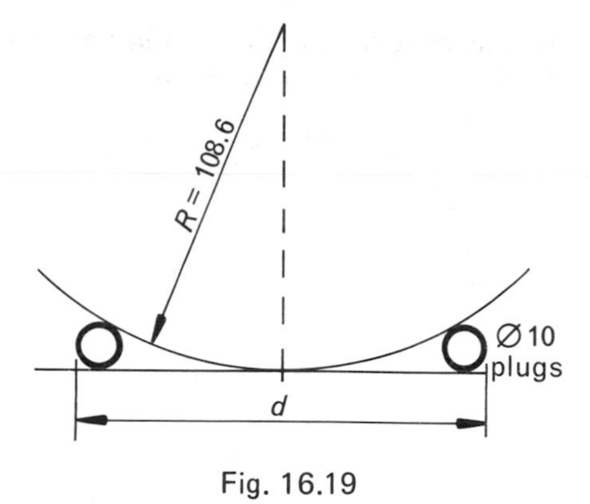

Fig. 16.19

8) Three holes are drilled as shown in Fig. 16.20, their centres lie on a circle 62 mm diameter. What is the dimension a?

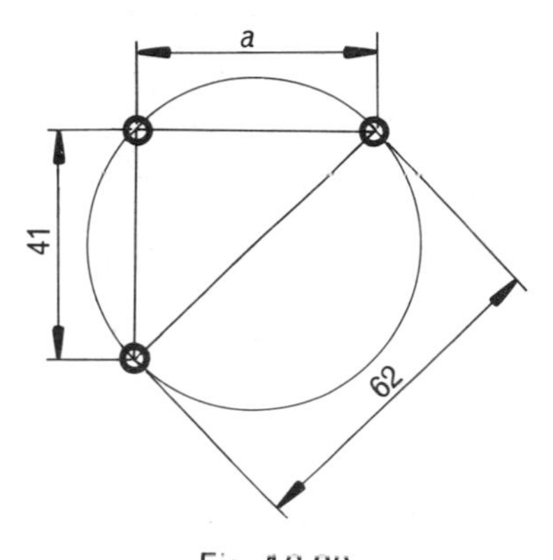

Fig. 16.20

9) Fig. 16.21 shows a component which is to be made from round bar. If the diameter of the bar is 58 mm, find the dimension x.

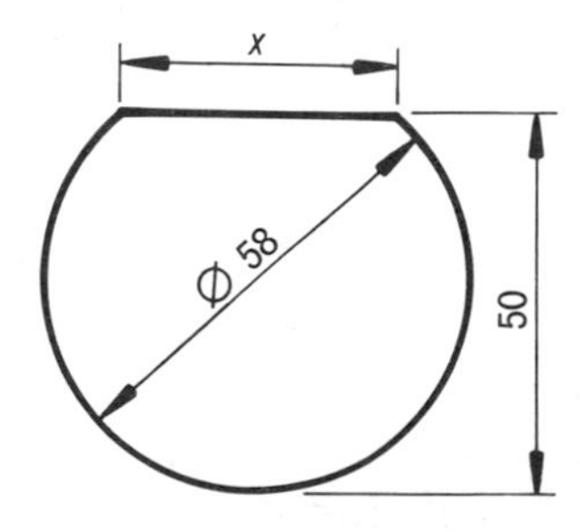

Fig. 16.21

10) In Fig. 16.22 find the dimension marked x.

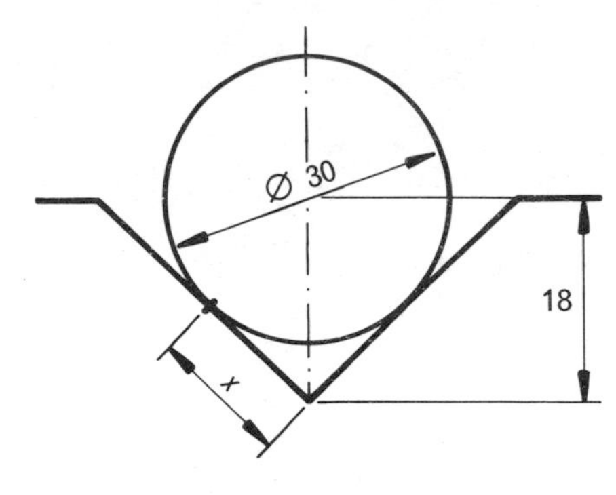

Fig. 16.22

SUMMARY

a) The circumference of a circle is found from the formula:

$$C = \pi d \quad \text{or} \quad C = 2\pi r$$

where d is the diameter and r is the radius.

b) The angle which an arc of a circle subtends at the centre of a circle is twice the angle which the arc subtends at the circumference.

c) If a triangle is inscribed in a semi-circle the angle opposite the diameter is a right angle.

d) A tangent to a circle is at right angles to a radius drawn to the point of tangency.

e) If two circles touch internally or externally then the line which joins their centres also passes through the point of tangency.

Self-Test 12

In the following questions state the letter (or letters) corresponding to the correct answer (or answers).

1) In Fig. 16.23 the line AB is called:

a a tangent b a chord c a diameter d a radius

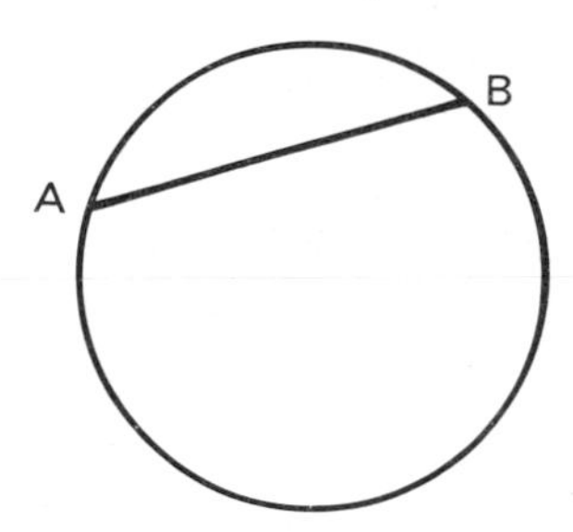

Fig. 16.23

2) The circumference of a circle is 264 mm. Hence the diameter is:

a 42 mm b 168 mm c 84 mm d none of these

3) In Fig. 16.24, O is the centre of the circle and angle BOC = 128°. Therefore the angle ABC is:

a 64° b 128° c 116° d 58°

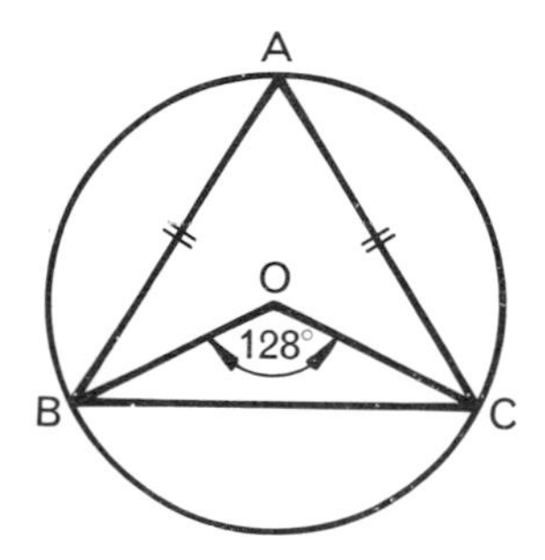

Fig. 16.24

4) Fig. 16.25 shows a belt passing ovcr two pullcys whose diameters are 200 mm and 100 mm respectively. If their centres are 180 mm apart then the length of a straight portion of the belt (AB in the diagram) is:

a 180 mm b 173 mm c 187 mm d 230 mm

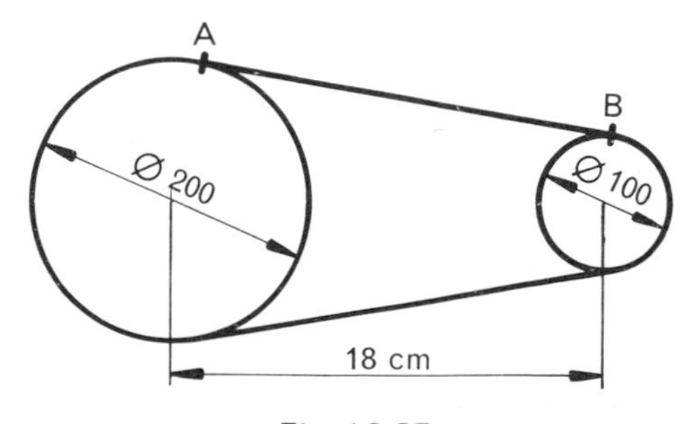

Fig. 16.25

5) In Fig. 16.26 the distance h is:

a 250 mm b 300 mm c 262 mm d 412 mm

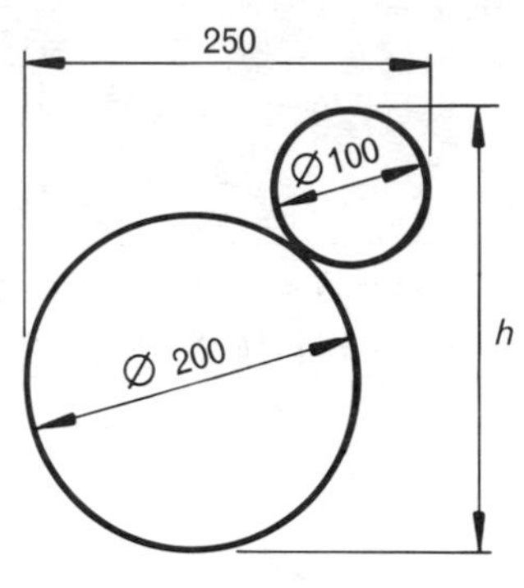

Fig. 16.26

17. AREA AND VOLUME

fter reaching the end of this chapter you should be able to:

1. *Define the properties of the following plane figures relating to sides, angles and diagonals: rectangle, square, parallelogram, rhombus, trapezium.*
2. *Use the properties of plane figures listed to solve practical problems.*
3. *Use given formulae to calculate areas of triangle, square, rectangle, parallelogram, circle, semi-circle, trapezium.*
4. *Use given formulae to calculate the volume of: cube, prism, cylinder.*
5. *Identify similar shapes and know that their areas are proportional to the squares of corresponding linear dimensions.*
6. *Identify similar solids and know that the volumes of similar solids are proportional to the cubes of corresponding linear dimensions.*

QUADRILATERALS

A quadrilateral is any four-sided figure (Fig. 17.1). Since it can be split up into two triangles the sum of its angles is 360°.

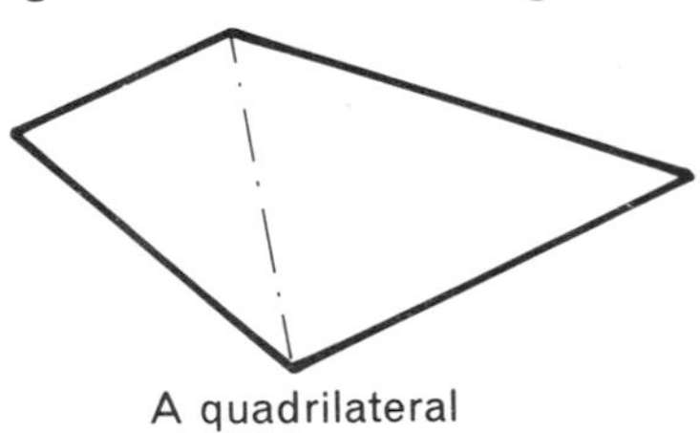

A quadrilateral

Fig. 17.1

PARALLELOGRAM

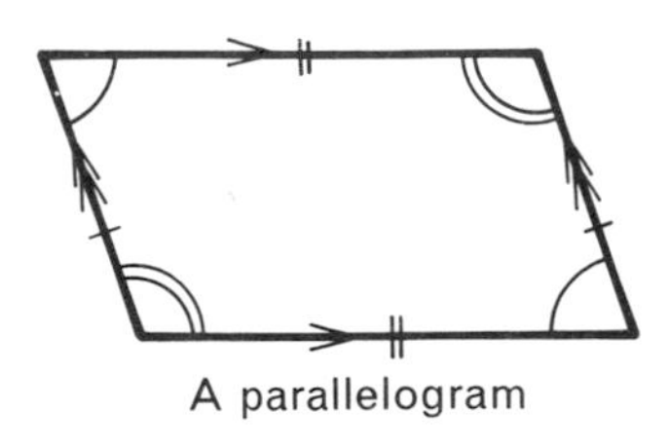

A parallelogram

Fig. 17.2

A parallelogram (Fig. 17.2) has both pairs of opposite sides parallel. It has the following properties:

(a) The sides which are opposite to each other are equal in length.
(b) The angles which are opposite to each other are equal.
(c) The diagonals bisect each other.
(d) The diagonals bisect the parallelogram so that two congruent triangles are formed.

RECTANGLE

A rectangle is a parallelogram with each of its angles equal to 90°. A rectangle has all the properties of a parallelogram but, in addition, the diagonals are equal in length.

RHOMBUS

A rhombus is a parallelogram with all its sides equal in length (Fig. 17.3). It has all the properties of a parallelogram, but in addition, it has the following properties:

(a) The diagonals bisect at right angles.
(b) The diagonal bisects the angle through which it passes.

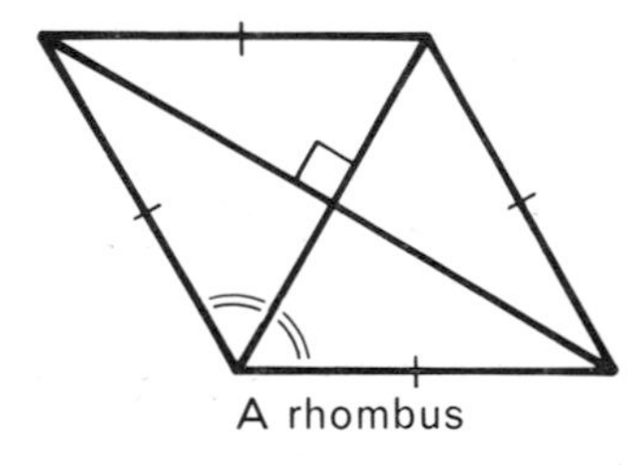
A rhombus

Fig. 17.3

SQUARE

A square is a rectangle with all its sides equal in length. It has all the properties of a parallelogram, rectangle and rhombus.

TRAPEZIUM

A trapezium (Fig. 17.4) is a quadrilateral with one pair of sides parallel.

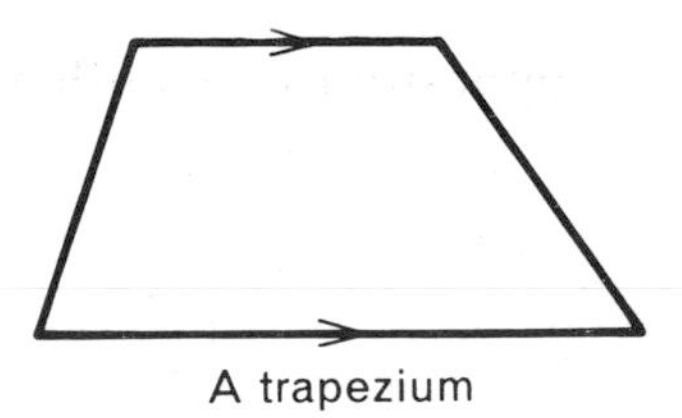

A trapezium

Fig. 17.4

EXAMPLE 17.1

Fig. 17.5 shows part of a structure. The angles A, B and C are 75°, 25° and 15° respectively. Find the angle α.

Since the sum of the angles of a quadrilateral is 360°,

$$\begin{aligned} D &= 360° - (A+B+C) \\ &= 360° - (75° + 25° + 15°) \\ &= 360° - 115° \\ &= 245° \\ \alpha &= 360° - D \\ &= 360° - 245° \\ &= 115° \end{aligned}$$

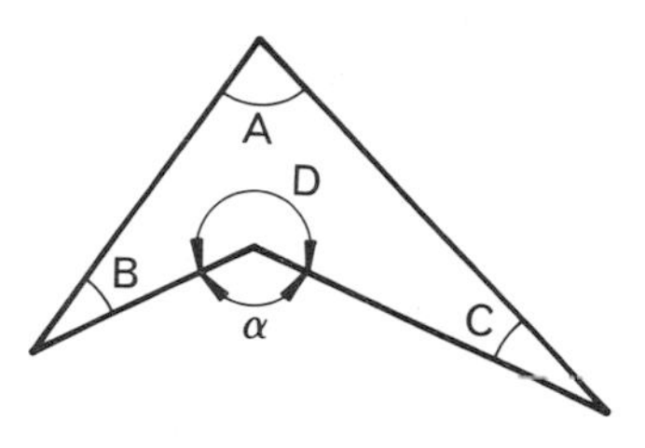

Fig. 17.5

Exercise 17.1

1) Fig. 17.6 shows part of a structure. Find the angle marked α.

2) Fig. 17.7 shows a template. What is the angle marked β?

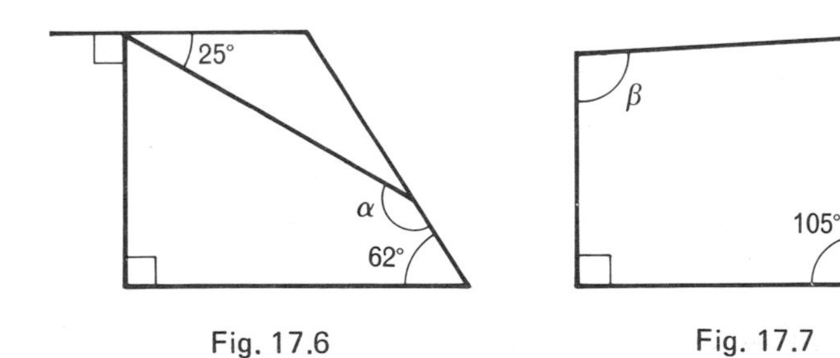

Fig. 17.6

Fig. 17.7

3) In Fig. 17.8, ABCD is a parallelogram. Calculate the angles α and β.

4) In a marking out problem (Fig. 17.9) the angle θ is required. Find its size.

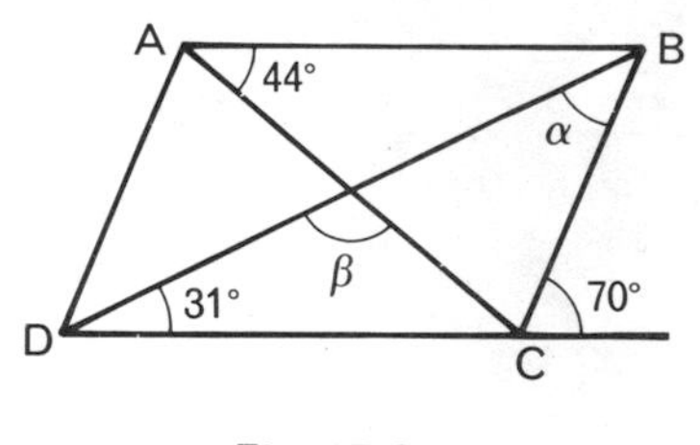

Fig. 17.8

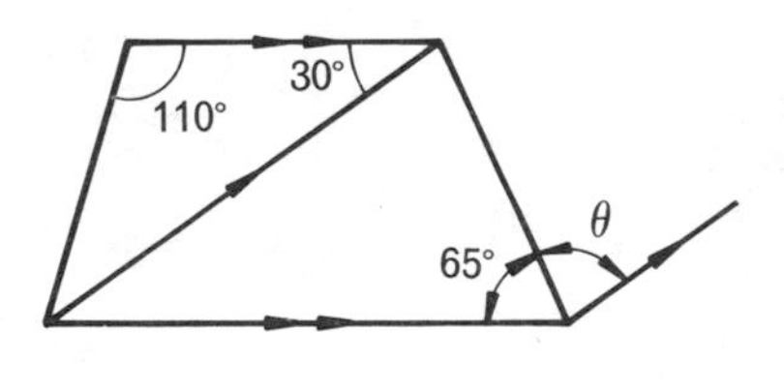

Fig. 17.9

5) In a drawing office problem (Fig. 17.10), the angles α and β are needed. Calculate their values.

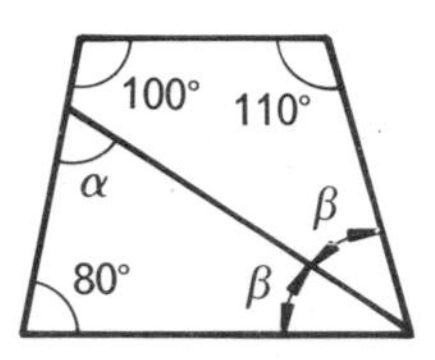

Fig. 17.10

6) ABCD is a template in the form of a quadrilateral. $\angle A = 86^\circ$, $\angle C = 110^\circ$ and $\angle D = 40^\circ$. $\angle ABC$ is bisected to cut the side AD at E. Determine $\angle AEB$.

7) PQRS is a square. T is a point on the diagonal PR such that PT = PQ. The line through T, perpendicular to PR, cuts QR at X. If QX = 50 mm find XT and TR.

8) A template is in the form of a quadrilateral. One of its angles equals 60°. What is the size of each of the other three angles if they are all equal?

9) Fig. 17.11 shows part of a drawing. For marking out, the angle θ is required. What is its size?

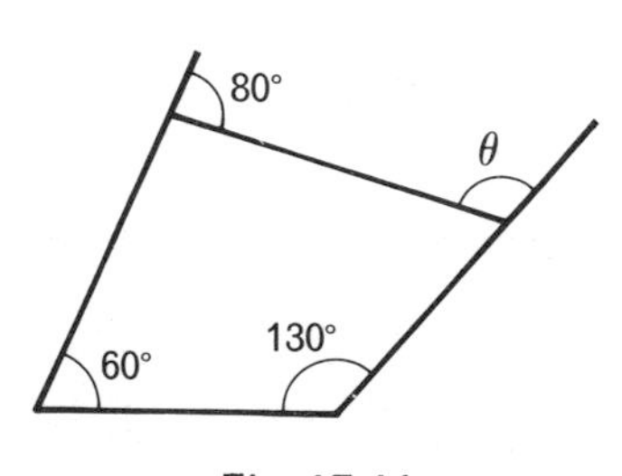

Fig. 17.11

10) Fig. 17.12 shows a trapezium ABCD which is part of a structure. Determine the angles ADB and ABD.

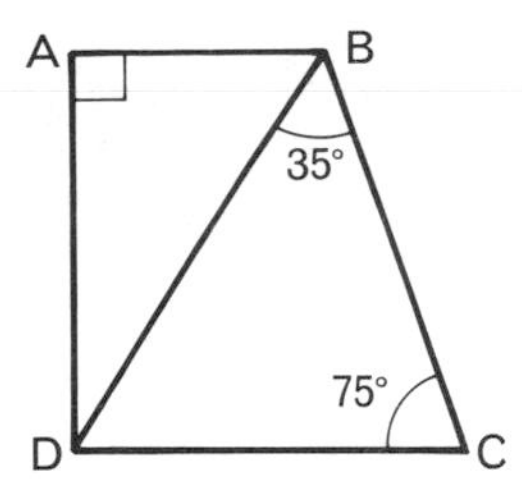

Fig. 17.12

UNITS OF LENGTH

The standard unit of length is the metre (abbreviation: m). The metre is split up into smaller units as follows:

$$\begin{aligned} 1 \text{ metre (m)} &= 10 \text{ decimetres (dm)} \\ &= 100 \text{ centimetres (cm)} \\ &= 1000 \text{ millimetres (mm)} \end{aligned}$$

UNITS OF AREA

The area of a plane figure is measured by seeing how many square units it contains. 1 square metre is the area contained inside a square which has a side of 1 metre (Fig. 17.13). Similarly 1 square centimetre is the area inside a square whose side is 1 cm and 1 square millimetre is the area inside a square whose side is 1 mm.

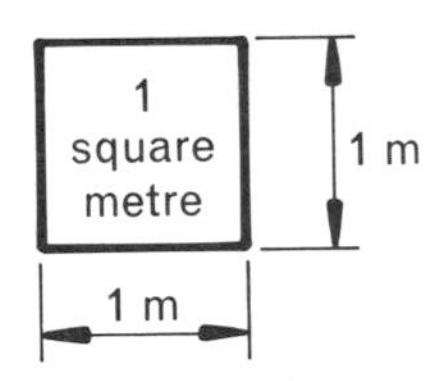

Fig. 17.13

The standard abbreviations for units of area are:

$$\begin{aligned} \text{square metre} &= \text{m}^2 \\ \text{square centimetre} &= \text{cm}^2 \\ \text{square millimetre} &= \text{mm}^2 \end{aligned}$$

THE RECTANGLE

The rectangle (Fig. 17.14) has been divided into 4 rows of 2 squares, each square having an area of $1\,m^2$. The rectangle, therefore, has an area of $4 \times 2\,m^2 = 8\,m^2$. All that we have done to find the area is to multiply the length by the breadth. The same rule will apply to any rectangle. Hence:

Area of rectangle = length × breadth

If we let A = the area of the rectangle

l = the length of the rectangle

b = the breadth of the rectangle

then $A = lb$

In using this formula the units of l and b must have the same units, that is they both must be in metres, centimetres or millimetres.

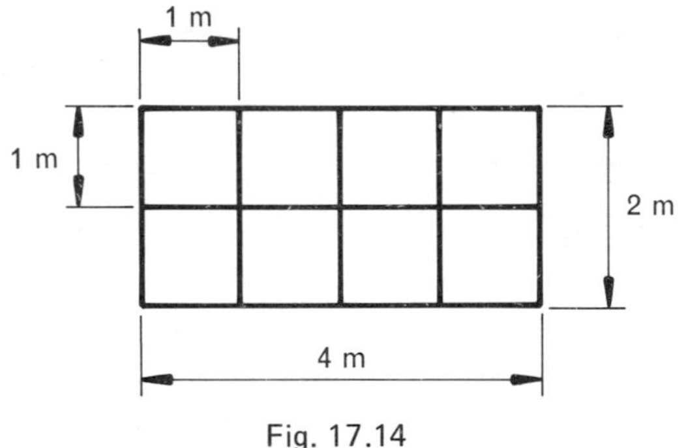

Fig. 17.14

EXAMPLE 17.2

A rectangle measures 5.2 m by 6.3 m. What is its area?

We are given that $l = 5.2$ m and $b = 6.3$ m. Hence the area is:

$$A = 5.2 \times 6.3 = 32.8\,m^2$$

EXAMPLE 17.3

Find the area of a piece of sheet metal measuring 1840 mm by 730 mm. Express the answer in square metres.

In problems of this type it is best to express each of the dimensions in metres before attempting to find the area. Thus:

$$1840\,mm = 1.84\,m \quad \text{and} \quad 730\,mm = 0.73\,m$$

The area of sheet metal is then:

$$A = 1.84 \times 0.73 = 1.34 \text{ m}^2$$

EXAMPLE 17.4

An office 9.3 m long and 7.6 m wide is to be carpeted so as to leave a surround 50 cm wide as shown in Fig. 17.15. Find the area of the surround.

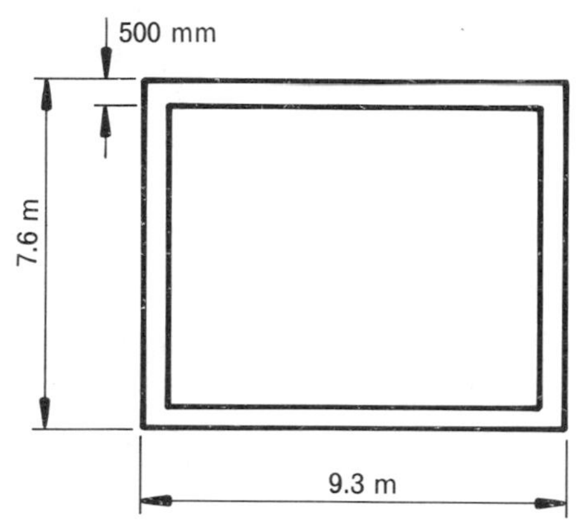

Fig. 17.15

The easiest way of solving this problem is to find the area of the office and subtract from it the area of the carpet.

$$\text{Area of office} = 9.3 \times 7.6 = 70.7 \text{ m}^2$$

$$\text{Area of carpet} = 8.3 \times 6.6 = 54.8 \text{ m}^2$$

$$\text{Area of surround} = 70.7 - 54.8 = 15.9 \text{ m}^2$$

The areas of many shapes can be found by splitting the shape up into rectangles and finding the area of each rectangle separately. The area of the shape is then found by adding the areas of the separate rectangles together.

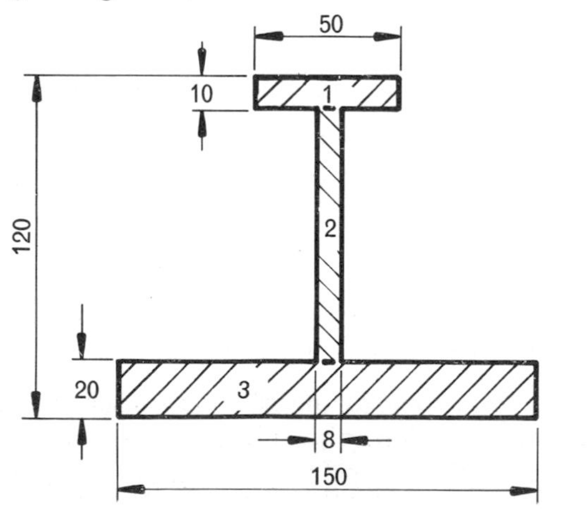

Fig. 17.16

EXAMPLE 17.5

Find the area of the I-section shown in Fig. 17.16 (p. 257).

The I-section can be divided up into three rectangles as shown in the diagram.

$$\begin{aligned}\text{Area of I-section} &= \text{area of } 1 + \text{area of } 2 + \text{area of } 3\\ &= (50\times 10)\,\text{mm}^2 + (90\times 8)\,\text{mm}^2\\ &\quad + (150\times 20)\,\text{mm}^2\\ &= 500\,\text{mm}^2 + 720\,\text{mm}^2 + 3000\,\text{mm}^2\\ &= 4220\,\text{mm}^2\end{aligned}$$

EXAMPLE 17.6

A rectangle has an area of $60\,\text{m}^2$ and its breadth is 12 m. What is its length?

We are given that $A = 60$ and $b = 12$. Hence:

$$60 = 12\times l$$

$$l = \frac{60}{12} = 5$$

Therefore the length of the rectangle is 5 m.

EXAMPLE 17.7

A piece of wood has an area of $1.8\,\text{m}^2$ and a width of 300 mm. What is its length?

First we convert 300 mm into metres, that is 300 mm = 0.3 m. We now have $A = 1.8$ and $b = 0.3$. Substituting these values in the formula for the area of a rectangle we have:

$$1.8 = 0.3\times l$$

$$l = \frac{1.8}{0.3} = 6$$

Hence the length of the piece of wood is 6 m.

THE SQUARE

A square is a rectangle with all its sides equal in length. Hence:

$$\boxed{\text{Area of square} = \text{side}\times\text{side} = \text{side}^2}$$

We can express this as a formula by letting A represent the area of the square and l the length of the side. Thus:

$$A = l^2$$

EXAMPLE 17.8

A square has an area of 2025 mm². What is the length of its side?

$$\text{Length of side} = \sqrt{\text{area}} = \sqrt{2025} = 45\,\text{mm}$$

Exercise 17.2

1) Find the areas of the following rectangles:

(a) 7 m by 8 m (b) 20 mm by 11 mm (c) 18 m by 35 m.

2) A piece of wood is 3.7 m long and 280 mm wide. What is its area in square metres?

3) A rectangular piece of metal is 1980 mm long and 880 mm wide. What is its area in square metres?

4) A room 5.8 m long and 4.9 m wide is to be covered with vinyl. What area of vinyl is needed?

5) What is the total area of the walls of a room which is 6.7 m long, 5.7 m wide and 2.5 m high?

6) A rectangular lawn is 32 m long and 23 m wide. A path 1.5 m wide is made around the lawn. What is the area of the path?

7) An office 8.5 m long and 6.3 m wide is to be carpeted to leave a surround 600 mm wide around the carpet. What is:

(a) the area of the office?

(b) the area of the carpet?

(c) the area of the surround?

8) Find the areas of the sections shown in Fig. 17.17.

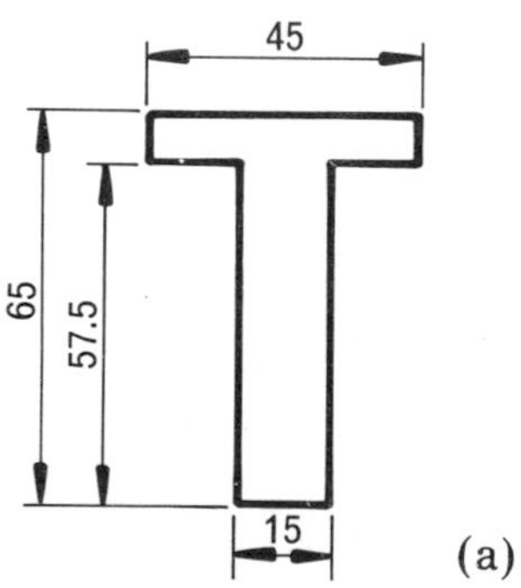

(a)

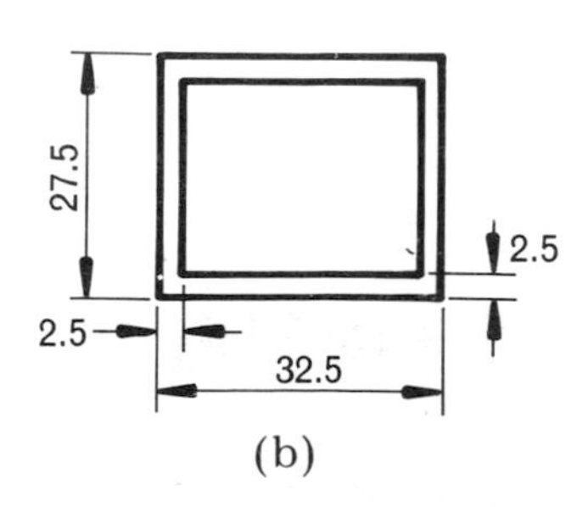

(b)

Fig. 17.17 *(continued on page 260)*

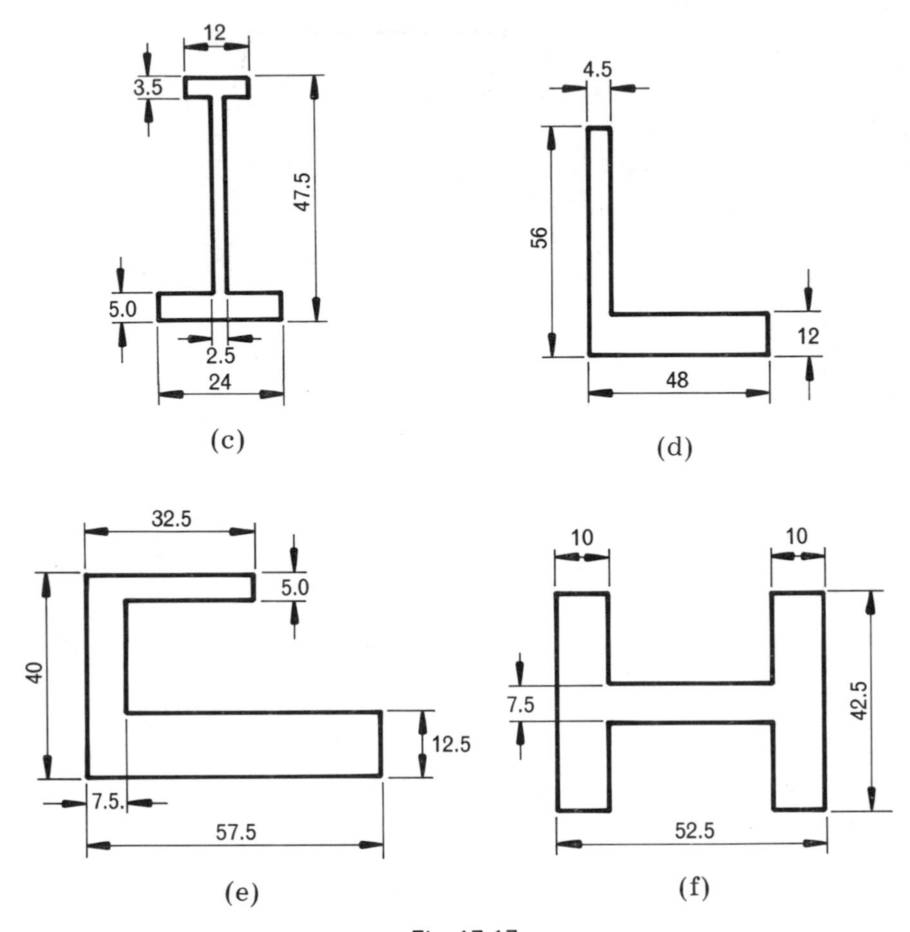

Fig. 17.17

THE PARALLELOGRAM

A parallelogram is, in effect, a rectangle pushed out of square as shown in Fig. 17.18, where the equivalent rectangle is shown dotted. Hence,

Area of parallelogram = length of base × vertical height

$$A = bh$$

where A = area, b = length of base and h = vertical height (or altitude).

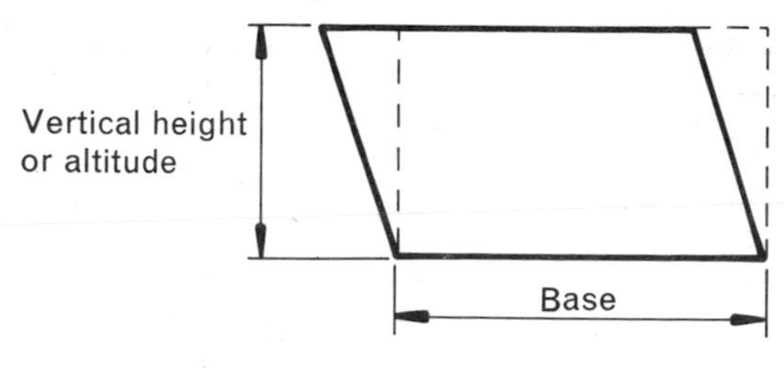

Fig. 17.18

EXAMPLE 17.9

Find the area of a parallelogram whose base is 150 mm long and whose altitude is 80 mm.

$$\text{Area} = 150 \times 80 = 12\,000\,\text{mm}^2$$

EXAMPLE 17.10

A parallelogram has an area of $36\,\text{m}^2$. If its base is 9 m, find its altitude.

We are given that $A = 36$ and that $b = 9$. Hence:

$$36 = 9 \times h \quad \text{or} \quad h = \frac{36}{9} = 4$$

Hence the altitude is 4 m.

Exercise 17.3

1) Find the area of a parallelogram whose base is 7 m long and whose vertical height is 8 m.

2) What is the area of a parallelogram whose base is 70 mm long and whose altitude is 650 mm? Give the answer in square metres.

3) The area of a parallelogram is $64\,\text{m}^2$. Its base is 16 m long. Calculate its altitude.

4) A parallelogram has an area of $25.92\,\text{mm}^2$. Its altitude is 3.6 mm. Find its length of base.

5) Fig. 17.19 shows a steel section. Find its area in square millimetres.

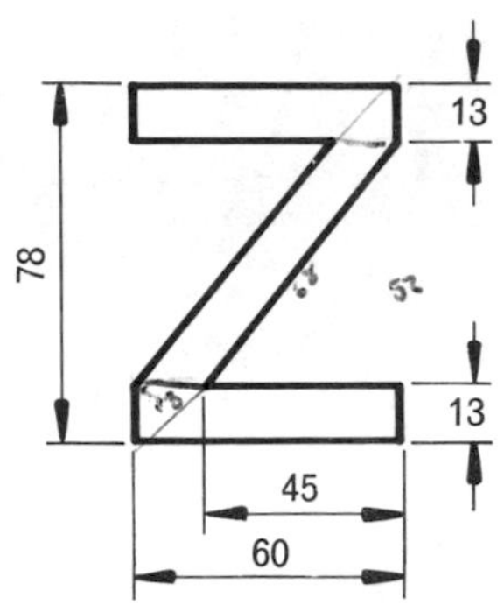

Fig. 17.19

AREA OF A TRAPEZIUM

A trapezium is a plane figure, bounded by four straight lines, which has one pair of parallel sides (Fig. 17.20).

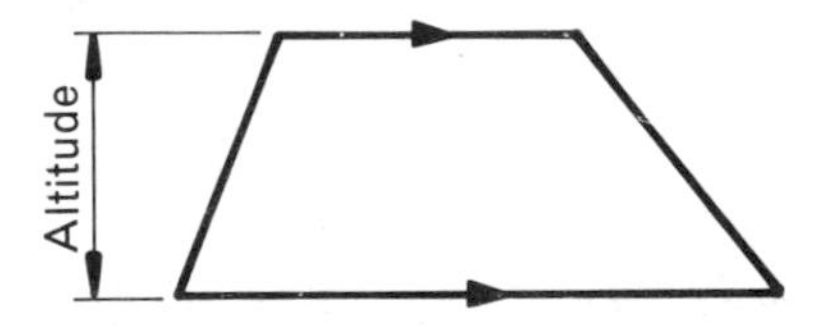

Fig. 17.20

$$\begin{pmatrix}\text{Area of}\\ \text{trapezium}\end{pmatrix} = \frac{1}{2} \times (\text{sum of lengths of parallel sides}) \times (\text{altitude})$$

EXAMPLE 17.11

The parallel sides of a trapezium are 12 m and 16 m long. The distance between the parallel sides is 9 m. What is the area of the trapezium?

Area of trapezium $= \frac{1}{2} \times (12 + 16) \times 9 = \frac{1}{2} \times 28 \times 9 = 126 \text{ m}^2$

EXAMPLE 17.12

The area of a trapezium is 220 mm^2 and the parallel sides are 26 mm and 14 mm long. Find the distance between the parallel sides.

$\frac{1}{2}$ the sum of the parallel sides $= \frac{1}{2} \times (26 + 14) = \frac{1}{2} \times 40 = 20$

Distance between the parallel sides $= 220 \div 20 = 11 \text{ mm}$

Exercise 17.4

1) Find the area of a trapezium whose parallel sides are 7 m and 9 m long and whose altitude is 5 m.

2) The parallel sides of a trapezium are 15 m and 9.8 m long. If the distance between the parallel sides is 7.6 m, what is the area of the trapezium?

3) The area of a trapezium is $500\ \text{mm}^2$ and its parallel sides are 35 mm and 65 mm long. Find the altitude of the trapezium.

4) Find the area of a trapezium whose parallel sides are 75 mm and 82 mm and whose vertical height is 39 mm. Give the answer in square centimetres.

5) Find the area of the trapezium shown in Fig. 17.21.

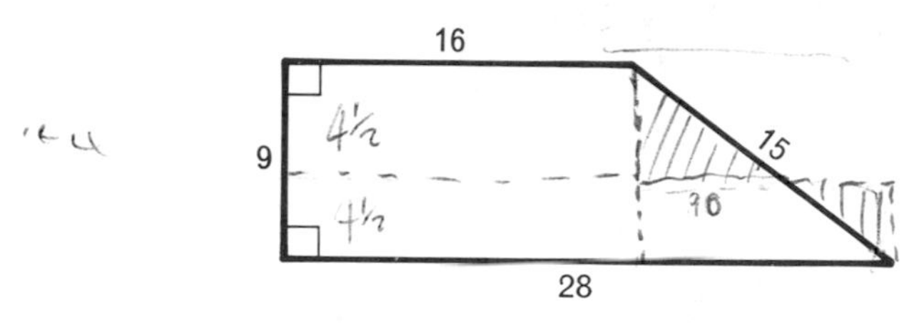

Fig. 17.21

AREA OF A TRIANGLE

The diagonal of the parallelogram shown in Fig. 17.22 splits the parallelogram into two equal triangles. Hence:

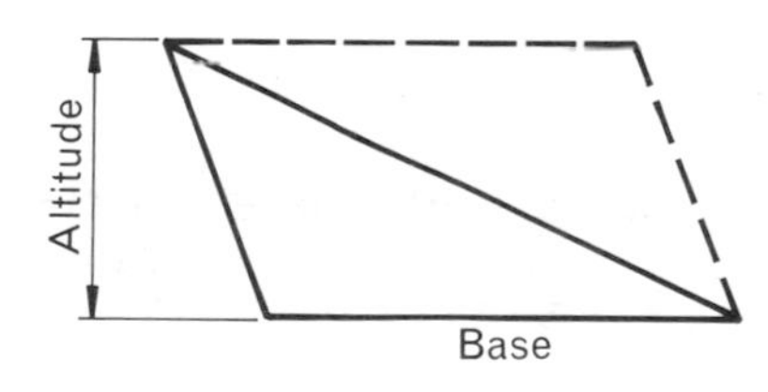

Fig. 17.22

Area of triangle $= \frac{1}{2} \times$ base $\times$ vertical height

Sometimes the vertical height is called the altitude and:

Area of triangle $= \frac{1}{2} \times$ base $\times$ altitude

As a formula the statement becomes:

$A = \frac{1}{2}bh$ where b = the base and h = the altitude

EXAMPLE 17.13

A triangle has a base 5 m long and a vertical height of 12 m. Calculate its area.

$$\text{Area of triangle} = \tfrac{1}{2} \times \text{base} \times \text{height} = \tfrac{1}{2} \times 5 \times 12 = 30\ \text{m}^2$$

EXAMPLE 17.14

An isosceles triangle has a base 118 mm long and the two equal sides are each 143 mm long. Calculate the area of the triangle.

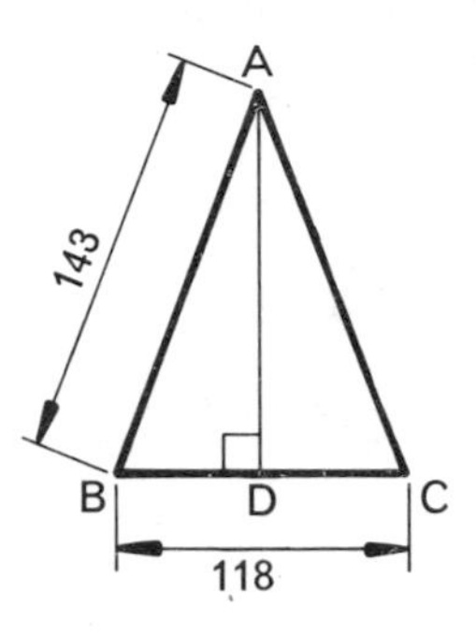

Fig. 17.23

We must first calculate the altitude of the triangle. Referring to Fig. 17.23 we see that, by Pythagoras' Theorem,

$$AD^2 = AB^2 - BD^2 = 143^2 - 59^2 = 2045 - 348 = 1697$$

$$AD = \sqrt{1697} = 130.3\ \text{mm}$$

$$\text{Area of triangle} = \tfrac{1}{2} \times \text{base} \times \text{height} = \tfrac{1}{2} \times 118 \times 130.3$$
$$= 7690\ \text{mm}^2$$

Area of Triangle (given three sides)

When we are given the lengths of three sides of a triangle we can find its area by using the formula given below:

$$\boxed{A = \sqrt{s \times (s-a) \times (s-b) \times (s-c)}}$$

where s stands for half of the perimeter of the triangle and a, b and c are the lengths of the sides of the triangle.

EXAMPLE 17.15

The sides of a triangle are 13 m, 8 m and 7 m long. Calculate the area of the triangle.

$$s = \tfrac{1}{2} \text{ perimeter} = \tfrac{1}{2} \times (13+8+7) = \tfrac{1}{2} \times 28 = 14 \text{ m}$$

$$a = 13 \text{ m}, \quad b = 8 \text{ m} \quad \text{and} \quad c = 7 \text{ m}$$

$$\text{Area of triangle} = \sqrt{14 \times (14-13) \times (14-8) \times (14-7)}$$

$$= \sqrt{14 \times 1 \times 6 \times 7} = \sqrt{588} = 24.2 \text{ m}^2$$

Exercise 17.5

1) Find the area of a triangle whose base is 180 mm and whose altitude is 120 mm.

2) Find the area of a triangle whose base is 7.5 m and whose altitude is 5.9 m.

3) A triangle has sides 4 m, 7 m and 9 m. What is its area?

4) A triangle has sides 37 mm, 52 mm and 63 mm long. What it its area?

5) An isosceles triangle has a base 12.4 m long and equal sides each 16.3 m long. Calculate the area of the triangle.

AREA OF A CIRCLE

It can be shown that:

$$\text{Area of circle} = \pi \times \text{radius}^2 = \pi r^2$$

Also since $\quad \text{Radius} = \dfrac{\text{diameter}}{2}$

$$\text{Area of circle} = \pi \left(\frac{\text{diameter}}{2}\right)^2 = \frac{\pi d^2}{4}$$

EXAMPLE 17.16

Find the area of a circle whose radius is 30 mm.

$$\text{Area of circle} = \pi \times 30^2 = \pi \times 900 = 2830 \text{ mm}^2$$

EXAMPLE 17.17

Find the area of a circle whose diameter is 28 m.

$$\text{Area of circle} = \frac{\pi \times 28^2}{4} = 616 \text{ m}^2$$

EXAMPLE 17.18

Find the area of the annulus shown in Fig. 17.24.

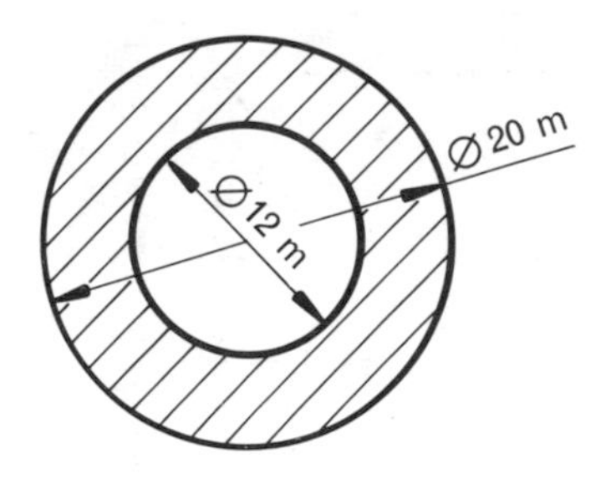

Fig. 17.24

Now Area of outer circle $= \dfrac{\pi \times 20^2}{4} = 314\ \text{m}^2$

and Area of inner circle $= \dfrac{\pi \times 12^2}{4} = 113\ \text{m}^2$

$\therefore$ Area of annulus $= 314 - 113 = 201\ \text{m}^2$

Exercise 17.6 ✓

Find the areas of the following circles:

1) 14 m radius
2) 350 mm radius
3) 2.82 m radius
4) 42 mm diameter
5) 7.78 m diameter
6) 197.6 mm diameter

7) An annulus has an inside radius of 6 m and an outside radius of 9 m. Calculate its area.

8) A copper pipe has a bore of 32 mm and an outside diameter of 42 mm. Find the area of its cross-section.

AREAS OF COMPOSITE FIGURES

Many shapes are composted of straight lines and arcs of circles. The areas of such shapes are found by splitting up the shape into figures such as rectangles, triangles, etc. and sectors of circles.

EXAMPLE 17.19

A template has the shape shown in Fig. 17.25. Finds its area.

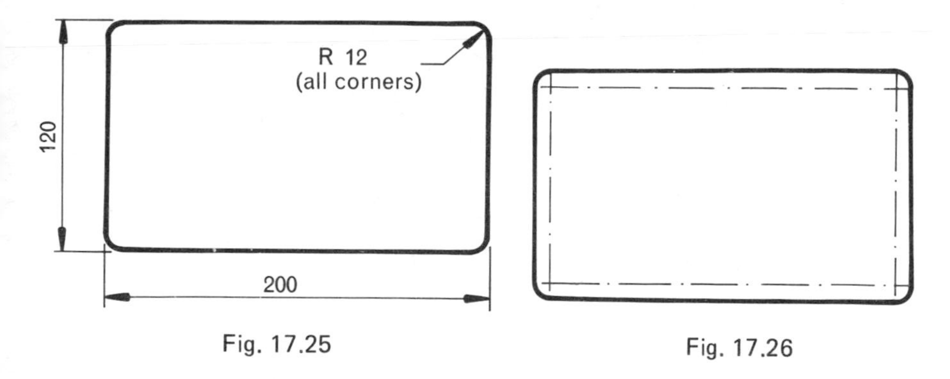

Fig. 17.25

Fig. 17.26

The template can be split up into 4 quarter circles and 5 rectangles as shown in Fig. 17.26.

$$\text{Area of 4 quarter circles} = 4 \times \tfrac{1}{4} \times \pi \times 12^2 = \pi \times 12^2 = 452\,\text{mm}^2$$

$$\text{Area of 5 rectangles} = 2 \times 12 \times 96 + 2 \times 12 \times 176 + 96 \times 176$$

$$= 2304 + 4224 + 16\,896 = 23\,424\,\text{mm}^2$$

$$\text{Area of template} = 452 + 23\,424 = 23\,900\,\text{mm}^2$$

to 3 significant figures

Exercise 17.7

Find the areas of the shaded portions of the figures shown in Fig. 17.27.

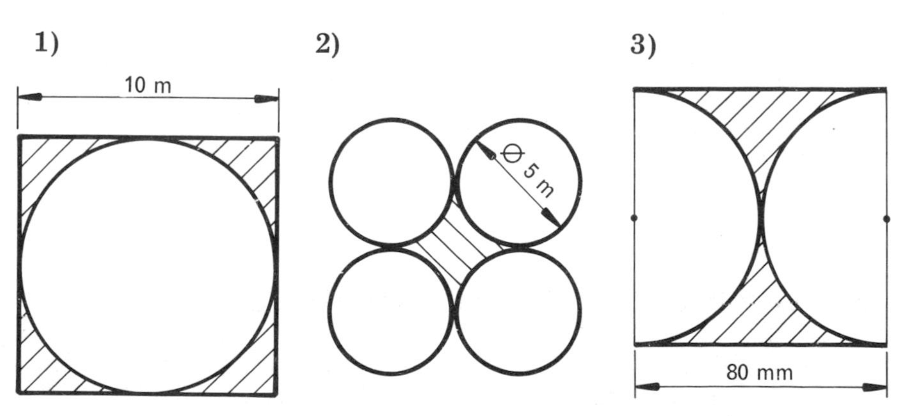

Fig. 17.27 (continued on page 268)

4) 5)

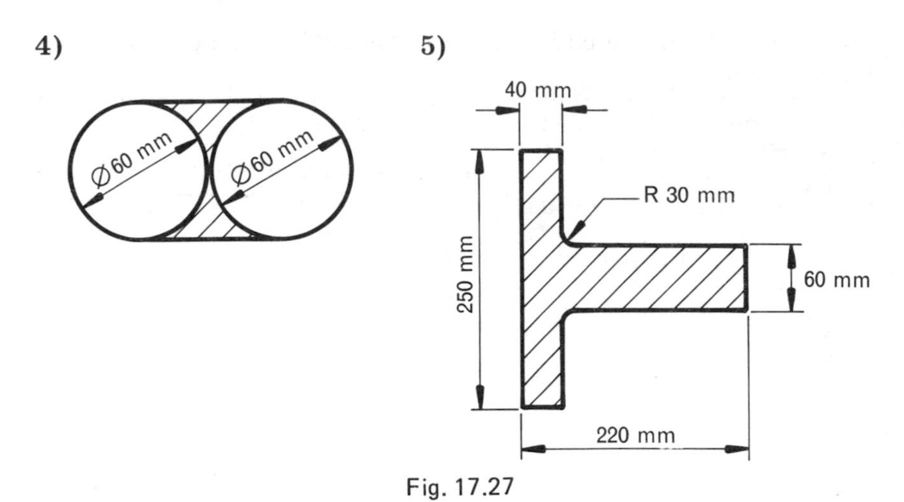

Fig. 17.27

THE UNIT OF VOLUME

We measure volume by seeing how many cubic units an object contains. For example, 1 cubic metre (abbreviation: m^3) is the volume contained in a cube whose edge is 1 m (Fig. 17.28).

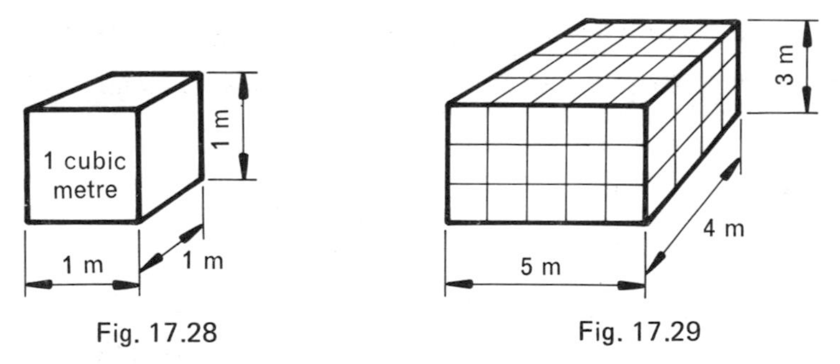

Fig. 17.28 Fig. 17.29

We can divide up the rectangular solid shown in Fig. 17.29 into three layers of small cubes each having a volume of $1\,m^3$. There are 5×4 cubes in each layer and, therefore, the total number of cubes is $5 \times 4 \times 3 = 60$. The solid therefore has a volume of $60\,m^3$.

All that we have done in order to find the volume is to multiply the length by the breadth by the height. This rule applies to any rectangular solid and hence:

Volume of a rectangular solid = length × breadth × height

$$V = lbh$$

Since the area of the end of a rectangular solid = length × breadth we can write:

Volume of rectangular solid = area of end × height

This statement is true for any solid which has the same shape (i.e. cross-section) throughout its length. For solids of this type:

Volume of solid = cross-sectional area × length

$$V = Al$$

EXAMPLE 17.20

Find the volume of a rectangular block which is 3.2 m long, 2.8 m wide and 2 m high.

$$\text{Volume} = \text{length} \times \text{breadth} \times \text{height} = 3.2 \times 2.8 \times 2 = 17.9\,\text{m}^3$$

EXAMPLE 17.21

Calculate the volume of a rectangular electric oven which is 1200 mm long, 820 mm wide and 520 mm high. Give the answer in cubic metres.

It is best to express the dimensions of the oven in metres before attempting to find the volume. Thus:

Length = 1.20 m, breadth = 0.82 m and height = 0.52 m

$$\begin{aligned} \text{Volume} &= \text{length} \times \text{breadth} \times \text{height} = 1.20 \times 0.82 \times 0.52 \\ &= 0.512\,\text{m}^3 \end{aligned}$$

EXAMPLE 17.22

Find the volume of the steel bar shown in Fig. 17.30.

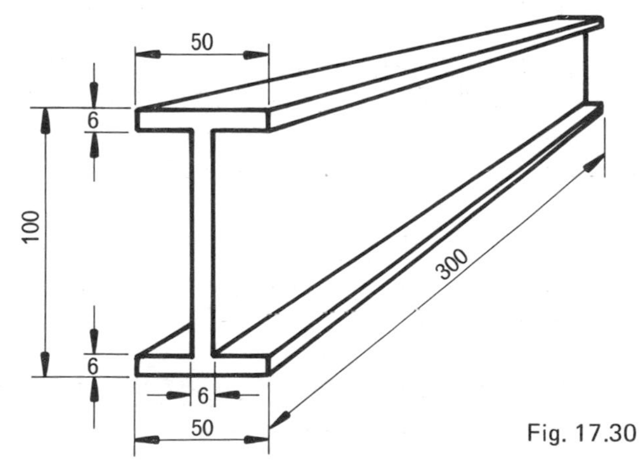

Fig. 17.30

Since the bar has the same cross-section throughout its length

$$\text{Area of cross-section} = 50 \times 6 + 50 \times 6 + 88 \times 6$$
$$= 300 + 300 + 528 = 1128\,\text{mm}^2$$
$$\therefore \quad \text{Volume} = \text{area of cross-section} \times \text{length}$$
$$= 1128 \times 300 = 338\,000\,\text{mm}^3$$

EXAMPLE 17.23

A steel bar has the cross-section shown in Fig. 17.31. If it is 9 m long calculate its volume.

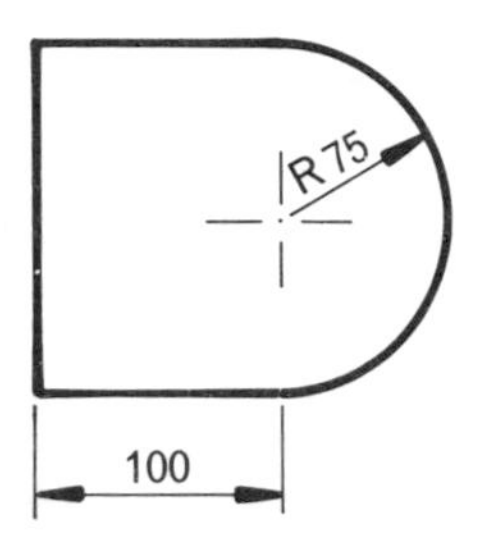

Fig. 17.31

Because the length is given in metres it will be convenient to find the volume in cubic metres.

$$\text{Area of cross-section} = 0.100 \times 0.150 + \tfrac{1}{2} \times \pi \times 0.075^2$$
$$= 0.015 + 0.008\,8 = 0.023\,8\,\text{m}^2$$
$$\therefore \quad \text{Volume} = \text{area of cross-section} \times \text{length}$$
$$= 0.023\,8 \times 9 = 0.214\,\text{m}^3$$

VOLUME OF A CYLINDER

A cylinder (Fig. 17.32) has a constant cross-section which is a circle.

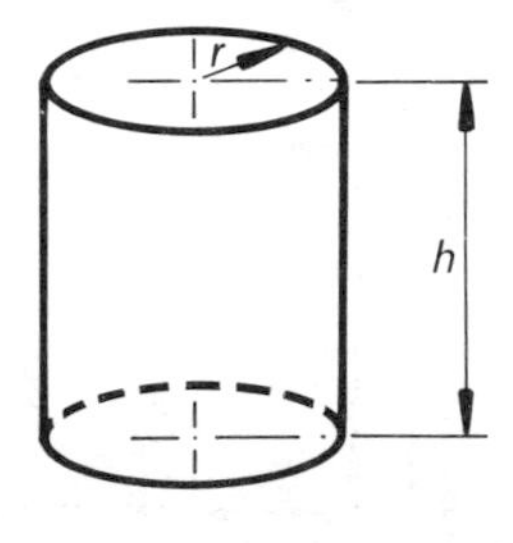

Fig. 17.32

Hence:

Volume of cylinder $= \pi \times \text{radius}^2 \times \text{length (or height)}$

or $$V = \pi r^2 h$$

EXAMPLE 17.24

Find the volume of a cylinder which has a radius of 1.4 m and a height of 1.2 m.

$$\text{Volume} = \pi \times 1.4^2 \times 1.2 = 7.39\,\text{m}^3$$

EXAMPLE 17.25

A pipe has the dimensions shown in Fig. 17.33. Calculate its volume.

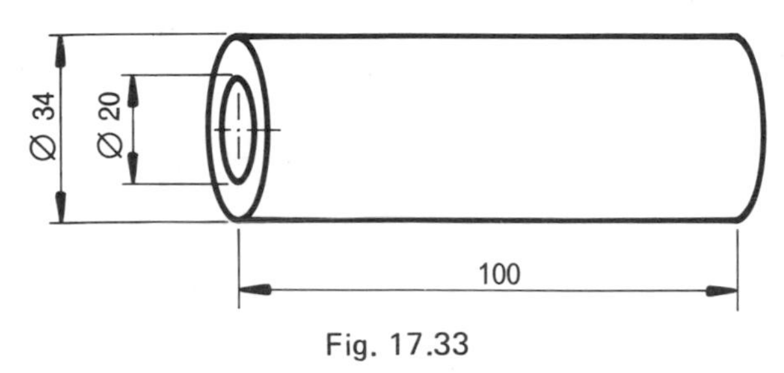

Fig. 17.33

$$\text{Cross-sectional area} = \pi \times 17^2 - \pi \times 10^2 = 593.8\,\text{mm}^2$$

$$\therefore \quad \text{Volume} = \text{cross-sectional area} \times \text{length}$$

$$= 593.8 \times 100$$

$$= 59\,400\,\text{mm}^3$$

Exercise 17.8

1) Find the volume of a rectangular block 8 m long, 5 m wide and 3.5 m high.

2) The diagram (Fig. 17.34) shows the cross-section of a steel bar. If it is 250 mm long, calculate its volume.

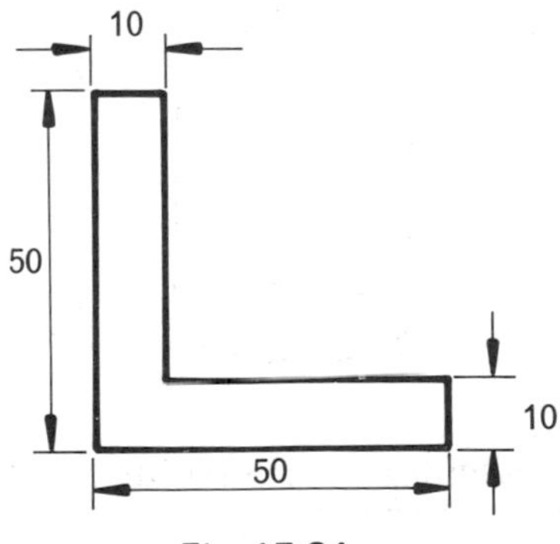

Fig. 17.34

3) Find the volume of a cylinder whose radius is 70 mm and whose height is 500 mm.

4) A hole 40 mm diameter is drilled in a plate 25 mm thick. What volume of metal is removed from the plate?

5) A steel component has the cross-section shown in Fig. 17.35. If it is 80 mm long calculate its volume.

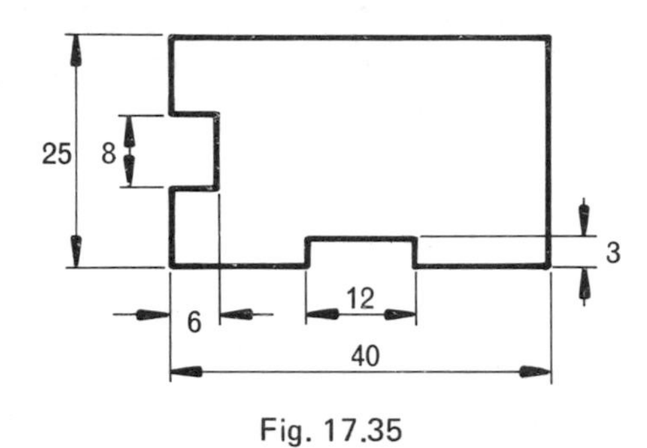

Fig. 17.35

6) Calculate the volume of a metal tube whose bore is 50 mm and whose thickness is 8 mm if it is 6 m long. Give the answer in cubic metres.

7) Fig. 17.36 shows a washer which is 2 mm thick. Calculate its volume.

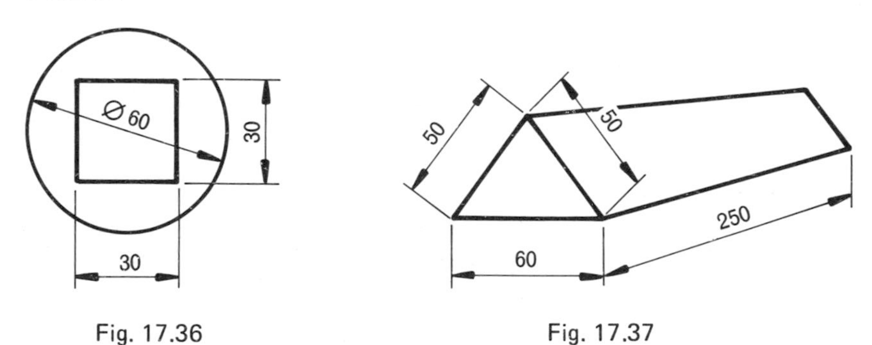

Fig. 17.36

Fig. 17.37

8) Fig. 17.37 shows a triangular prism. Calculate its volume.

9) A tent has a triangular cross-section whose base is 3 m and whose height is 2.2 m. If it is 7 m long, what is the volume inside the tent?

10) A pipe is 8 m long. It has a bore of 80 mm and an outside diameter of 100 mm. Calculate the volume of the pipe in cubic metres.

SIMILAR SHAPES

Two plane figures are similar if they are equi-angular and the ratio of adjacent sides is the same. Thus the parallelograms shown in Fig. 17.38 are similar because they are equi-angular and $\frac{AD}{WZ} = \frac{CD}{YZ} = \frac{1}{3}$.

> The ratio of the areas of similar shapes is equal to the ratio of the squares on corresponding sides.

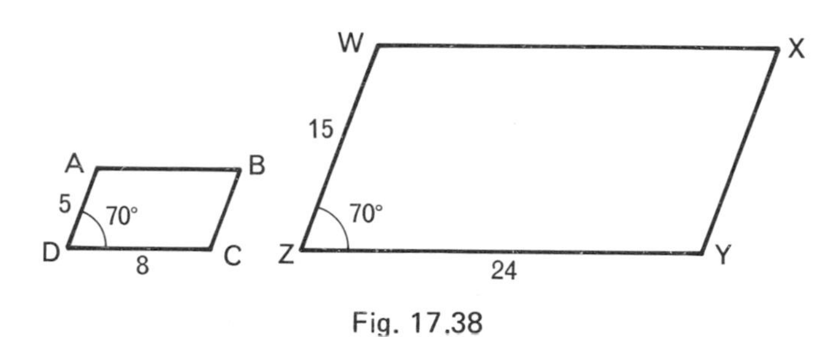

Fig. 17.38

Thus for the parallelograms of Fig. 17.38, AD and WZ are corresponding sides and hence

$$\frac{\text{Area WXYZ}}{\text{Area ABCD}} = \frac{WZ^2}{AD^2} = \frac{3^2}{1^2} = \frac{9}{1}$$

EXAMPLE 17.26

Find the area of triangle XYZ given that the area of triangle ABC is $12\,\text{m}^2$ (see Fig. 17.39).

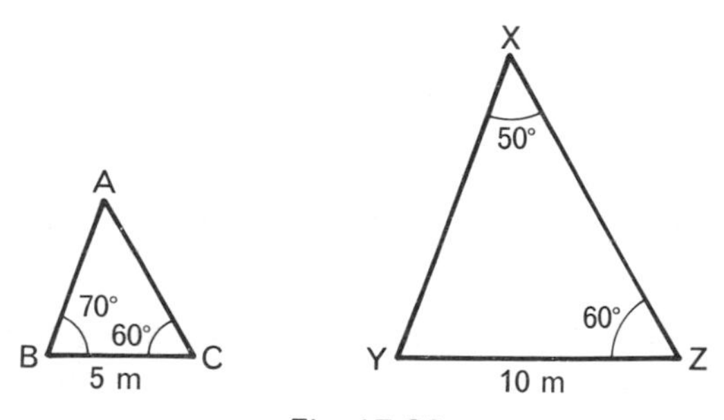

Fig. 17.39

In triangle XYZ, $\angle Y = 80°$ and in triangle ABC, $\angle A = 50°$. Hence the two triangles are equi-angular; that is, they are similar. Thus BC and YZ correspond, hence:

$$\frac{\text{Area } \Delta\text{XYZ}}{\text{Area } \Delta\text{ABC}} = \frac{\text{YZ}^2}{\text{BC}^2}$$

thus
$$\frac{\text{Area } \Delta\text{XYZ}}{12} = \frac{10^2}{5^2} = \frac{100}{25} = 4$$

$\therefore$
$$\text{Area } \Delta\text{XYZ} = 12 \times 4 = 48\,\text{m}^2$$

SIMILAR SOLIDS

Two solids are similar if the ratios of their corresponding linear dimensions are equal. The two cones shown in Fig. 17.40 are similar if

$$\frac{h_1}{h_2} = \frac{r_1}{r_2}$$

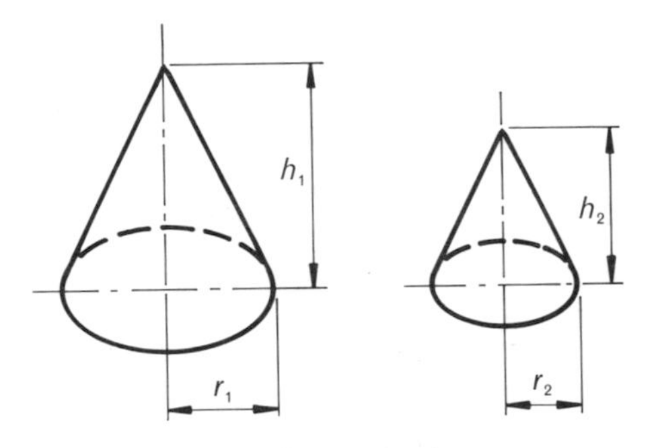

Fig. 17.40

The two cylinders shown in Fig. 17.41 are similar since

$$\frac{150}{75} = \frac{100}{50}$$

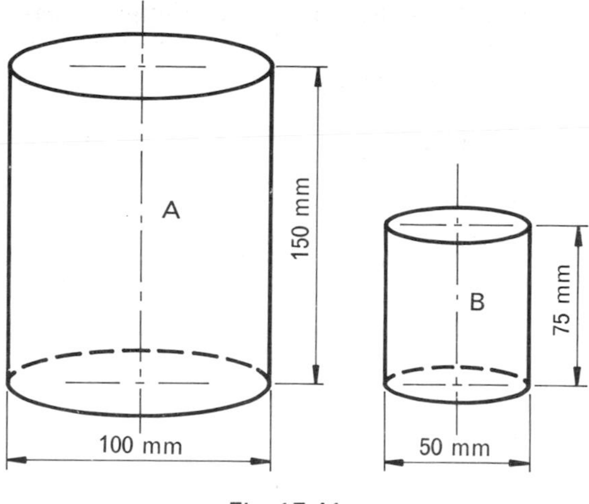

Fig. 17.41

> The volumes of similar solids are proportional to the cubes of their corresponding linear dimensions.

(1) If two spheres of radii r_1 and r_2 have volumes V_1 and V_2 respectively, then

$$\frac{V_1}{V_2} = \frac{r_1^3}{r_2^3}$$

(2) The ratio of the volumes of the two cones shown in Fig. 17.40 is

$$\frac{V_1}{V_2} = \frac{r_1^3}{r_2^3} = \frac{h_1^3}{h_2^3}$$

(3) The ratio of the volumes of the two cylinders shown in Fig. 17.41 is

$$\frac{V_1}{V_2} = \frac{100^3}{50^3} = \frac{150^3}{75^3} = \frac{8}{1}$$

EXAMPLE 17.27

The volume of a cone of height 135 mm is 1090 mm^3. Find the volume of a similar cone whose height is 72 mm.

Let V_1 and V_2 be the volumes of the two cones. Then:

$$\frac{V_2}{V_1} = \frac{{h_2}^3}{{h_1}^3}$$

Thus $$\frac{V_2}{1090} = \frac{72^3}{135^3}$$

$\therefore$ $$V_2 = 1090 \times \frac{72^3}{135^3} = 165\,\text{mm}^3$$

Exercise 17.9

1) In Fig. 17.42, the triangles ABC and EFG are similar. If the area of ΔABC is $8\,\text{m}^2$, calculate the area of ΔEFG.

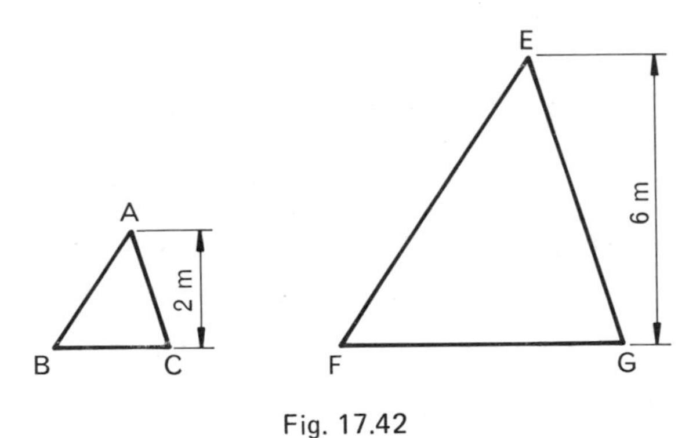

Fig. 17.42

2) In Fig. 17.43, the area of triangle XYZ is $9\,\text{m}^2$. What is the area of ΔABC?

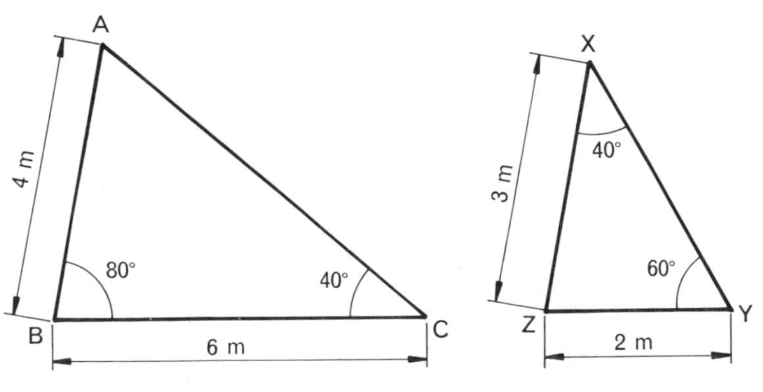

Fig. 17.43

3) A rectangle ABCD has a length of 30 mm and an area of 600 mm^2. A similar rectangle WXYZ has a length of 20 mm. What is the area of WXYZ?

4) A trapezium ABCD (Fig. 17.44) has an area of 12 300 mm^2. What is the area of the trapezium PQRS?

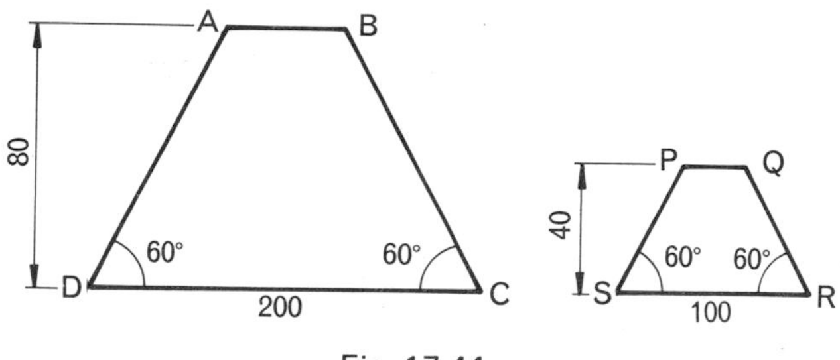

Fig. 17.44

5) A rhombus ABCD (Fig. 17.45) has an area of 18 m^2. The rhombus EFGH has an area of 72 m^2. Calculate the length EF.

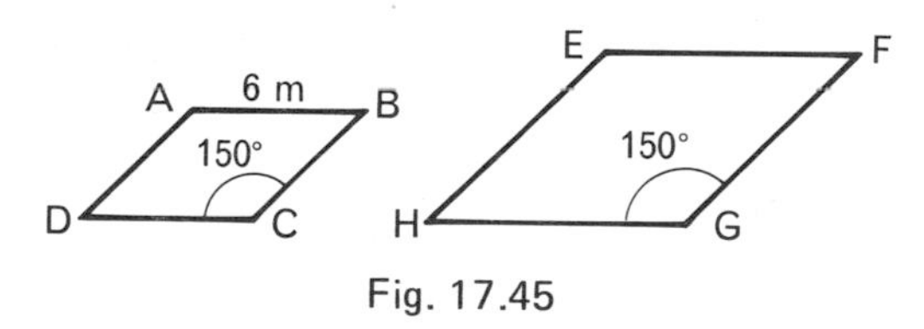

Fig. 17.45

6) In a scale model of a building the linear dimensions are $\frac{1}{10}$th of the actual dimensions. If the volume of the model is 2 m^3, what is the volume of the building?

7) Fig. 17.46 represents three similar rectangular boxes.

(a) The dimensions of box A are 16 mm by 35 mm by 53 mm. Calculate its volume.

(b) In box C the dimensions are double those of box A. Calculate the volume of box C.

(c) The ratio of the volume of box B to the volume of box A is 27 : 8. Calculate the dimensions of box B.

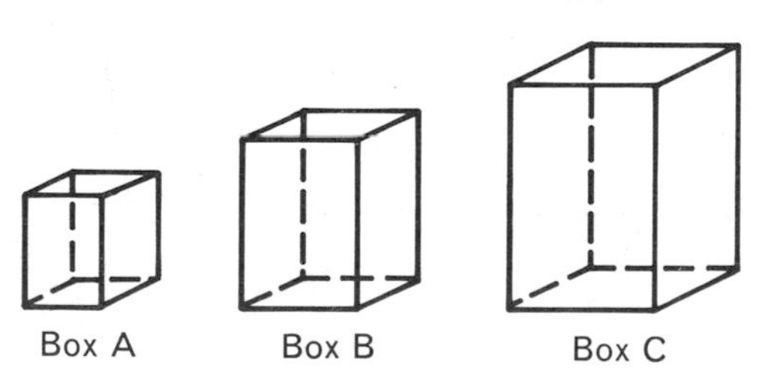

Fig. 17.46

8) Fig. 17.47 shows three similar cylinders.

(a) If the volume of B is 25.1 m^3, calculate the volume of A.

(b) If the volume of C is 84.7 m^3, what is its height?

9) A ball-bearing has a diameter of 15 mm and its volume is 14 000 mm^3. Calculate the diameter of a ball bearing whose volume is 112 000 mm^3.

10) Fig. 17.48 shows a rain water gutter whose capacity, when full, is 61.6 litres. What is the capacity of a gutter of the same length whose diameter is 180 mm?

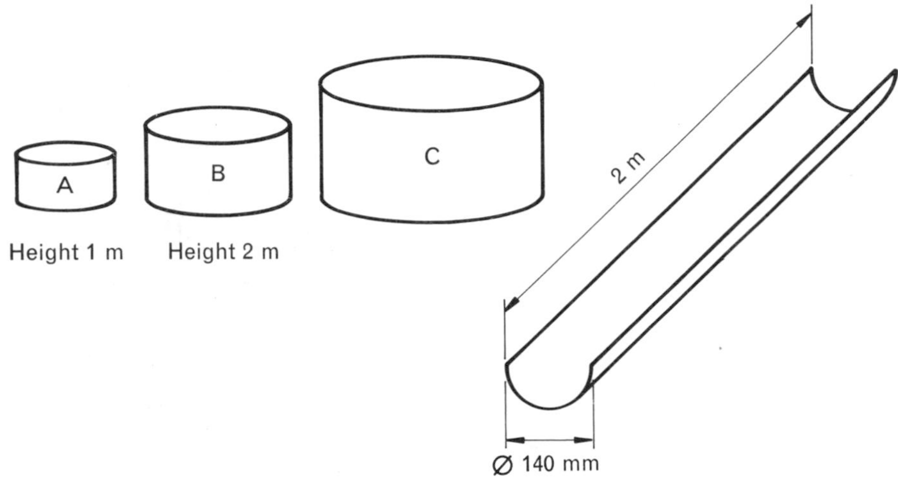

Fig. 17.47

Fig. 17.48

SUMMARY

Area of rectangle = (length) × (breadth)

Area of square = $(\text{length of side})^2$

Area of parallelogram = (base) × (altitude)

Area of trapezium = $\frac{1}{2}$ (sum of lengths of parallel sides) × (altitude)

Area of triangle = $\frac{1}{2}$ (base) × (altitude)

Area of triangle = $\sqrt{s(s-a)(s-b)(s-c)}$

where $s = \frac{1}{2}(a+b+c)$

$$\text{Area of circle} = \pi\,(\text{radius})^2 \quad \text{or} \quad \frac{\pi\,(\text{diameter})^2}{4}$$

$$\text{Volume of cube} = (\text{length of edge})^3$$

$$\left(\begin{array}{c}\text{Volume of rectangular}\\ \text{solid (or prism)}\end{array}\right) = \left(\begin{array}{c}\text{area of uniform}\\ \text{cross-section}\end{array}\right) \times (\text{length})$$

$$\text{Volume of cylinder} = \pi\,(\text{radius})^2 \times (\text{height or length})$$

For similar shapes:

$$\text{Ratio of areas} = (\text{ratio of corresponding lengths})^2$$

For similar solids:

$$\text{Ratio of volumes} = (\text{ratio of corresponding lengths})^3$$

Self-Test 13

1) A quadrilateral has one pair of sides parallel. It is therefore a:

a square b parallelogram c trapezium d rectangle

2) A quadrilateral has diagonals which bisect at right angles. It is therefore a:

a trapezium b parallelogram c square d rectangle

3) A rectangle has a length of 80 mm and a width of 30 mm. Its area is:

a $240\ \text{mm}^2$ b $2400\ \text{mm}^2$ c $0.24\ \text{m}^2$ d $2.4\ \text{m}^2$

4) A triangle has an altitude of 100 mm and a base of 50 mm. Its area is:

a $2500\ \text{mm}^2$ b $5000\ \text{mm}^2$ c $0.025\ \text{m}^2$ d $0.050\ \text{m}^2$

5) A parallelogram has a base 100 mm long and a vertical height of 50 mm. Its area is:

a $0.025\ \text{m}^2$ b $0.050\ \text{m}^2$ c $2500\ \text{mm}^2$ d $5000\ \text{mm}^2$

6) A trapezium has parallel sides whose lengths are 18 mm and 22 mm. The distance between the parallel sides is 10 mm. Hence the area of the trapezium is:

a $400\ \text{mm}^2$ b $200\ \text{mm}^2$ c $3960\ \text{mm}^2$ d $495\ \text{mm}^2$

7) The area of a circle is given by:

a $\dfrac{\pi(\text{radius})^2}{2}$ b $\pi(\text{diameter})^2$ c $\pi(\text{radius})^2$ d $\dfrac{\pi(\text{diameter})^2}{4}$

8) The cross-sectional area of a pipe of outside diameter 8 and inside diameter 4 is:

a $\pi(4^2-2^2)$ b $\pi 8^2-\pi 4^2$ c $\pi(8^2-4^2)$ d $\dfrac{\pi 8^2-\pi 4^2}{4}$

9) The volume of a cylinder of diameter 40 mm and height 20 mm is:

a $\pi\times 40^2\times 20$ b $\pi\times 40\times 20^2$ c $\dfrac{\pi\times 40^2\times 20}{4}$ d $\pi\times 20^2\times 10$

10) The volume of the prism shown in Fig. 17.49 is:

a $1.5\times 2\times 3$ b $\pi\times 1.5\times 2\times 3$
c $2\times 1.5\times 2\times 3$ d $\frac{1}{2}\times 1.5\times 2\times 3$

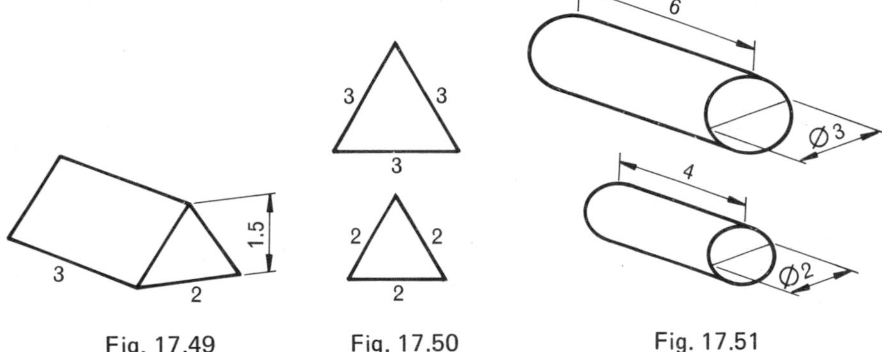

Fig. 17.49 Fig. 17.50 Fig. 17.51

11) The ratio of the areas of the triangles shown in Fig. 17.50 is:

a $\dfrac{2}{3}$ b $\dfrac{2^2}{3^2}$ c $\left(\dfrac{2}{3}\right)^2$ d $\dfrac{2^3}{3^3}$

12) The ratio of the volumes of the cylinders shown in Fig. 17.51 is:

a $\dfrac{\pi(3^3-2^2)}{6-4}$ b $\left(\dfrac{\pi}{4}3^2-\dfrac{\pi}{4}2^2\right)(6-4)$

c $\left(\dfrac{3}{2}\right)^3$ d $\dfrac{6^3}{4^3}$

18. TRIGONOMETRY

fter reaching the end of this chapter you should be able to:

1. *Sketch a right-angled triangle from given data.*
2. *Define sine, cosine and tangent ratios for acute angles.*
3. *Use tables or scientific calculator to find the three trigonometrical ratios for an acute angle.*
4. *Determine from tables or calculator an acute angle given its sine, cosine, tangent.*
5. *Calculate the fractional or surd forms of the three trigonometrical ratios for angles of 30°, 45° and 60°.*
6. *State the relationships: $\cos\theta = \sin(90-\theta)$ and $\sin\theta = \cos(90-\theta)$.*
7. *Solve problems involving the trigonometrical ratios and/or Pythagoras.*

THE TRIGONOMETRICAL RATIOS

Consider any angle θ which is bounded by the lines OA and OB as shown in Fig. 18.1. Take any point P on the boundary line OB. From P draw line PM perpendicular to OA to meet it at the point M.

Then, the ratio $\dfrac{MP}{OP}$ is called the sine of $\angle AOB$,

the ratio $\dfrac{OM}{OP}$ is called the cosine of $\angle AOB$, and

the ratio $\dfrac{MP}{OM}$ is called the tangent of $\angle AOB$

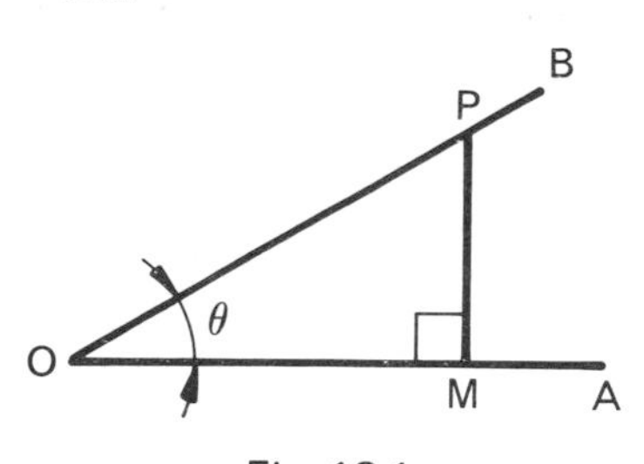

Fig. 18.1

THE SINE OF AN ANGLE

The abbreviation 'sin' is usually used for sine. Consider right-angled triangle (Fig. 18.2).

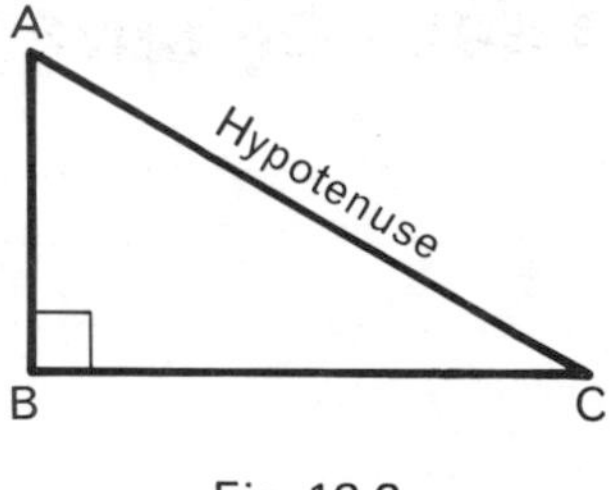

Fig. 18.2

Now the sine of angle $= \dfrac{\text{side opposite the angle}}{\text{hypotenuse}}$

Hence $\sin A = \dfrac{BC}{AC}$

and $\sin C = \dfrac{AB}{AC}$

EXAMPLE 19.1

Find by drawing a suitable triangle the value of sin 30°.

Draw the lines AX and AY which intersect at A so that the angle $\angle YAX = 30°$ as shown in Fig. 18.3. Along AY measure off AC equal to 1 unit (say 100 mm) and from C draw CB perpendicular to AX. Measure CB which will be found to be 0.5 units (50 mm in this case).

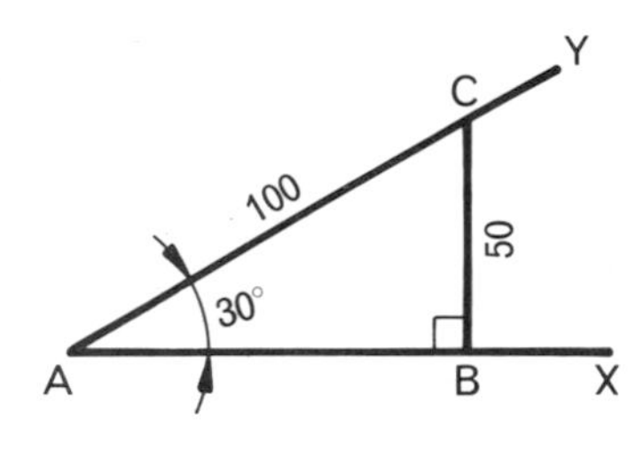

Fig. 18.3

Therefore $\sin 30° = \dfrac{5}{10} = 0.5$

Although it is possible to find the sines of angles by drawing this is inconvenient and not very accurate. Tables of sines have been calculated which allow us to find the sine of any angle.

READING THE TABLE OF SINES OF ANGLES

(1) *To find* $\sin 12°$. The sine of an angle with an exact number of degrees is shown in the column headed 0. Thus $\sin 12° = 0.2079$

(2) *To find* $\sin 12°36'$. The value will be found under the column headed 36′. Thus $\sin 12°36' = 0.2181$

(3) *To find* $\sin 12°40'$. If the number of minutes is not an exact multiple of 6 we use the table of mean differences. Now $12°36' = 0.2181$ and 40′ is 4′ more than 36′. Looking in the mean difference column headed 4 we find the value 11. This is *added* on to the sine of 12°36′ and we have $\sin 12°40' = 0.2181 + 0.0011 = 0.2192$

(4) *To find the angle whose sine is* 0.1711. Look in the table of sines to find the nearest number *lower* than 0.1711. This is found to be 0.1702 which corresponds to an angle of 9°48′. Now 0.1702 is 0.0009 less than 0.1711 so we look in the mean difference table in the row marked 9° and find 9 in the column headed 3′. The angle whose sine is 0.1711 is then $9°48' + 3' = 9°51'$ or $\sin 9°51' = 0.1711$

INVERSE NOTATION

The phrase **The angle whose sine is 0.1711** is often written as **arc sin 0.1711** or **inv sin 0.1711** or $\mathbf{\sin^{-1} 0.1711}$

Thus if $\sin\theta = 0.1771$

Then $\theta = \text{arc}\sin 0.1711$

or $\theta = \text{inv}\sin 0.1711$

or $\theta = \sin^{-1} 0.1711$

SINES FROM A SCIENTIFIC CALCULATOR

On most calculators there is a sliding switch which may be positioned at either RAD (radians), DEG (degrees) or GRAD (grades — a grade being one hundredth of a right angle, more used on the continent). Here we wish to work in degrees and so we set the sliding switch to the DEG position.

(1) To find sin 12°

The sequence of operations is:

[AC] [1] [2] [sin] The display gives 0.207 912

Thus $\sin 12° = 0.2079$ correct to four decimal places.

(2) To find sin 12°36′

We must first convert the 36 minutes to decimals of a degree. Some calculators have a special facility for this conversion as your instruction booklet will indicate—if not we may use the fact that $12°36' = \left(12 + \frac{36}{60}\right)^{\circ}$ and the sequence of operations would be:

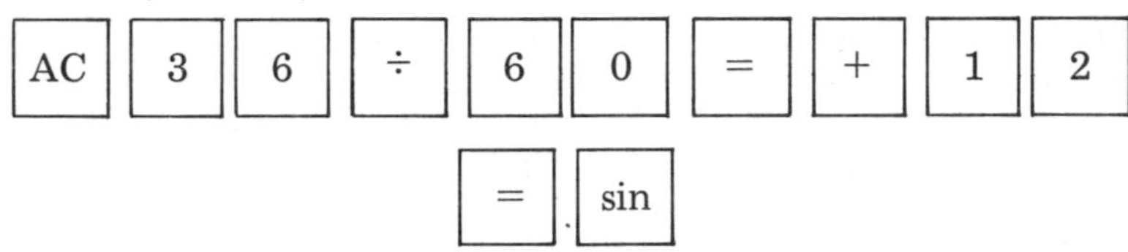

The display gives 0.218 143

Thus $\sin 12°36' = 0.2181$ correct to four decimal places.

(3) To find the angle whose sine is 0.1711

One of the phrases inv, arc or $\sin^{-1}$ will be present on your calculator keyboard — consult your instruction booklet. We will assume here that the phrase 'inv' appears on a separate key.

Thus the sequence of operations for finding inv sin 0.1711 is:

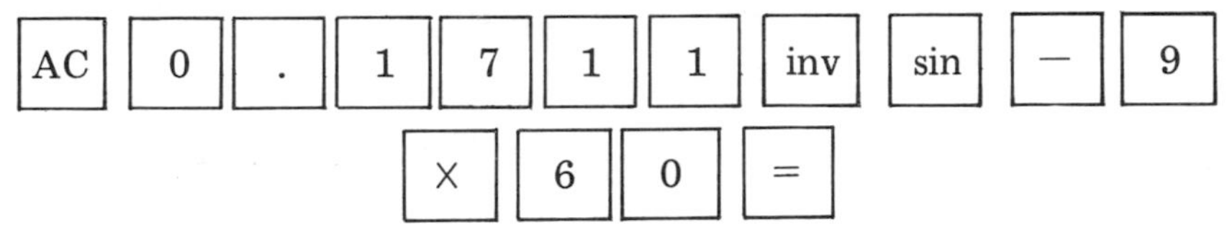

The display here is 9.851 78 which represents the required angle in degrees. The subsequent operations subtract the whole number part, 9, leaving the decimal part which is then multiplied by 60 to find the number of minutes.

The final display is 51.1069

Thus inv sin 0.1711 = 9°51′ correct to the nearest minute.

EXAMPLE 18.2

a) Find the length of AB in Fig. 18.4.

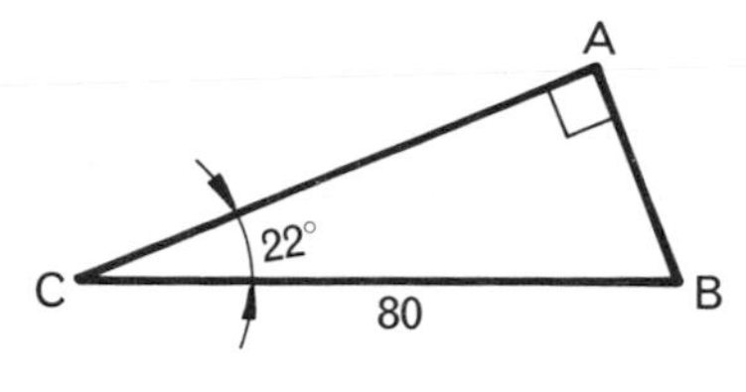

Fig. 18.4

AB is the side opposite $\angle$ACB, BC is the hypotenuse since it is opposite the right angle.

$$\therefore \qquad \frac{AB}{BC} = \sin 22^\circ$$

$$\text{or} \qquad AB = BC \times \sin 22^\circ = 80 \times 0.3746 = 30.0\text{ mm}$$

b) Find the length of AB in Fig. 18.5.

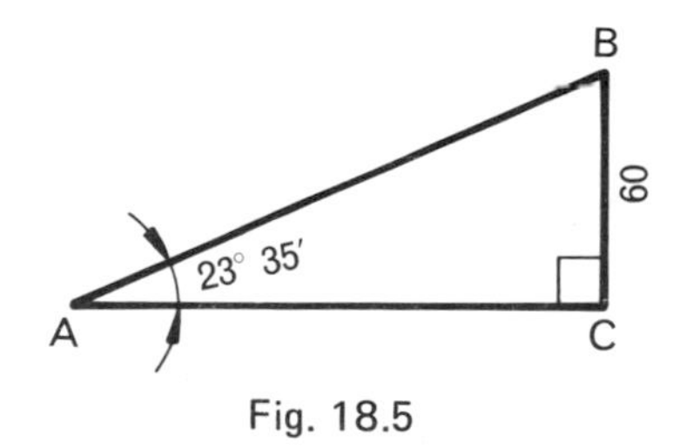

Fig. 18.5

BC is the side opposite $\angle$BAC and AB is the hypotenuse.

$$\therefore \qquad \frac{BC}{AB} = \sin 23^\circ 35'$$

$$\text{or} \qquad AB = \frac{BC}{\sin 23^\circ 35'} = \frac{60}{0.4000} = 150\text{ mm}$$

c) Find the angles CAB and ABC in ΔABC which is shown in Fig. 18.6.

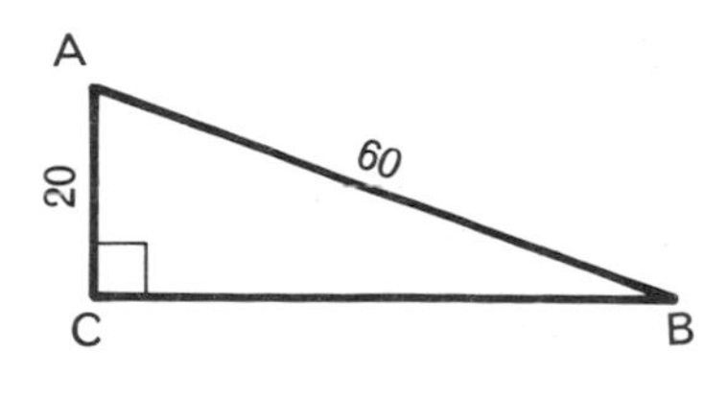

Fig. 18.6

Now $\sin B = \frac{AC}{AB} = \frac{20}{60} = 0.3333$

From the sine tables or calculator:

$$\angle B = \text{inv}\sin 0.3333 = 19°28'$$

$$\therefore \quad \angle A = 90° - 19°28' = 70°32'$$

Exercise 18.1

1) Find, by drawing, the sines of the following angles:

(a) 30° (b) 45° (c) 68°

2) Find, by drawing, the angles whose sines are:

(a) $\frac{1}{3}$ (b) $\frac{3}{4}$ (c) 0.72

3) Using the tables or calculator find the values of:

(a) sin 12° (b) sin 18°12′ (c) sin 74°42′
(d) sin 7°23′ (e) sin 87°35′ (f) sin 0°11′

4) Using the tables or calculator find the angles whose sines are:

(a) 0.1564 (b) 0.9135 (c) 0.9880 (d) 0.0802
(e) 0.9814 (f) 0.7395 (g) 0.0500 (h) 0.2700

5) Find the lengths of the sides marked x in Fig. 18.7 the triangles being right-angled.

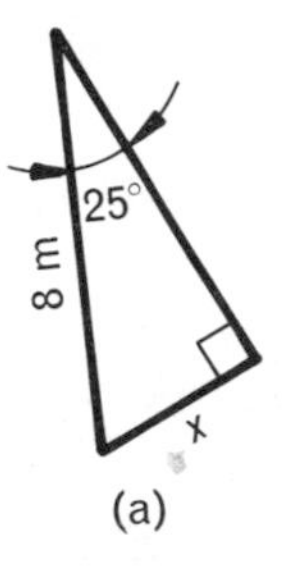

(a)

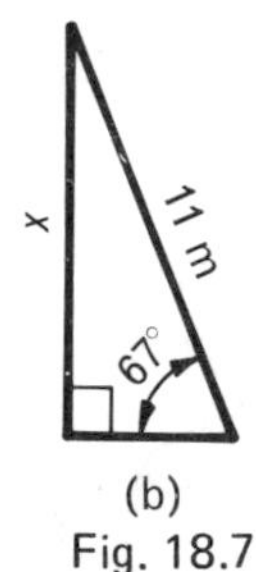

(b)

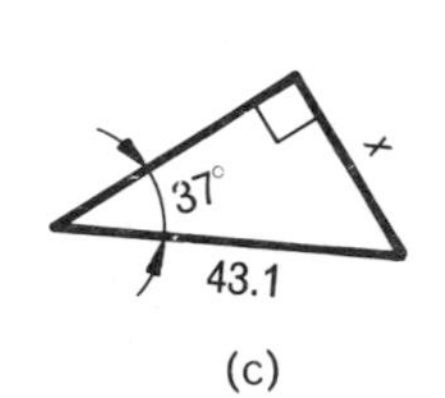

(c)

Fig. 18.7

6) Find the angles marked θ in Fig. 18.8, the triangles being right-angled.

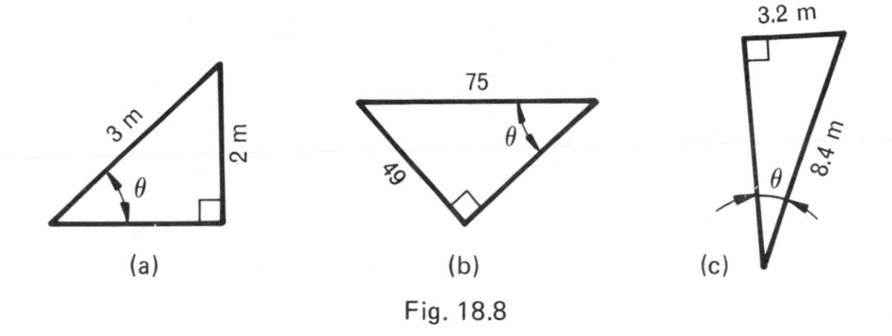

Fig. 18.8

7) In $\triangle ABC$, $\angle C = 90°$, $\angle B = 23°17'$ and $AC = 112$ mm. Find AB.

8) In $\triangle ABC$, $\angle B = 90°$, $\angle A = 67°28'$ and $AC = 0.86$ m. Find BC.

9) An equilateral triangle has an altitude of 187 mm. Find the length of the equal sides.

10) Find the altitude of an isosceles triangle whose vertex angle is 38° and whose equal sides are 7.9 m long.

11) The equal sides of an isosceles triangle are each 270 mm and the altitude is 190 mm. Find the angles of the triangle.

THE COSINE OF AN ANGLE

In any right-angled triangle (Fig. 18.9):

$$\text{the cosine of an angle} = \frac{\text{side adjacent to the angle}}{\text{hypotenuse}}$$

Thus $$\cos A = \frac{AB}{AC}$$

and $$\cos C = \frac{BC}{AC}$$

The abbreviation 'cos' is usually used for cosine.

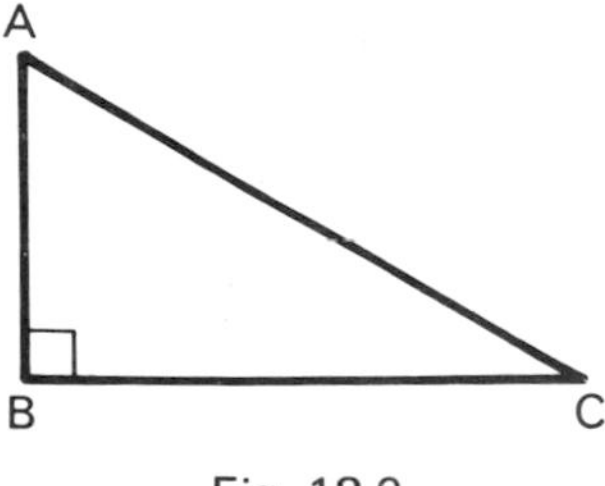

Fig. 18.9

The cosine of an angle may be found by drawing, the construction being similar to that used for the sine of an angle. However, tables of cosines are available and these are used in a similar way to the table of sines except that the mean differences are now *subtracted*. Cosines may also be found using a scientific calculator.

EXAMPLE 18.3

a) Find the length of the side BC in Fig. 18.10.

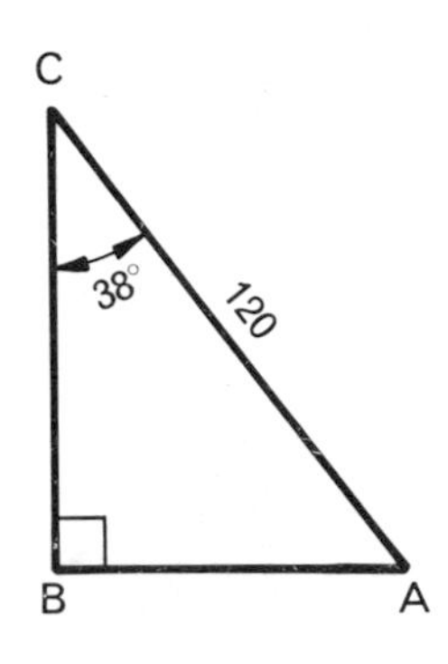

Fig. 18.10

BC is the side adjacent to ∠BCA and AC is the hypotenuse.

$$\therefore \qquad \frac{BC}{AC} = \cos 38^\circ$$

or $\qquad BC = AC \times \cos 38^\circ = 120 \times 0.7880 = 94.6 \text{ mm}$

b) Find the length of the side AC in Fig. 18.11.

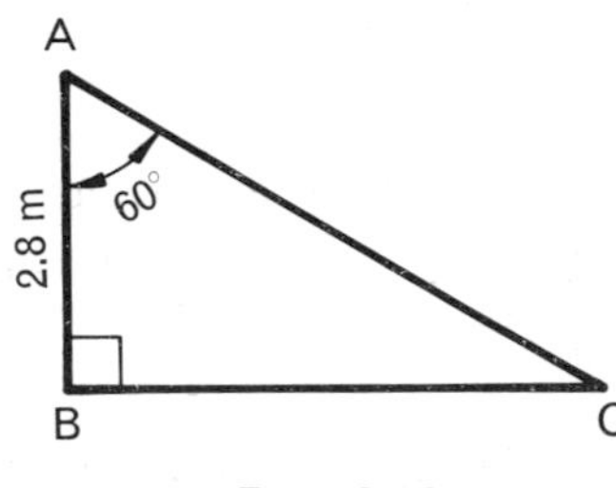

Fig. 18.11

AB is the side adjacent to ∠BAC and AC is the hypotenuse.

$$\therefore \qquad \frac{AB}{AC} = \cos 60^\circ$$

or $$AC = \frac{AB}{\cos 60^\circ} = \frac{2.8}{0.5000} = 5.60\,\text{m}$$

c) Find the angle θ shown in Fig. 18.12.

Since ΔABC is isosceles the perpendicular AD bisects the base BC and hence $BD = 15\,\text{mm}$.

Now $$\cos\theta = \frac{BD}{AB} = \frac{15}{50} = 0.3$$

or $$\theta = \text{inv}\cos 0.3 = 72^\circ 32'$$

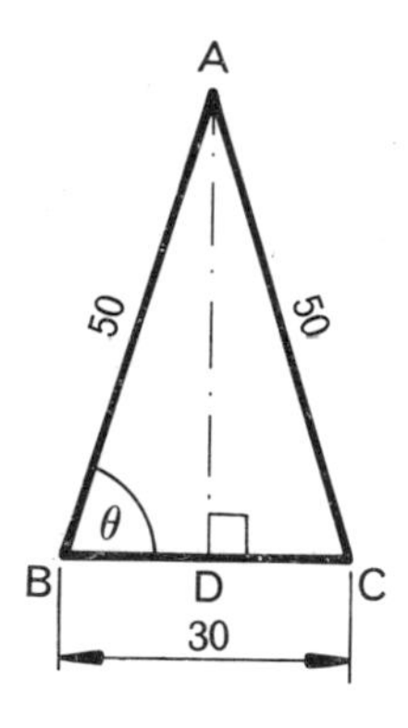

Fig. 18.12

Exercise 18.2

1) Use the tables or calculator to find the values of:

(a) $\cos 15^\circ$ (b) $\cos 24^\circ 18'$ (c) $\cos 78^\circ 24'$ (d) $\cos 0^\circ 11'$
(e) $\cos 73^\circ 22'$ (f) $\cos 39^\circ 59'$

2) Use the tables or calculator to find the angles whose cosine are:

(a) 0.9135 (b) 0.3420 (c) 0.9673 (d) 0.4289
(e) 0.9586 (f) 0.0084 (g) 0.2611 (h) 0.4700

3) Find the lengths of the sides marked x in Fig. 18.13, the triangles being right-angled.

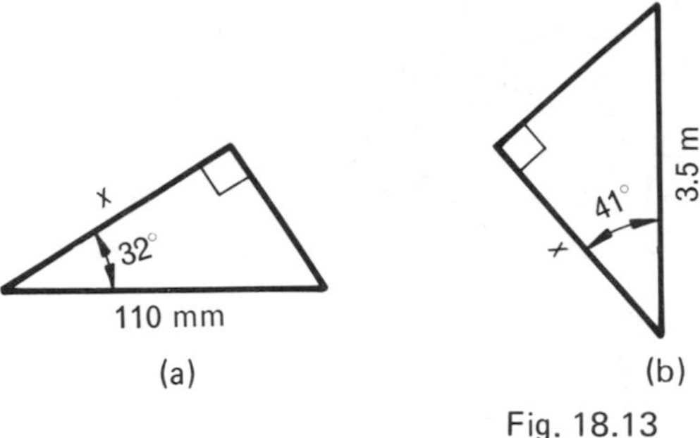

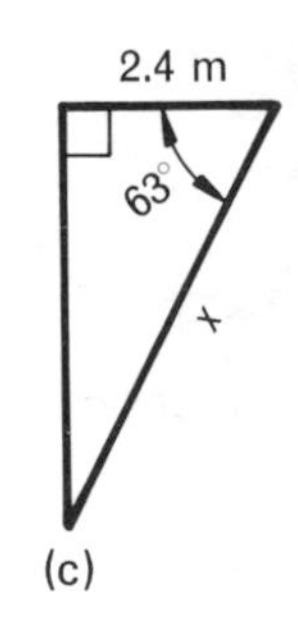

Fig. 18.13

4) Find the angles marked θ in Fig. 18.14 the triangles being right-angled.

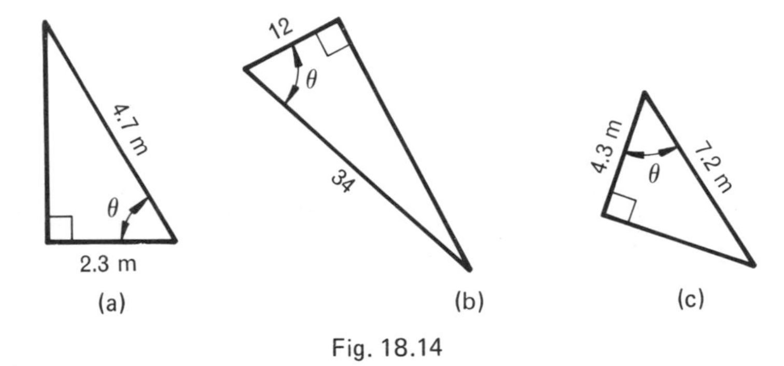

Fig. 18.14

5) An isosceles triangle has a base of 34 mm and the equal sides are each 42 mm long. Find the angles of the triangle and also its altitude.

6) In $\triangle ABC$, $\angle C = 90°$, $\angle B = 33°27'$ and BC = 24 mm. Find AB.

7) In $\triangle ABC$, $\angle B = 90°$, $\angle A = 62°45'$ and AC = 4.3 m. Find AB.

8) In Fig. 18.15, calculate $\angle BAC$ and the length BC.

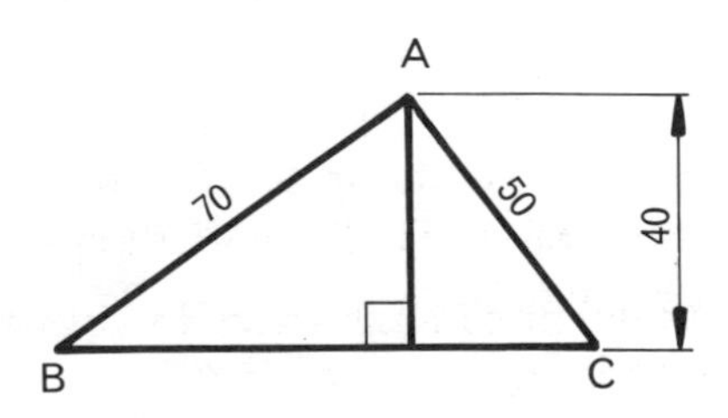

Fig. 18.15

9) In Fig. 18.16 calculate BD, AD, AC and BC.

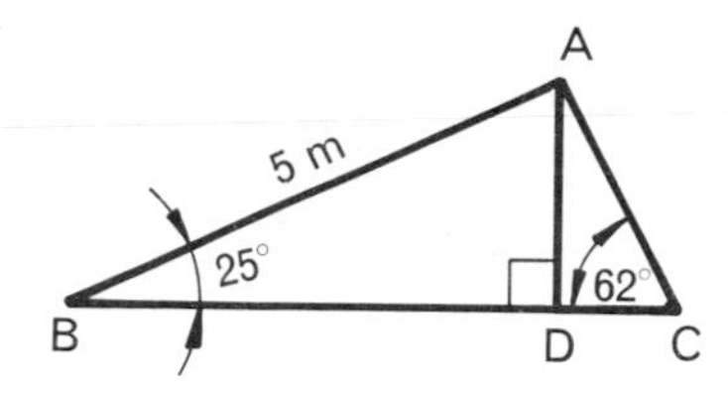

Fig. 18.16

10) Lay out accurately the following angles by using their cosines:
(a) 39° (b) 70°6′ (c) 18°11′

THE TANGENT OF AN ANGLE

In any right-angled triangle (Fig. 18.17),

$$\text{the tangent of an angle} = \frac{\text{side opposite to the angle}}{\text{side adjacent to the angle}}$$

Thus $$\tan A = \frac{BC}{AB}$$

and $$\tan C = \frac{AB}{BC}$$

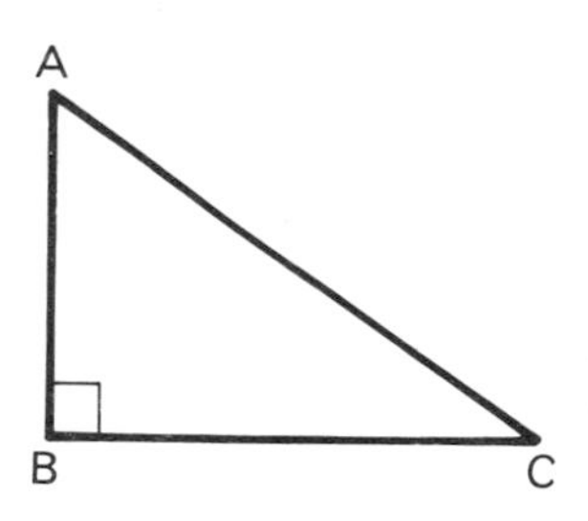

Fig. 18.17

The abbreviation 'tan' is usually used for tangent.

Values of the tangents of angles may be found using the tables or a scientific calculator.

EXAMPLE 18.4

a) Find the length of the side AB in Fig. 18.18.

AB is the side opposite $\angle C$ and AC is the side adjacent to $\angle C$.

Hence, $$\frac{AB}{AC} = \tan \angle C$$

or $$\frac{AB}{AC} = \tan 42^\circ$$

$\therefore$ $$AB = AC \times \tan 42^\circ = 40 \times 0.9004 = 36.0\,\text{mm}$$

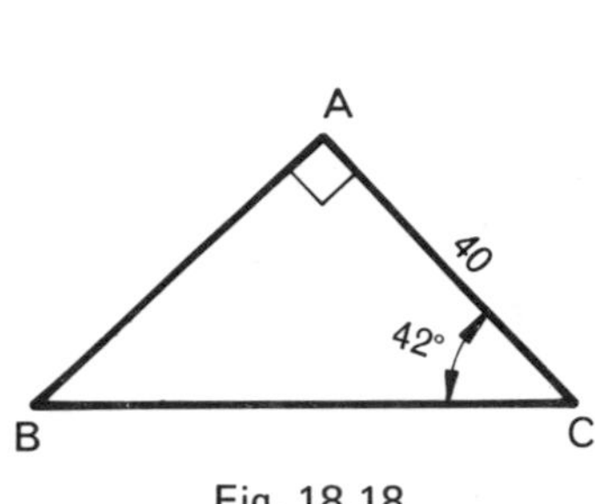

Fig. 18.18

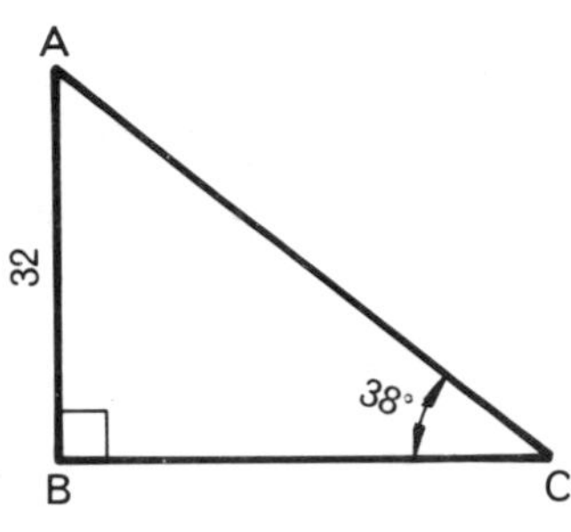

Fig. 18.19

b) Find the length of the side BC in Fig. 18.19.

There are two ways of doing this problem.

(i) Now $$\frac{AB}{BC} = \tan 38^\circ \quad \text{or} \quad BC = \frac{AB}{\tan 38^\circ}$$

$\therefore$ $$BC = \frac{32}{0.7813} = 41.0\,\text{mm}$$

(ii) Since $$\angle C = 38^\circ, \angle A = 90^\circ - 38^\circ = 52^\circ$$

now $$\frac{BC}{AB} = \tan A \quad \text{or} \quad BC = AB \times \tan A$$

$\therefore$ $$BC = 32 \times \tan 52^\circ = 41.0\,\text{mm}$$

c) Find the angle θ shown in Fig. 18.20.

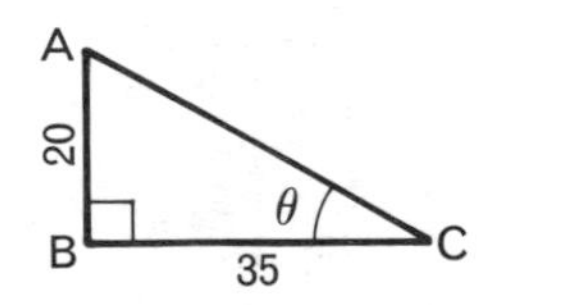

Fig. 18.20

Now $\quad \tan\theta = \dfrac{AB}{BC} = \dfrac{20}{35} = 0.5714$

or $\quad \theta = \text{inv}\tan 0.5714 = 29°45'$

Exercise 18.3

1) Use tables or calculator find the values of:

(a) tan 18° (b) tan 32°24′ (c) tan 53°42′
(d) tan 39°27′ (e) tan 11°20′ (f) tan 69°23′

2) Use tables or calculator to find the angles whose tangents are:

(a) 0.4452 (b) 3.2709 (c) C.0769 (d) 0.3977
(e) 0.3568 (f) 0.8263 (g) 1.9251 (h) 0.0163

3) Find the lengths of the sides marked y in Fig. 18.21, the triangles being right-angled.

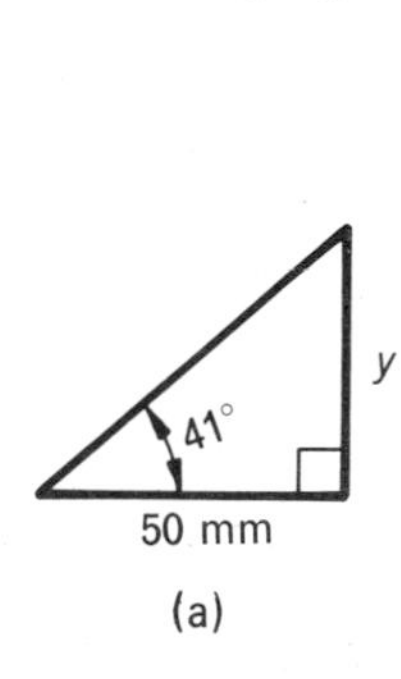

(a)

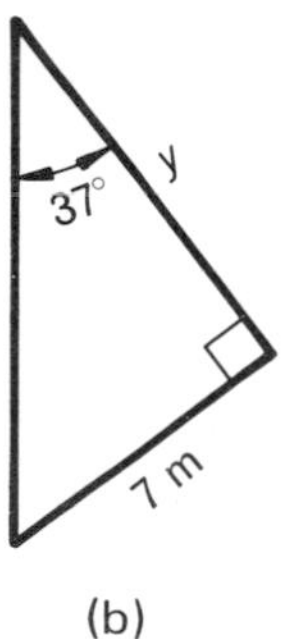

(b)

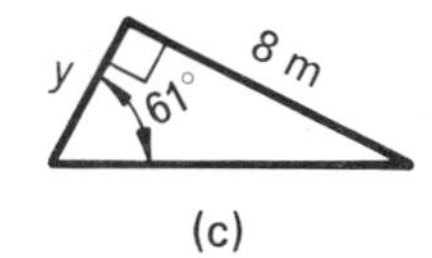

(c)

Fig. 18.21

4) Find the angles marked α in Fig. 18.22, the triangles being right-angled.

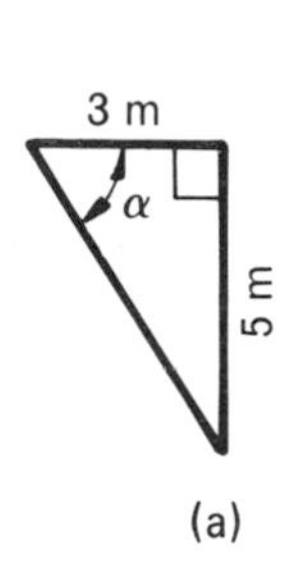

(a)

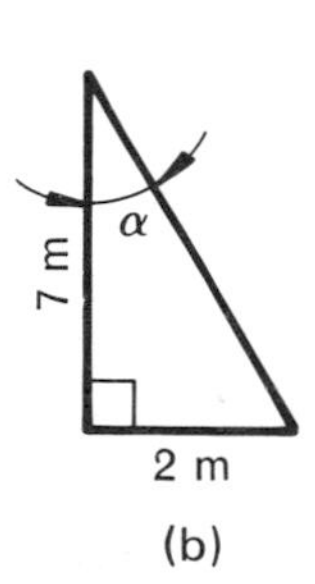

(b)

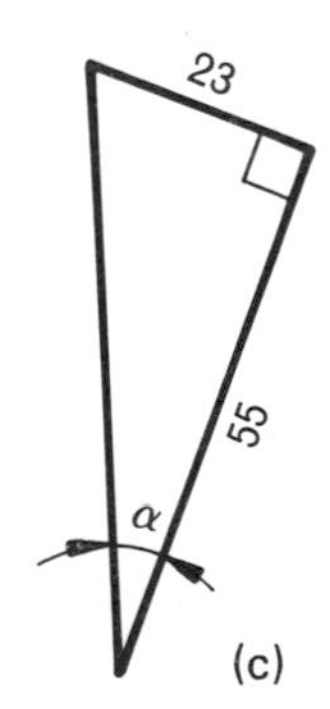

(c)

Fig. 18.22

5) An isosceles triangle has a base 100 mm long and the two equal angles are each 57°. Calculate the altitude of the triangle.

6) In $\triangle ABC$, $\angle B = 90°$, $\angle C = 49°$ and $AB = 32$ mm. Find BC.

7) In $\triangle ABC$, $\angle A = 12°23'$, $\angle B = 90°$ and $BC = 7.31$ m. Find AB.

8) Calculate the distance x in Fig. 18.23.

9) Calculate the distance d in Fig. 18.24.

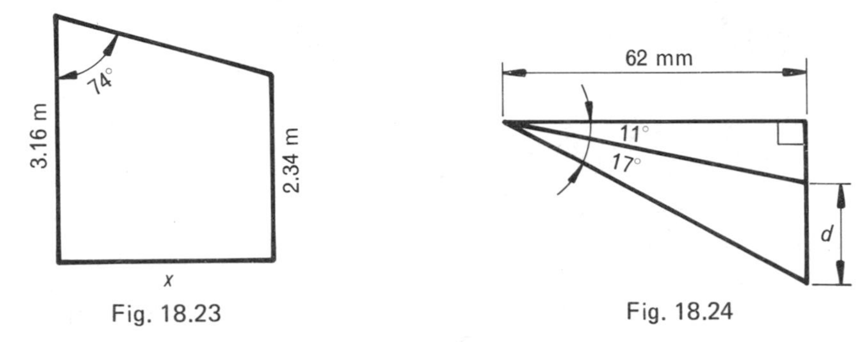

Fig. 18.23

Fig. 18.24

10) Lay out accurately the following angles using their tangents:
(a) 18°16′ (b) 37°11′ (c) 68°19′

TRIGONOMETRICAL RATIOS FOR 30°, 60° and 45°

Ratios for 30° and 60°

Fig. 18.25 shows an equilateral triangle ABC with each of the sides equal to 2 units. From C draw the perpendicular CD which bisects the base AB and also bisects $\angle C$.

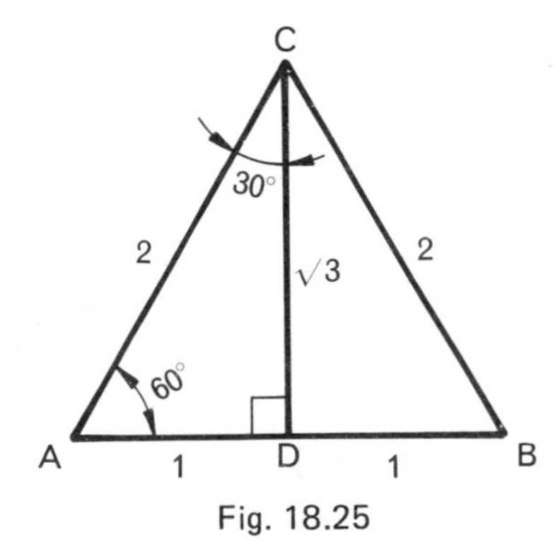

Fig. 18.25

In ΔACD,

$$CD^2 = AC^2 - AD^2 = 2^2 - 1^2 = 3$$

$$\therefore \quad CD = \sqrt{3}$$

Since all the angles of ΔABC are 60° and $\angle$ACD = 30°,

$$\sin 60^\circ = \frac{\sqrt{3}}{2} \qquad \sin 30^\circ = \frac{1}{2}$$

$$\cos 60^\circ = \frac{1}{2} \qquad \cos 30^\circ = \frac{\sqrt{3}}{2}$$

$$\tan 60^\circ = \frac{\sqrt{3}}{1} = \sqrt{3} \qquad \tan 30^\circ = \frac{1}{\sqrt{3}} = \frac{\sqrt{3}}{3}$$

Ratios for 45°

Fig. 18.26 shows a right-angled isosceles triangle ABC with the equal sides each 1 unit in length. The equal angles are each 45°. Now,

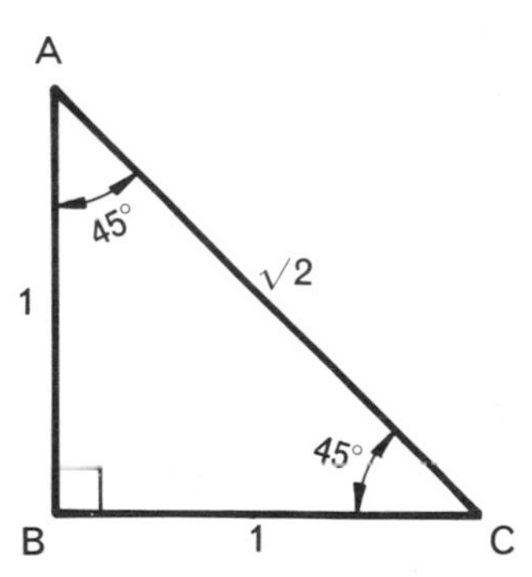

Fig. 18.26

$$AC^2 = AB^2 + BC^2 = 1^2 + 1^2 = 2$$

$$\therefore \quad AC = \sqrt{2}$$

$$\sin 45^\circ = \frac{1}{\sqrt{2}} = \frac{\sqrt{2}}{2}$$

$$\cos 45^\circ = \frac{1}{\sqrt{2}} = \frac{\sqrt{2}}{2}$$

$$\tan 45^\circ = \frac{1}{1} = 1$$

GIVEN ONE RATIO TO FIND THE OTHERS

The method is shown in the following example.

EXAMPLE 18.5

If $\cos A = 0.7$, find, without using tables, the values of $\sin A$ and $\tan A$.

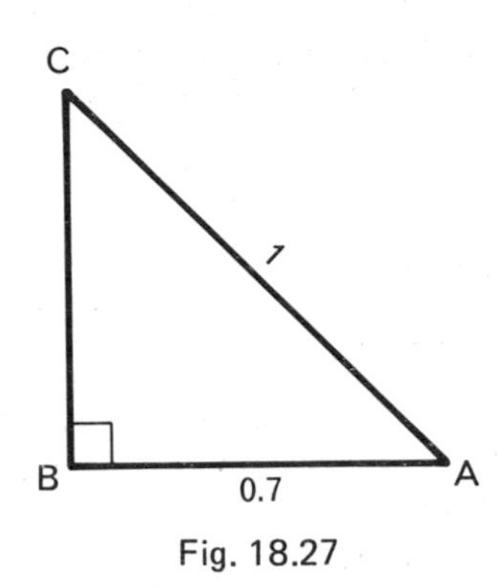

Fig. 18.27

In Fig. 18.27 if we make $AB = 0.7$ units and $AC = 1$ unit, then

$$\cos A = \frac{0.7}{1} = 0.7$$

By Pythagoras' theorem,

$$BC^2 = AC^2 - AB^2 = 1^2 - 0.7^2 = 0.51$$

$$\therefore \quad BC = \sqrt{0.51} = 0.7141$$

then $$\sin A = \frac{BC}{AC} = \frac{0.7141}{1} = 0.7141$$

and $$\tan A = \frac{BC}{AB} = \frac{0.7141}{0.7} = 1.020$$

COMPLEMENTARY ANGLES

Consider the triangle ABC shown in Fig. 18.28.

$$\sin A = \frac{a}{b} \qquad \cos C = \frac{a}{b}$$

Hence, $\sin A = \cos C = \cos(90° - A)$

Similarly $\cos A = \sin(90° - A)$

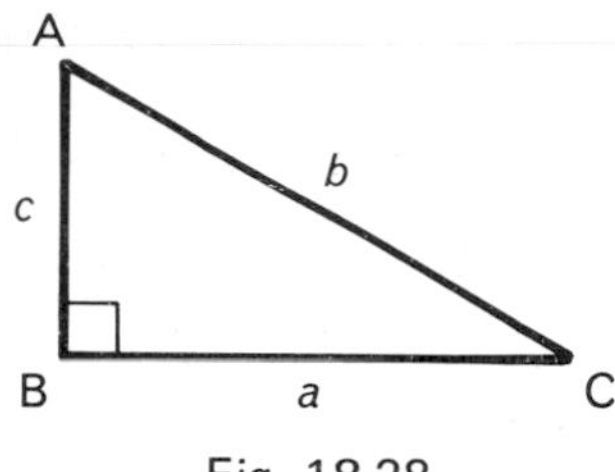

Fig. 18.28

Therefore, *the sine of an angle is equal to the cosine of its complementary angle and vice versa.*

Thus $\sin 26° = \cos 64° = 0.4384$

and $\cos 70° = \sin 20° = 0.3420$

Exercise 18.4

1) If $\sin A = 0.3171$ find the values of cos A and tan A without using tables or calculator.

2) If $\tan A = \frac{3}{4}$, find the values of sin A and cos A without using tables or calculator.

3) If $\cos A = \frac{12}{13}$, find without using tables or calculator the values of sin A and tan A.

4) Show that $\cos 60° + \cos 30° = \frac{1+\sqrt{3}}{2}$.

5) Show that $\sin 60° + \cos 30° = \sqrt{3}$.

6) Show that $\cos 45° + \sin 60° + \sin 30° = \frac{\sqrt{2}+\sqrt{3}+1}{2}$.

7) Given that $\sin 48° = 0.7431$ find the values of $\cos 42°$, without using tables or calculator.

8) If $\cos 63° = 0.4540$, what is the value of $\sin 27°$?

PRACTICAL TRIGONOMETRY PROBLEMS

EXAMPLE 18.6

a) The end of a round bar is to be machined as shown in Fig. 18.29. Calculate the length x.

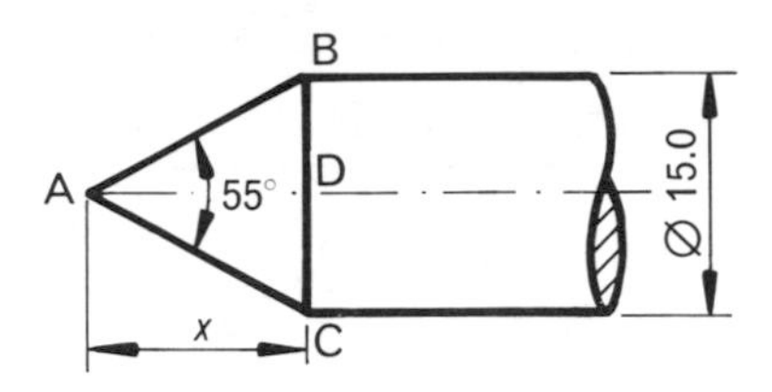

Fig. 18.29

ΔABC is isosceles. The perpendicular AD bisects BC and $\angle$BAC.

In ΔACD, $\angle CAD = 27°30'$

$\therefore$ $\angle ACD = 90° - 27°30' = 62°30'$

Now $CD = 7.5\,\text{mm}$

and $\dfrac{AD}{CD} = \tan \angle ACD$

or $AD = CD \times \tan \angle ACD = 7.5 \times \tan 62°30'$

$= 7.5 \times 1.921 = 14.4$

Hence the length x is 14.4 mm.

b) A flange has 10 holes equally spaced around a pitch circle of 65.00 mm diameter. Calculate the centre distance between a pair of holes.

The flange is shown in Fig. 18.30. Since AO = BO (radii) the triangle AOB is isosceles.

Then $\angle AOB = \dfrac{360°}{10} = 36°$

Draw the perpendicular OC as shown in the diagram, then

$$\angle AOC = \tfrac{1}{2} \times 36° = 18°$$

In ΔCOA, $\dfrac{AC}{AO} = \sin \angle AOC$

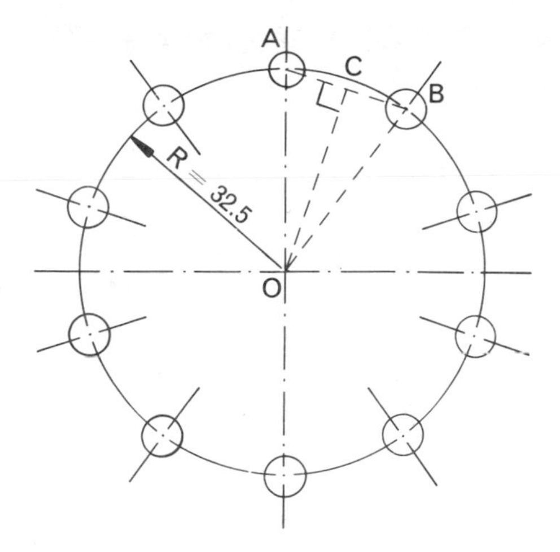

Fig. 18.30

or $\quad AC = AO \times \sin \angle AOC = 32.5 \sin 18^\circ$

$= 32.5 \times 0.3090 = 10.04$

Thus $\quad AB = 2 \times AC = 2 \times 10.04 = 20.08$

Hence the centre distance between a pair of adjacent holes is 20.08 mm.

c) The power developed in an a.c. circuit is given by the expression $P = VI \cos \phi$ where ϕ is the phase angle. Calculate the value of $\cos \phi$ given that $P = 500$, $V = 250$ and $I = 8$. Hence obtain the value of ϕ correct to the nearest minute.

Substituting the given values we have

$$500 = 250 \times 8 \times \cos \phi$$

$$\cos \phi = \frac{500}{250 \times 8} = 0.25$$

From the table of cosines (or using a scientific calculator),

$$\phi = 75^\circ 31'$$

Exercise 18.5

1) A vertical aerial has a wire stay 40 m long. The wire makes an angle of 55° with the ground. Find:

(a) the distance of the wire from the base of the aerial;

(b) the height of the aerial.

2) In Fig. 18.31, BC = 5 m, AC = 10 m and DC = 3 m. Calculate:
(a) the difference in length between AB and DB;
(b) the angle ABD.

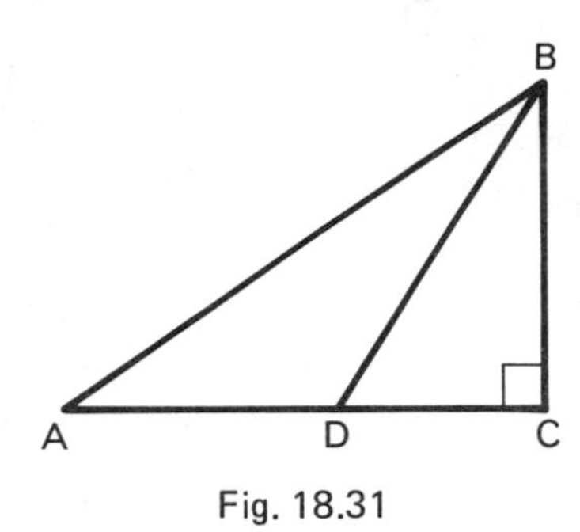

Fig. 18.31

3) The base of an isosceles triangle is 4 m long. The angles at the base are each 72°. Find the height of the triangle, the area and the length of its perimeter.

4) The current I in a tangent galvanometer is given by:

$$I = \frac{5HR \tan \theta}{\pi n}$$

If $H = 0.1853$, $R = 12$ and $n = 4$, calculate:
(a) the value of I when $\theta = 36°18'$
(b) the value of θ when $I = 0.26$

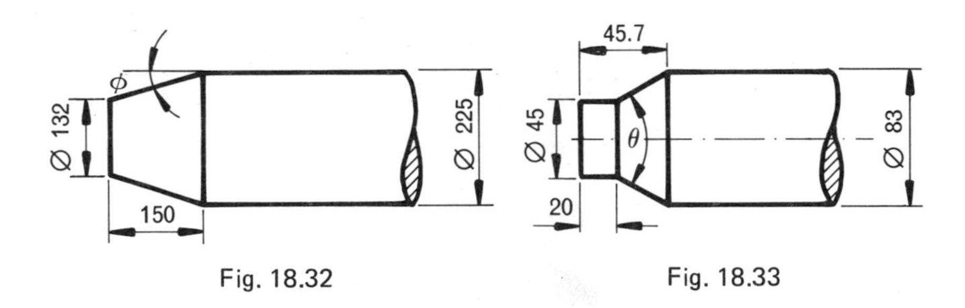

Fig. 18.32 Fig. 18.33

5) Find the angle ϕ for the spindle shown in Fig. 18.32.

6) Find the angle θ in Fig. 18.33.

7) A round bar has to be machined as shown in Fig. 18.34. Calculate the minimum length of bar required.

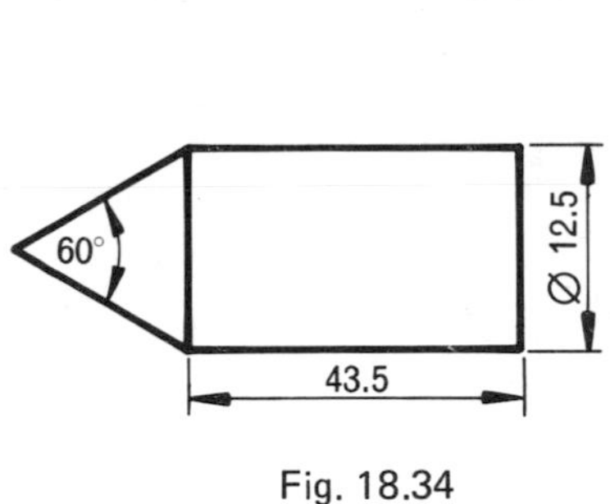

Fig. 18.34

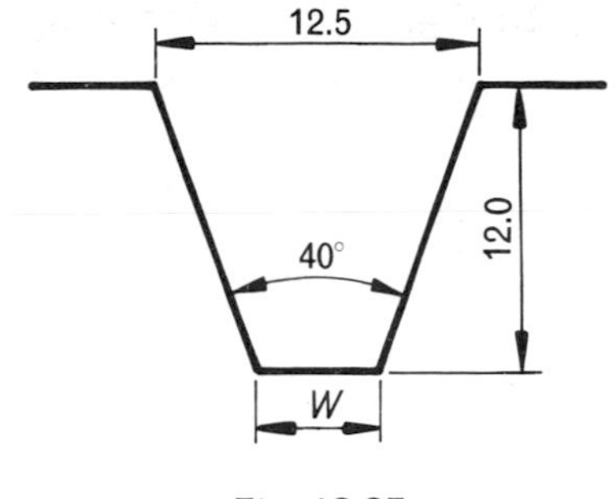

Fig. 18.35

8) 15 holes are equally spaced on a pitch circle 82.5 mm diameter. Calculate the length of chord joining the centres of two adjacent holes.

9) Fig. 18.35 shows a groove for a vee belt. Calculate the distance W.

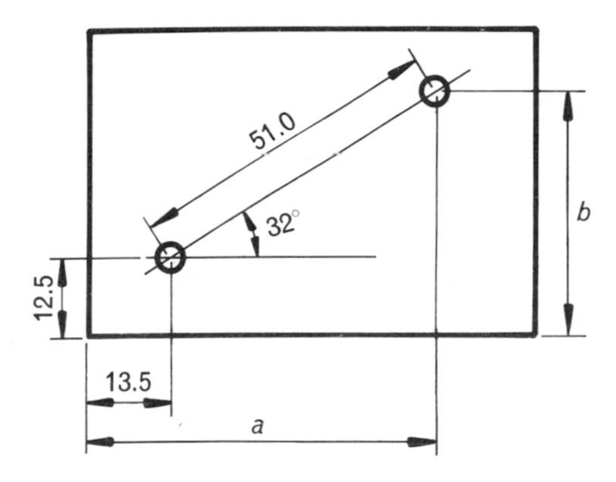

Fig. 18.36

10) Two holes are to be drilled in a plate as shown in Fig. 18.36. Calculate the coordinate dimensions a and b.

11) Fig. 18.37 represents four bushed holes in a drill jig. The lines AB and CD are parallel. Calculate:

(a) the centre distances AB, CD, BC and AC;

(b) the angle BAC.

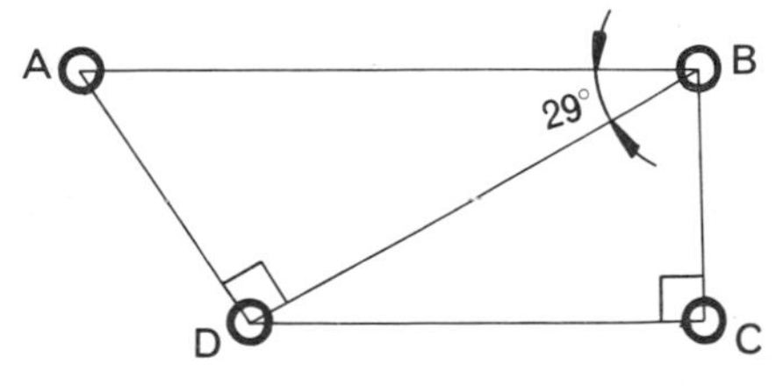

Fig. 18.37

12) A triangular template is shown in Fig. 18.38. Find the distance AB.

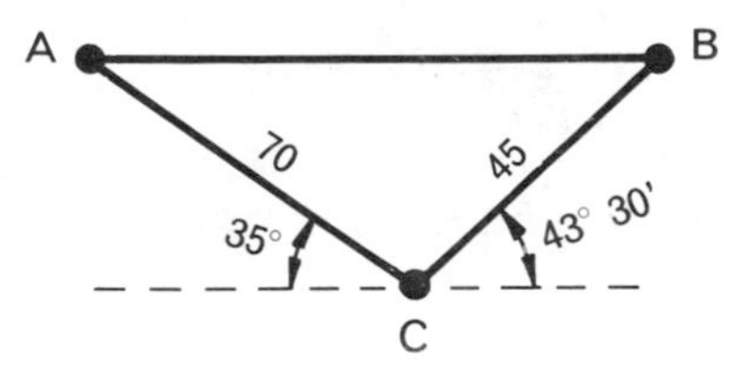

Fig. 18.38

SUMMARY

a)

$$\text{sine of an angle} = \frac{\text{opposite side}}{\text{hypotenuse}}$$

$$\text{cosine of an angle} = \frac{\text{adjacent side}}{\text{hypotenuse}}$$

$$\text{tangent of an angle} = \frac{\text{opposite side}}{\text{adjacent side}}$$

b)

$\sin 60° = \frac{\sqrt{3}}{2}$	$\sin 30° = \frac{1}{2}$	$\sin 45° = \frac{\sqrt{2}}{2}$
$\cos 60° = \frac{1}{2}$	$\cos 30° = \frac{\sqrt{3}}{2}$	$\cos 45° = \frac{\sqrt{2}}{2}$
$\tan 60° = \sqrt{3}$	$\tan 30° = \frac{\sqrt{3}}{3}$	$\tan 45° = 1$

c) The sine of an angle is equal to the cosine of its complement and vice versa. That is,

$$\sin\theta = \cos(90° - \theta) \qquad \cos\theta = \sin(90° - \theta)$$

Self-Test 14

1) In Fig. 18.39, $\sin x$ is equal to:

a $\frac{h}{p}$ **b** $\frac{h}{m}$ **c** $\frac{m}{p}$ **d** $\frac{p}{h}$

2) In Fig. 18.39, $\cos x$ is equal to:

a $\frac{h}{p}$ **b** $\frac{h}{m}$ **c** $\frac{m}{p}$ **d** $\frac{p}{h}$

3) In Fig. 18.39, $\tan x$ is equal to:

a $\frac{h}{p}$ **b** $\frac{h}{m}$ **c** $\frac{m}{p}$ **d** $\frac{p}{h}$

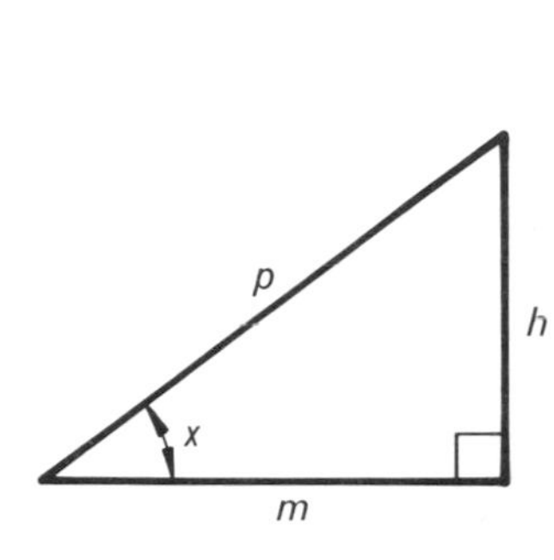

Fig. 18.39

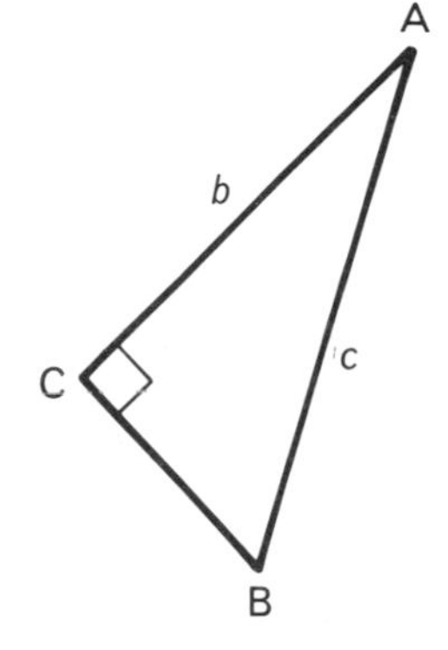

Fig. 18.40

4) In Fig. 18.40, sin A is equal to:

a $\frac{a}{b}$ **b** $\frac{a}{c}$ **c** $\frac{b}{c}$ **d** $\frac{c}{a}$

5) In Fig. 18.41, $\tan x$ is equal to:

a $\frac{q}{p}$ **b** $\frac{q}{r}$ **c** $\frac{p}{q}$ **d** $\frac{r}{q}$

6) In Fig. 18.42, $\cos y$ is equal to:

a $\frac{s}{t}$ **b** $\frac{r}{s}$ **c** $\frac{s}{r}$ **d** $\frac{t}{s}$

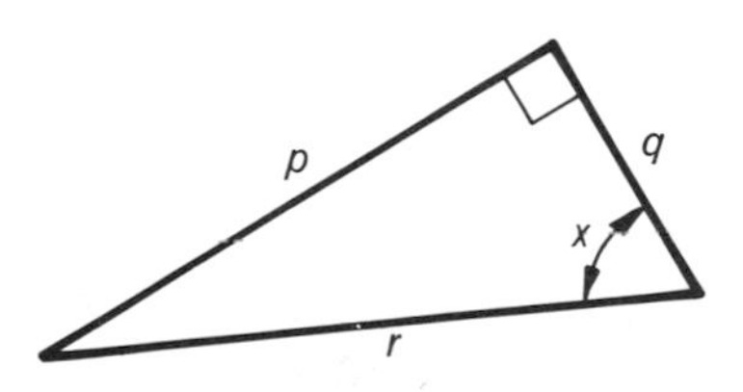

Fig. 18.41

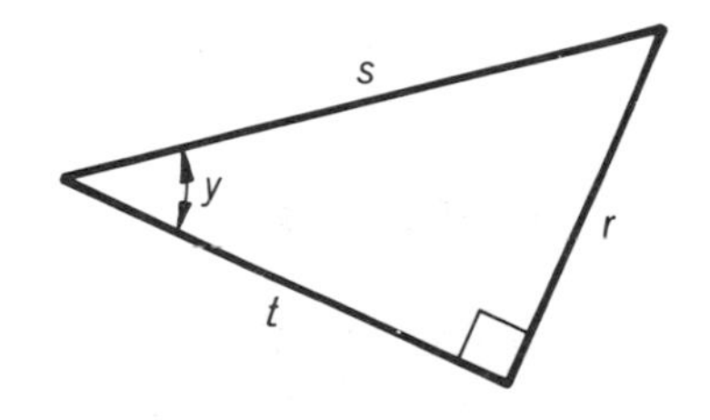

Fig. 18.42

7) The expression for the length AB (Fig. 18.43) is:

a $40 \tan 50^\circ$ b $40 \sin 50^\circ$ c $\dfrac{40}{\tan 50^\circ}$ d $\dfrac{40}{\sin 50^\circ}$

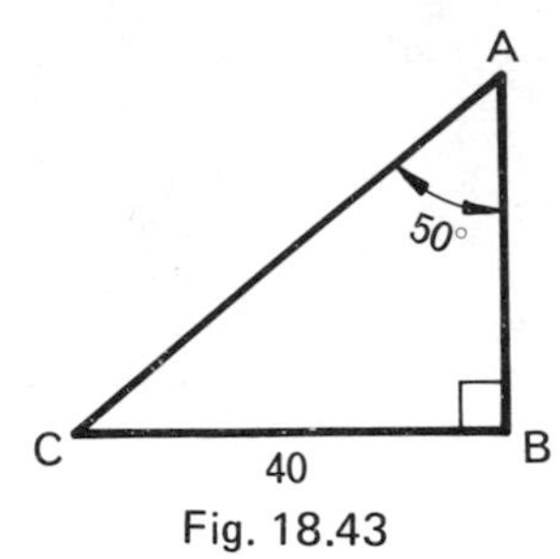

Fig. 18.43

8) The expression for the length AC (Fig. 18.43) is:

a $40 \sin 50^\circ$ b $40 \cos 50^\circ$ c $\dfrac{40}{\sin 50^\circ}$ d $\dfrac{40}{\cos 50^\circ}$

9) In Fig. 18.44, the expression for the side RN is:

a $(2 \sin 30^\circ + 3 \sin 60^\circ)$ m b $(2 \sin 30^\circ + 3 \cos 60^\circ)$ m

c $(2 \cos 30^\circ + 3 \cos 60^\circ)$ m d $(2 \cos 30^\circ + 3 \sin 60^\circ)$ m

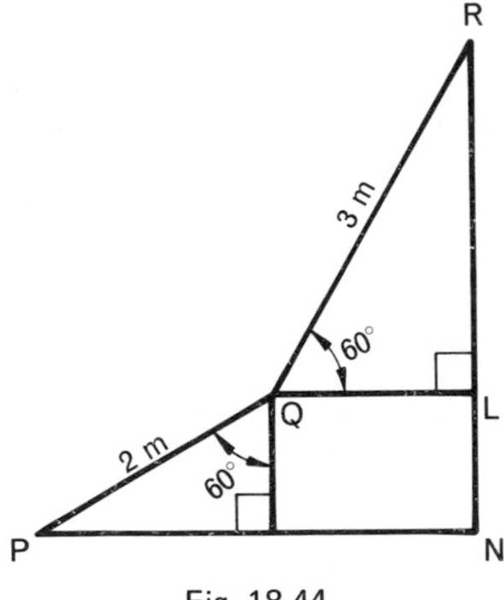

Fig. 18.44

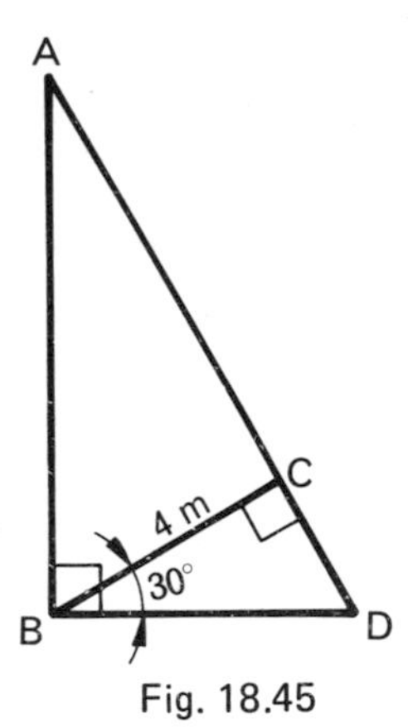

Fig. 18.45

10) In Fig. 18.45, an expression for the length AD is:

a $\dfrac{16\sqrt{3}}{3}$ m b $\dfrac{16}{\sqrt{3}}$ m c $8\sqrt{3}$ m d $\dfrac{40\sqrt{3}}{3}$ m

11) In Fig. 18.46, an expression for the angle θ is:

a $\tan \theta = \dfrac{30 - 10\sqrt{3}}{40}$ b $\tan \theta = \dfrac{40}{30 - 10\sqrt{3}}$

c $\tan\theta = \dfrac{20}{50-10\sqrt{3}}$

d $\tan\theta = \dfrac{50-10\sqrt{3}}{20}$

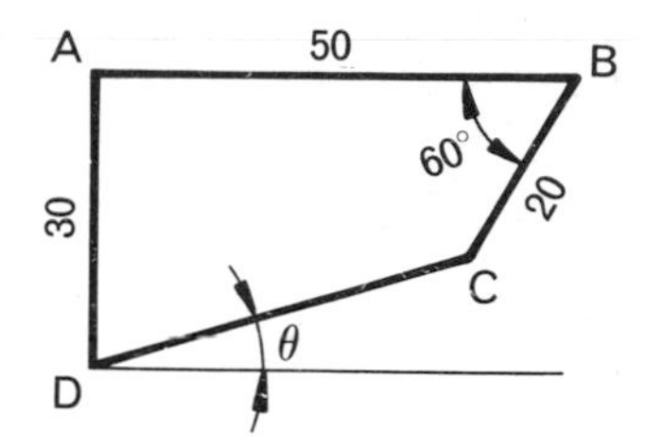

Fig. 18.46

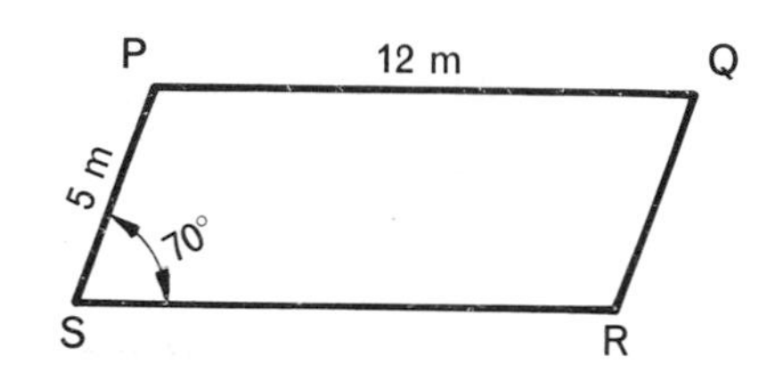

Fig. 18.47

12) The area of the parallelogram PQRS (Fig. 18.47) is:

a $30\sin 70°$ m^2 b $30\cos 70°$ m^2

c $60\sin 70°$ m^2 d $60\cos 70°$ m^2

19. STATISTICS

After reaching the end of this chapter you should be able to:

1. *Collect data from practical work in other subjects and from publications.*
2. *Distinguish between discrete and continuous data.*
3. *Distinguish between a sample and a population.*
4. *Determine the range and the appropriate density of the data and use this information to form appropriate groups (equal or unequal) to cover the set of data.*
5. *Define frequency and relative frequency.*
6. *Determine, using a tally count, the frequency and hence the relative frequency of objec in each group.*
7. *Represent the data using either the frequenci or relative frequencies by suitable ful labelled diagrams, e.g. bar charts, compone bar charts, pie charts, pictograms, etc.*
8. *Draw a labelled histogram and frequen polygon to represent a given set of data.*
9. *Calculate cumulative frequencies and dra an ogive.*
10. *Interpret descriptively data summarised tables and diagrams.*

RECORDING INFORMATION

Suppose that in a certain factory the number of persons employed on various jobs is as given in the following table:

TABLE 1

Type of personnel	*Number employed*	*Percentage*
Machinists	140	35
Fitters	120	30
Clerical staff	80	20
Labourers	40	10
Draughtsmen	20	5
Total	400	100

The information in the table can be represented pictorially in several ways:

1) *The pie chart* (Fig. 19.1) displays the proportions as angles (or sector areas), the complete circle representing the total number

employed. Thus for machinists the angle is $\frac{140}{400} \times 360 = 126°$ and for fitters $\frac{120}{400} \times 360 = 108°$ etc.

2) *The bar chart* (Fig. 19.2) relies on heights (or areas) to convey the proportions; the total height of the diagram represents 100%.

3) *The horizontal bar chart* (Fig. 19.3) gives a better comparison of the various types of personnel employed but it does not readily display the total number employed in the factory.

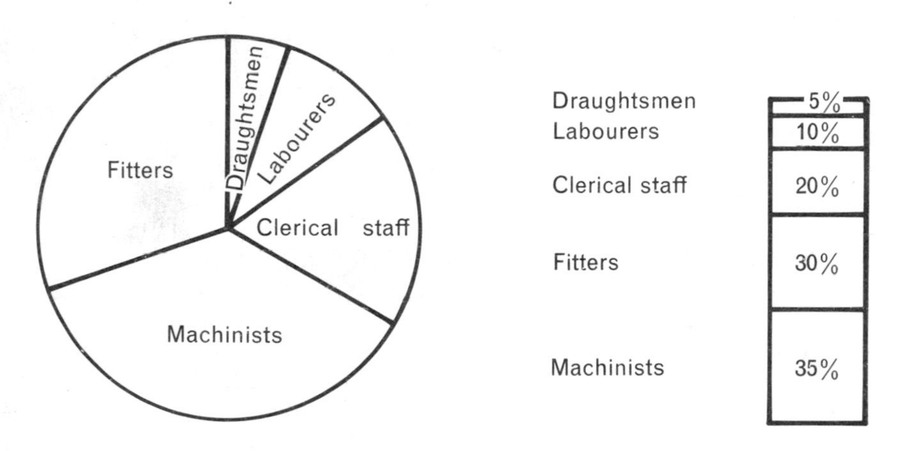

Fig. 19.1 The Pie Chart

Fig. 19.2 100% Bar Chart

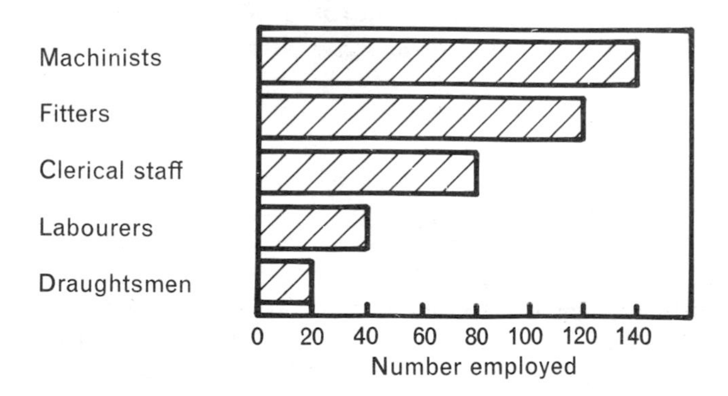

Fig. 19.3 Horizontal Bar Chart

4) *A vertical bar chart* (Fig. 19.4) is sometimes used instead of a horizontal bar chart.

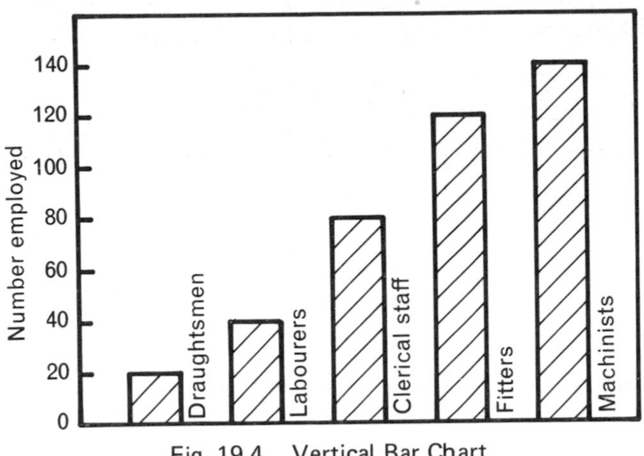

Fig. 19.4 Vertical Bar Chart

THE POPULATION

'In 1978, 14 million people watched the Cup Final on BBC television.' Statements like this are made every day but how could the BBC be so confident that their figure is correct?

Clearly they cannot send researchers to every household in the country to see how many people are watching. What they do is to select a *sample* from the total population and then use the results of this sample to estimate the number watching.

In this case the parent population is the population of the country. But in the case of the data of Table 1, the parent population is the total workforce in the factory. Again, suppose a factory produces one million ball-bearings. This is the parent population of ball-bearings.

SAMPLING

It is rarely possible to examine every item making up a parent population and recourse has to be made to sampling. For the information obtained to be of value the sample must be representative of the population as a whole. We might take a sample of 100 ball-bearings and measure their diameters. The results obtained would then be regarded as being representative of the population as a whole.

FREQUENCY DISTRIBUTIONS

Suppose we measure the diameters of a sample of 100 ball-bearings. We might get the following reading in millimetres:

TABLE 2

15.02	15.03	14.98	14.98	15.00	15.01	15.00
15.01	14.99	15.02	14.99	15.03	15.02	15.01
15.01	15.02	15.04	14.98	15.01	15.02	14.98
15.01	15.01	15.01	14.99	15.00	15.00	15.00
15.01	15.01	15.01	15.03	14.98	14.99	14.99
14.98	15.00	14.97	15.00	15.02	15.00	15.01
15.00	14.99	15.00	15.00	15.02	14.96	15.01
14.98	15.01	15.00	15.01	15.00	15.01	14.99
15.01	15.00	14.99	15.02	14.99	15.01	15.00
15.01	15.00	14.99	14.98	14.97	14.99	15.00
14.98	14.97	15.00	14.99	14.98	15.03	14.99
15.03	15.00	14.99	14.97	15.02	15.03	15.03
14.99	15.00	14.99	14.99	14.96	15.04	14.99
15.01	15.00	15.00	15.00	15.03	14.98	14.99
15.01	14.99					

These figures do not mean very much as they stand and so we re-arrange them into what is called a frequency distribution. To do this we collect all the 14.96 mm readings together, all the 14.97 mm readings and so on. A tally chart (Table 3) is the best way of doing this. Each time a measurement arises a tally mark is placed opposite the appropriate measurement. The fifth tally mark is usually made

TABLE 3

Measurement (mm)	*Number of bars with this measurement*	*Frequency*
14.96	//	2
14.97	////	4
14.98	卌 卌 /	11
14.99	卌 卌 卌 卌	20
15.00	卌 卌 卌 卌 ///	23
15.01	卌 卌 卌 卌 /	21
15.02	卌 ////	9
15.03	卌 ///	8
15.04	//	2

in an oblique direction thus tying the tally marks into bundles of five to make counting easier. When the tally marks are complete the marks are counted and the numerical value recorded in the column headed 'frequency'. The frequency is the number of times each measurement occurs. From Table 3 it will be seen that the measurement 14.96 occurs twice (that is, it has a frequency of 2). The measurement 14.97 occurs four times (a frequency of 4) and so on.

THE CLASS WIDTH

There are very many instruments which will measure accurately to 0.001 mm. If such an instrument is used to obtain the measurement given in Table 2 then in the group or class stated as 15.01 mm we would place all the ball-bearings with diameters between 15.005 and 15.015 mm. The diameter of 15.005 mm is called the *lower boundary* of the class and 15.015 mm is called the *upper class boundary*. The difference between the upper and lower class boundaries is called the *class width*. In this case the class width is $15.015 - 15.005 = 0.010$ mm.

THE HISTOGRAM

The frequency distribution of Table 2 becomes even more understandable if we draw a diagram to represent it. The best type of

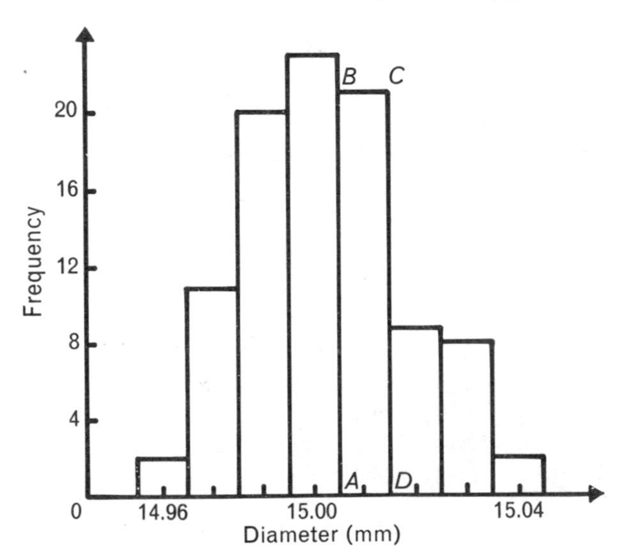

Fig. 19.5

diagram is the histogram (Fig. 19.5) which consists of a set of rectangles whose areas represent the frequencies.

If all the class widths are the same, which is usually the case, then the heights of the rectangles represent the frequencies. Note also that the left hand edge of each rectangle represents the lower boundary and the right hand edge represents the upper class boundary. Thus for the class 15.01 mm AB represents 15.005 and CD represents 15.015 mm which are the lower and upper class boundaries respectively.

On studying the histogram the pattern of the variation becomes clear, most of the measurements being grouped near the centre of the diagram with a few values more widely dispersed.

GROUPED DATA

When dealing with a large number of observations it is often useful to group them into classes or categories. We can then determine the number of items which belong to each class thus obtaining the class frequency.

EXAMPLE 19.1

The following gives the heights, in centimetres, of 100 men:

153	162	168	154	151	168	162	153	161
167	157	154	165	156	163	165	160	171
170	173	164	158	163	155	167	162	168
153	161	166	163	162	163	163	163	164
165	163	162	162	163	161	162	162	158
162	167	158	169	168	168	156	160	158
169	156	158	157	168	163	164	162	164
163	162	161	169	164	161	156	167	168
167	159	164	174	168	165	159	165	168
159	162	165	168	154	163	162	157	150
165	169	164	163	163	163	169	160	164
167								

Draw up a tally chart for the classes 150–154, 155–159, etc.

The tally chart is shown in Table 4.

TABLE 4

Class (cm)	*Tally*	*Frequency*
150–154	~~IIII~~ III	8
155–159	~~IIII~~ ~~IIII~~ ~~IIII~~ I	16
160–164	~~IIII~~ ~~IIII~~ ~~IIII~~ ~~IIII~~ ~~IIII~~ ~~IIII~~ ~~IIII~~ ~~IIII~~ III	43
165–169	~~IIII~~ ~~IIII~~ ~~IIII~~ ~~IIII~~ ~~IIII~~ IIII	29
170–174	IIII	4
	Total	100

The main advantage of grouping is that it produces a clear overall picture of the distribution. The first step in forming the frequency distribution is to determine the range of the data, i.e. the difference between the largest and smallest numbers in the data.

Thus for the data of Example 19.1, the smallest number is 150 and the largest is 174. Therefore the range is 24. We could therefore form 8 classes with a class width of 3, i.e. 150–153, 154–157, etc. or we could form 6 classes with a class width of 4 as has been done in Table 4.

The number of classes chosen depends upon the amount of original data. However, it should be borne in mind that too many groups will destroy the pattern of the data, whilst too few will destroy much of the detail contained in the original data. The number of classes is usually between 5 and 20.

Note that for the first class (150–154) the lower boundary is 149.5 cm and the upper boundary is 154.5, the class width being 5 cm. For the fourth class (165–169) the lower and upper boundaries are 164.5 and 169.5 cm respectively. The width of each class is the same and hence the frequencies of the various classes are represented by the heights of the rectangles in Fig. 19.6.

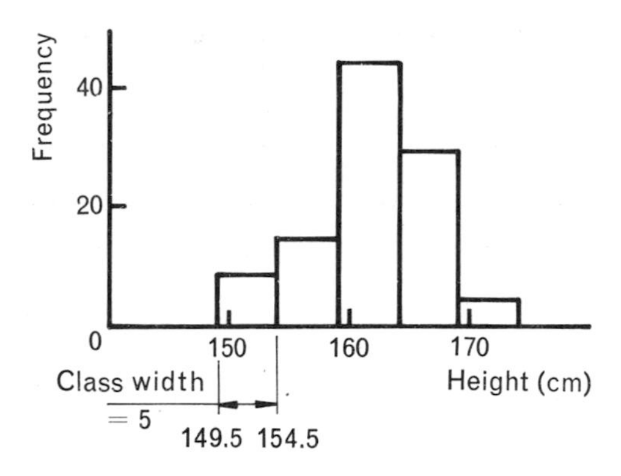

Fig. 19.6

It is usual to use class intervals of equal size but there are circumstances where this may not be appropriate.

EXAMPLE 19.2

The table below shows the age distribution for workers in a certain firm:

Age group	16–20	21–25	26–30	31–40	41–50	51–70
Number of workers	20	18	14	18	16	24

Draw a histogram of this information.

The first thing to notice is that the classes are not all of the same width. In drawing the histogram we must remember that it is the *areas of the rectangles* which give the frequencies of the various classes. Let us examine the widths of the various classes:

Age group	16–20	21–25	26–30	31–40	41–50	51–70
Class width	5	5	5	10	10	20

In drawing the histogram (Fig. 19.7) let us take 1 unit to represent a class width of 5 years. Then a class width of 10 years is represented by 2 units and a class width of 20 years by 4 units.

The heights of the rectangles then become:

Age group	16–21	21–25	26–30	31–40	41–50	51–70
Height of rectangle	$\frac{20}{1}=20$	$\frac{18}{1}=18$	$\frac{14}{1}=14$	$\frac{18}{2}=9$	$\frac{16}{2}=8$	$\frac{24}{4}=6$

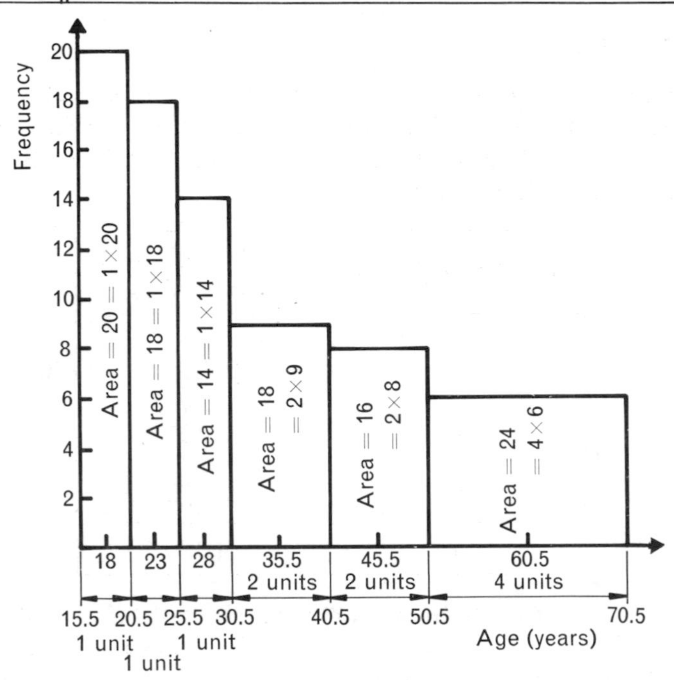

Fig. 19.7

DISCRETE AND CONTINUOUS VARIABLES

A variable which can take any value between two given values is called a *continuous variable.* Thus the height of an individual which can be 158 cm, 164.2 cm or 177.832 cm, depending upon the accuracy of measurement, is a continuous variable.

A variable which can only have certain values is called a *discrete variable.* Thus the number of children in a family can only take whole number values such as 0, 1, 2, 3, etc. It cannot be $2\frac{1}{2}$, $3\frac{1}{4}$, etc., and it is therefore a discrete variable. Note that the values of a discrete variable need not be whole numbers. The size of shoes is a discrete variable but it can be $4\frac{1}{2}$, 5, $5\frac{1}{2}$, 6, etc.

DISCRETE DISTRIBUTIONS

The histogram shown in Fig. 19.6 represents a distribution in which the variable is continuous. The data in Example 19.2 are discrete and we shall see now a discrete distribution is represented.

EXAMPLE 19.3

Five coins were tossed 100 times and after each toss the number of heads was recorded. The table below gives the number of tosses during which 0, 1, 2, 3, 4 and 5 heads were obtained. Represent these data in a suitable diagram.

Number of heads	*Number of tosses (frequency)*
0	4
1	15
2	34
3	29
4	16
5	2
Total	100

Since the data are discrete (there cannot be 2.3 or 3.6 heads), Fig. 19.8 seems the most natural diagram to use. This diagram is in the form of a vertical bar chart in which the bars have zero width. Fig. 19.9 shows the same data represented as a histogram.

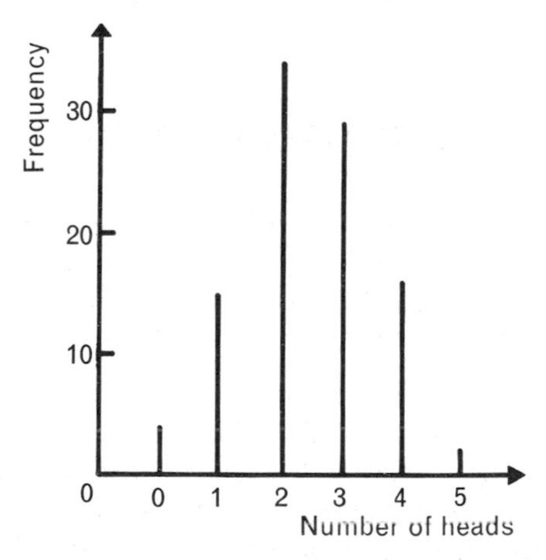

Fig. 19.8

Note that the area under the diagram gives the total frequency of 100 which is as it should be. Discrete data are often represented as a histogram as was done in Fig. 19.9, despite the fact that in doing this we are treating the data as though they were continuous.

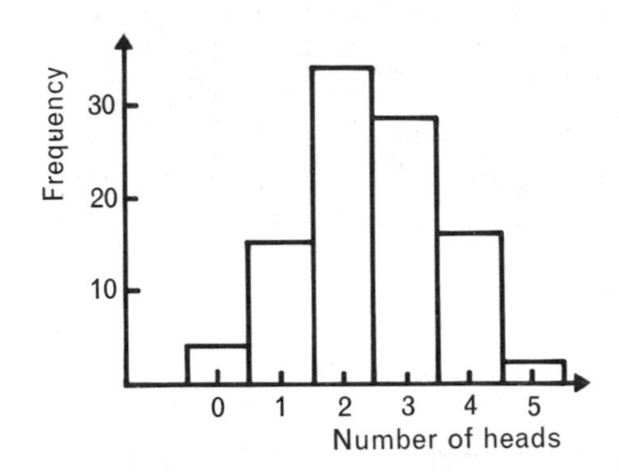

Fig. 19.9

THE FREQUENCY POLYGON

The frequency polygon is a second way of representing a frequency distribution. It is drawn by connecting the mid-points of the tops of the rectangles in a histogram by straight lines as shown in Example 19.4.

EXAMPLE 19.4

Draw a frequency polygon for the information given below which relates to the ages of people working in a factory:

Age (years)	15–19	20–24	25–29	30–34	35–39
Frequency	5	23	58	104	141

Age (years)	40–44	45–49	50–54	55–59
Frequency	98	43	19	6

The frequency polygon is drawn in Fig. 19.10. It is usual to add the extensions PQ and RS to the next lower and next higher class mid-points at the ends of the diagram. When this is done the area of the frequency polygon is equal to the area of the histogram.

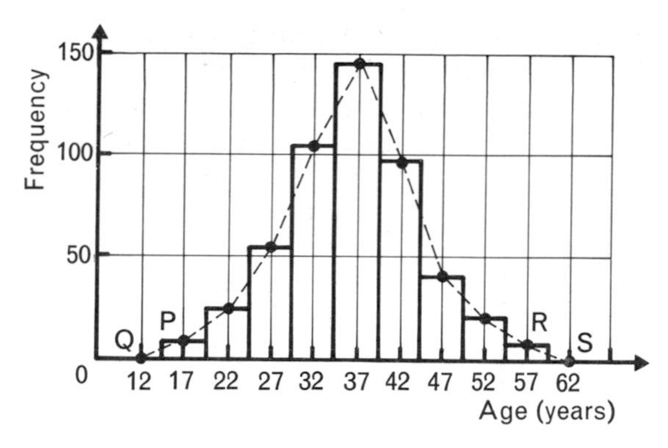

Fig. 19.10

CUMULATIVE FREQUENCY DISTRIBUTIONS

A cumulative frequency distribution is an alternative method of presenting a frequency distribution. The way in which a cumulative frequency distribution is obtained is shown in Example 19.5.

EXAMPLE 19.5

Obtain a cumulative frequency distribution for the information given in Table 4, which is repeated below.

Height (cm)	150–154	155–159	160–164	165–169	170–174
Frequency	8	16	43	29	4

The class boundaries are 149.5 to 154.5, 154.5 to 159.5, 159.5 to 164.5, 164.5 to 169.5 and 169.5 to 174.5. In obtaining the cumulative frequency distribution the lower boundary limit for each class is normally used.

Height (cm)	*Cumulative frequency*
less than 149.5	0
less than 154.5	8
less than 159.5	$8+16=24$
less than 164.5	$24+43=67$
less than 169.5	$67+29=96$
less than 174.5	$96+4=100$

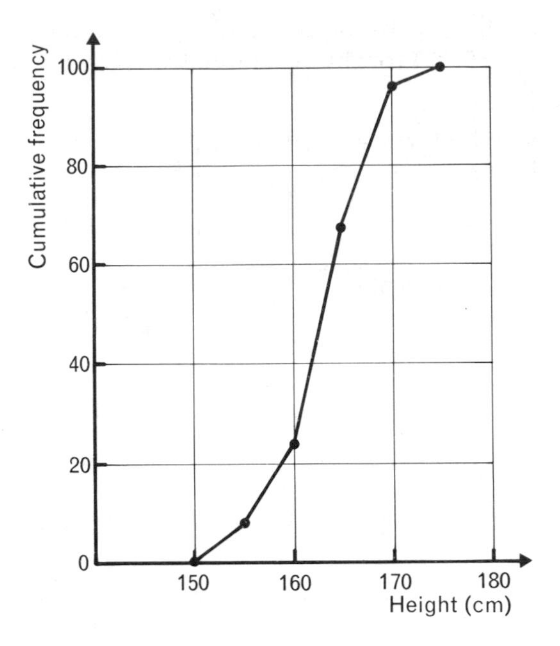

Fig. 19.11

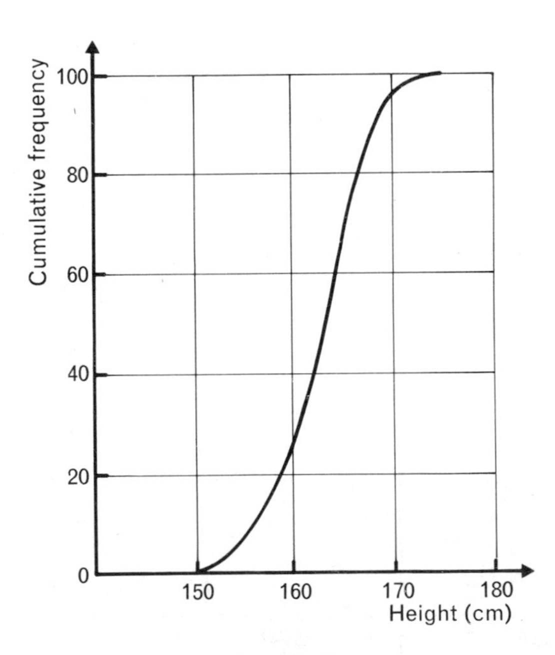

Fig. 19.12

The distribution may be represented by a cumulative frequency polygon (Fig. 19.11) or by a cumulative frequency curve (Fig. 19.12).

The cumulative frequency curve is often called an *ogive* after the architectural term for this shape.

RELATIVE FREQUENCY

In statistics many situations occur where the actual frequencies of a distribution are not as important as the relative frequencies. The relative frequency of a class is found as follows:

$$\text{Relative frequency} = \frac{\text{class frequency}}{\text{total frequency}}$$

EXAMPLE 19.6

Determine the relative frequencies for the distribution shown below which relates to the diameters of ball-bearings.

Diameter (mm)	*Frequency*
5.94–5.96	8
5.97–5.99	37
6.00–6.02	90
6.03–6.05	52
6.06–6.08	13

Diameter (mm)	*Frequency*	*Relative frequency*
5.94–5.96	8	$\frac{8}{200} = 0.040$
5.97–5.99	37	$\frac{37}{200} = 0.185$
6.00–6.02	90	$\frac{90}{200} = 0.450$
6.03–6.05	52	$\frac{52}{200} = 0.260$
6.06–6.08	13	$\frac{13}{200} = 0.065$
Total	200	1.000

RELATIVE PERCENTAGE FREQUENCY

Sometimes relative percentage frequencies are required instead of relative frequencies.

$$\text{Relative percentage frequency} = \frac{\text{class frequency}}{\text{total frequency}} \times 100$$

For Example 19.6, the percentage relative frequencies are as follows:

Diameter (mm)	*Frequency*	*Relative % frequency*
5.94–5.96	8	$\frac{8}{200} \times 100 = 4.0$
5.97–5.99	37	$\frac{37}{200} \times 100 = 18.5$
6.00–6.02	90	$\frac{90}{200} \times 100 = 45.0$
6.03–6.05	52	$\frac{52}{200} \times 100 = 26.0$
6.06–6.08	13	$\frac{13}{200} \times 100 = 6.5$
Total	200	100.0

Relative frequencies or relative percentage frequencies may be used to construct histograms, etc. The histogram shown in Fig. 19.13 represents the relative percentage frequencies calculated above.

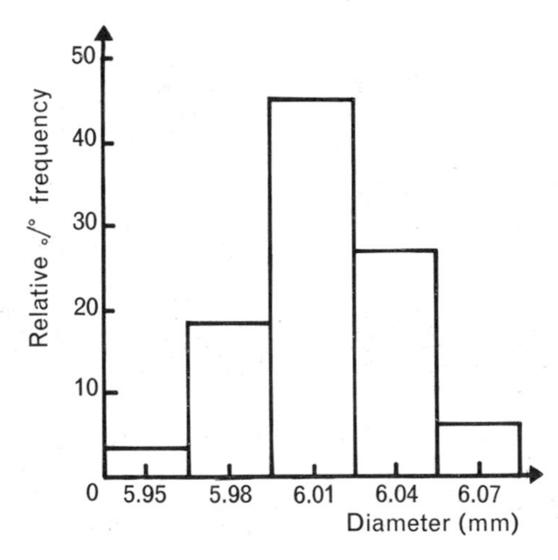

Fig. 19.13

REPRESENTING FREQUENCIES BY PICTOGRAMS

Pictures are often used to represent values and when these are used the resulting diagram is called a *pictogram.*

EXAMPLE 19.7

The table below shows the output of bicycles for the years 1970 to 1974.

Year	1970	1971	1972	1973	1974
Output	2000	4000	7000	8500	9000

Source: *Industrial News*

Represent these data in the form of a pictogram.

The pictogram is shown in Fig. 19.14 and it will be seen that each bicycle represents an output of 2000 bicycles. Part of a symbol as shown in 1972, 1973 and 1974 is used to represent a fraction of 2000 but clearly this is not a very precise way of representing a frequency distribution.

It is important that the diagram is labelled correctly and that the source of the information is stated.

Output of bicycles by Thomas & Co.

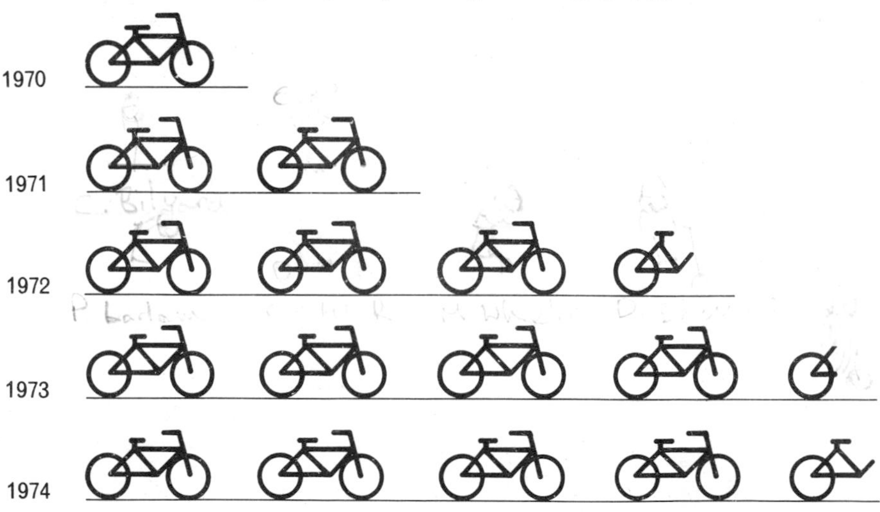

Source: *Industrial News*

Fig. 19.14

REPRESENTING FREQUENCIES BY MEANS OF BAR CHARTS

A simpler and more accurate way of representing a frequency distribution is by using a vertical bar chart as shown in Example 19.8.

EXAMPLE 19.8

The table show the output of motor-cycles by Richards & Co. for the years 1970 to 1976 inclusive. Represent this information in the form of a vertical bar chart.

Year	1970	1971	1972	1973	1974	1975	1976
Output	8000	7300	6200	5600	5700	5400	5900

The bar chart is shown in Fig. 19.15.

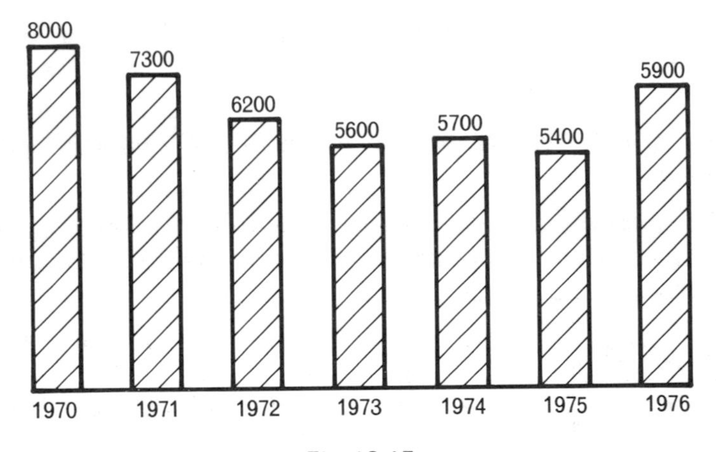

Fig. 19.15

Source: *Business Statistics*

Exercise 19.1

1) A building contractor surveying his labour force finds that 35% are engaged on factories, 40% are engaged on house building and 25% are engaged on public works (schools, hospitals, etc.).

(a) Draw a pie chart of this information.

(b) Present the information in the form of a single bar chart.

2) A firm finds that each pound received from sales is spent in the following way:

Raw materials	£0.38
Wages and salaries	£0.29
Machinery etc.	£0.08
Advertising etc.	£0.15
Profit	£0.10

Construct a pie chart of this information.

3) A department store has monthly takings of £40 000. It is divided between the various departments as follows:

Men's clothing	£10 000
Women's clothing	£15 000
Hardware	£5 000
Electrical	£8 000
Stationery	£2 000

Draw a pie chart of this information.

4) Draw a horizontal bar chart for the information given in question 3.

5) An industrial organisation gives an aptitude test to all applicants for employment. The results of 150 people taking the test were:

Score (out of 10)	1	2	3	4	5	6	7	8	9	10
Frequency	6	12	15	21	35	24	20	10	6	1

Draw a histogram of this information.

6) The diameters of 40 steel bars were measured (in millimetres) with the following results:

24.98	24.96	24.97	24.98	24.99	25.02	24.99	25.01
25.03	25.01	25.00	25.02	25.00	25.02	25.01	25.02
25.01	24.97	24.98	25.01	25.03	25.05	24.95	24.98
24.99	24.99	25.02	24.97	25.04	25.00	24.97	25.04
25.00	25.00	24.99	25.01	25.03	25.03	25.02	25.01

Draw up a frequency table and then draw a histogram to represent the frequency distribution.

7) The data below relates to the marks of a class of 30 students in an end of term test. The marks are out of 10.

4 3 8 8 9 7 7 6 5 6 7 8 4 6 4 8 7 6 7 8 5 5 7 9 6 9 5 7 6 9

Draw up a frequency distribution table and from that draw a histogram.

8) The lengths of 100 pieces of wood were measured with the following results:

Length (cm)	29.5	29.6	29.7	29.8	29.9	30.0	30.1	30.2	30.3
Frequency	2	4	11	18	31	22	8	3	1

Draw a histogram of this information.

9) 200 candidates sat an examination. The following table shows the frequency distribution obtained.

Mark	0–10	11–20	21–30	31–40	41–50
Frequency	4	12	20	25	38

Mark	51–60	61–70	71–80	81–90	91–100
Frequency	43	30	16	8	0

Draw a histogram of this information.

10) The table below gives the forging temperatures for various metals.

Metal	*Forging temperature °C*
High speed steel	850–1090
High carbon steel	770–1120
Medium carbon steel	770–1290
Wrought iron	860–1340
High tensile brass	600– 800
Brass (60/40)	600– 800
Copper	450–1000

Represent this information in a horizontal bar chart.

11) The figures below relate to the value of exports from the UK to various trade areas in a certain year.

Area	*Value of exports (£ millions)*
EEC	78
EFTA	52
Commonwealth	110
USA	35
Latin America	12
Soviet bloc	5
Others	82

Draw a vertical bar chart to represent this information.

12) The sales of motor cars by Mortimer & Co. were as follows:

Year	1	2	3	4	5
Sales	2000	2500	3200	2700	3000

Represent this information in a pictogram.

13) The table below shows the distribution of marks gained by students in an examination.

Mark	0–10	11–20	21–30	31–40	41–50	51–60
Frequency	1	4	12	47	64	42

Calculate the relative frequency percentage for each class and draw a histogram using these percentages.

14) The table below gives the number of houses completed in the South West area of England.

Year	1	2	3	4	5
Number of houses completed (thousands)	81	69	73	84	80

Represent this information by means of a pictogram.

15) The information below gives the production of tyres by the Treadwear Tyre Company for the first six months of a certain year.

Month	*Jan.*	*Feb.*	*March*	*April*	*May*	*June*
Production (thousands)	40	43	39	38	37	45

Represent this information in the form of a pictogram and in the form of a vertical bar chart. Which diagram gives the best information?

16)

45	93	35	56	16	50	63	30	86	65	57	39	44
75	25	45	74	93	84	25	77	28	54	50	12	85
55	34	50	57	55	48	78	15	27	79	68	26	66
80	91	62	67	52	50	75	96	36	83	20	45	71
63	51	40	46	61	62	67	57	53	45	51	40	46
31	54	67	66	52	49	54	55	52	56	59	38	52
43	55	51	47	54	56	56	42	53	40	51	58	52
27	56	42	86	50	31	61	33	36				

Draw up a tally chart for the classes 0–9, 10–19, 20–29, . . . , 90–99 and hence form a frequency distribution and draw a histogram.

17) The lengths of 100 pieces of wood were measured with the following results:

Length (mm)	*Frequency*
295	2
296	4
297	11
298	18
299	31
300	22
301	8
302	3
303	1

Draw a frequency polygon.

18) Fig. 19.16 shows a frequency polygon. Draw the corresponding histogram.

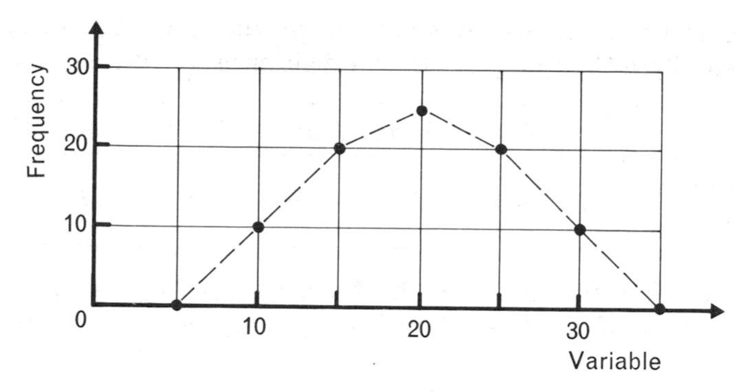

Fig. 19.16

19) The diameters of 200 ball-bearings were measured with the results shown below:

Diameter (mm)	*Frequency*
5.94–5.96	8
5.97–5.99	37
6.00–6.02	90
6.03–6.05	52
6.06–6.08	13

Draw a frequency polygon.

20) A firm decided that all its 900 workers should sit an intelligence test. The following results were obtained.

Range of mark	*Frequency*
0–9	5
10–19	33
20–29	82
30–39	148
40–49	270
50–59	215
60–69	87
70–79	44
80–89	12
90–99	4

Construct a cumulative frequency distribution and hence draw an ogive.

21) The figures below are measurements of the noise level (in dBA units) of 36 areas in a factory during an industrial operation.

93	90	98	88	103	92	89	82	86
87	91	89	85	95	86	86	94	85
103	92	88	102	99	85	98	100	95
98	105	86	100	92	96	91	87	88

(a) Display these data in the form of a grouped frequency table using intervals 81–85, 86–90, etc.

(b) Draw a cumulative frequency curve for your grouped distribution.

22) The table below shows the heights of workers in a factory.

Height (m)	*Frequency*	*Height (m)*	*Frequency*
1.60	1	1.64	10
1.61	5	1.65	6
1.62	10	1.66	2
1.63	16		

Draw a cumulative frequency curve to represent this information.

23) Draw a histogram to represent the following distribution:

Height (m)	0–9	10–19	20–29	30–49	50–69
Frequency	40	32	48	56	60

24) Fig. 19.17 shows a histogram for the lifetime of electric light bulbs. Draw up a frequency distribution.

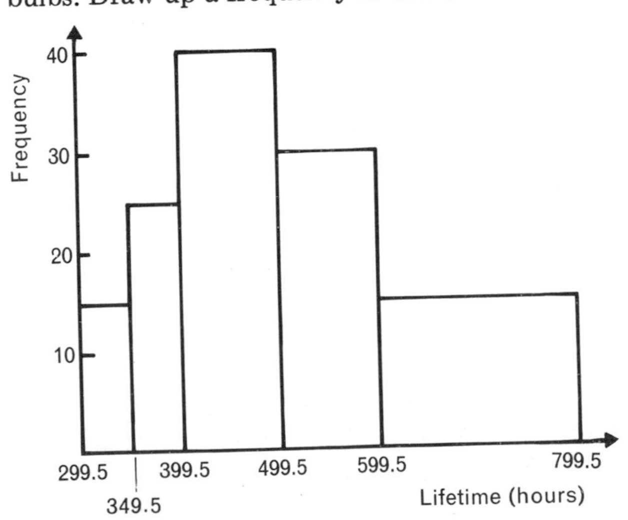

Fig. 19.17

25) Draw a histogram for the following information:

Length (mm)	56–58	59–61	62–67	68–73	74–85	86–97
Frequency	6	8	16	20	12	8

ANSWERS

ANSWERS TO CHAPTER 1

Exercise 1.1

1) 457 2) 9536 3) 7777
4) 3008 5) 705 6) 30 028
7) 5090 8) 4904 9) 125 906
10) 3 800 007 11) 95 827 000
12) 300 000 009
13) two hundred and twenty five
14) eight thousand, three hundred and twenty one
15) three thousand and seventeen
16) three thousand, nine hundred and sixty
17) one thousand, eight hundred and seven
18) twenty thousand and four
19) seventeen thousand
20) one hundred and ninety eight thousand, three hundred and seventy six
21) two hundred thousand and five
22) seven million, three hundred and sixty five thousand, two hundred and thirty five
23) twenty seven million, three hundred and nine

Exercise 1.2

1) 351 2) 4570
3) 58 190 4) 8 579 649
5) 126 331 6) 3104 ohm
7) 238 mm 8) 3308 mm

Exercise 1.3

1) 32 2) 335
3) 14 4) 1558
5) 9226 6) 154 mm
7) 129 kg 8) 759 ℓ

Exercise 1.4

1) 11 2) 32 3) 36
4) 18 5) 22 mm 6) 30 mm

Exercise 1.5

1) 928 2) 9334
3) 1 010 829 4) 4 483 887
5) 1 022 656 6) 15 000 g
7) 272 mm 8) 4752 kg

Exercise 1.6

1) 246 2) 56
3) 433 remainder 3
4) 1842 remainder 1
5) 624 remainder 5

Exercise 1.7

1) 546 remainder 4 2) 1264
3) 309 remainder 1
4) 909 remainder 2
5) 903 remainder 1 6) 1701
7) 59 817 8) 5923 9) 7
10) 83 kg 11) 18 mm
12) 17 ohms

Exercise 1.8

1) (a) 2×12; 4×6; 8×3; 24×1
   (b) 2×28; 4×14; 7×8; 56×1
   (c) 2×21, 3×14, 6×7, 42×1
2) 2, 3, 4, 6, 12; 12, 18, 24
3) 12, 15, 18, 21, 24, 27, 30, 33, 36, 39
4) (a) $2 \times 2 \times 2 \times 3$
   (b) $2 \times 2 \times 3 \times 3$
   (c) $2 \times 2 \times 2 \times 7$
   (d) $2 \times 2 \times 3 \times 11$
5) 23 and 29

Exercise 1.9

1) 24 2) 60 3) 12
4) 24 5) 40 6) 100
7) 160 8) 120 9) 420
10) 5040

Exercise 1.10

1) 4 2) 12 3) 5
4) 13 5) 6 6) 14

Exercise 1.11

1) 13 2) 10 3) 57
4) 7 5) 35 6) 74
7) 15 8) 13 9) 45
10) 20 11) 754 mm
12) 203 13) 90 14) 20
15) 2014 mm

Self-Test 1

1) b 2) a 3) e 4) a
5) (a) 22 676 (b) 22 527
(c) 15 891
6) (a) b (b) c (c) a
7) (a) 1105 (b) 1316
(c) 6116 (d) 261
(e) 114 (f) 903
8) (a) 114 786 (b) 7 625 868
(c) 37 883 967 (d) 23 114 250
(e) 56 770 371 (f) 57 566 124
9) (a) 11 587 (b) 539
(c) 48 (d) 18
10) (a) 17 (b) 12 (c) 6
(d) 72
11) (a) 4, 3 (b) 5 (c) 3, 9
(d) 4 (e) 11 (f) 4, 11
12) (a) b (b) a (c) b
(d) c (e) a
13) true 14) false 15) false
16) true 17) false 18) true
19) false 20) true

ANSWERS TO CHAPTER 2

Exercise 2.1

1) $\frac{21}{28}$ 2) $\frac{12}{20}$ 3) $\frac{25}{30}$
4) $\frac{7}{63}$ 5) $\frac{8}{12}$ 6) $\frac{4}{24}$
7) $\frac{24}{64}$ 8) $\frac{25}{35}$

Exercise 2.2

1) $\frac{1}{2}$ 2) $\frac{3}{5}$ 3) $\frac{1}{8}$
4) $\frac{3}{5}$ 5) $\frac{7}{8}$ 6) $\frac{3}{4}$
7) $\frac{5}{7}$ 8) $\frac{18}{35}$ 9) $\frac{2}{3}$
10) $\frac{2}{3}$

Exercise 2.3

1) $\frac{19}{8}$ 2) $\frac{51}{10}$ 3) $\frac{26}{3}$
4) $\frac{127}{20}$ 5) $\frac{31}{7}$

Exercise 2.4

1) $\frac{1}{2}$, $\frac{7}{12}$, $\frac{2}{3}$ and $\frac{5}{6}$ 2) $\frac{3}{4}$, $\frac{6}{7}$, $\frac{7}{8}$ and $\frac{9}{10}$
3) $\frac{11}{20}$, $\frac{3}{5}$, $\frac{7}{10}$ and $\frac{13}{16}$ 4) $\frac{3}{5}$, $\frac{5}{8}$, $\frac{13}{20}$ and $\frac{3}{4}$
5) $\frac{9}{14}$, $\frac{11}{16}$, $\frac{7}{10}$ and $\frac{3}{4}$ 6) $\frac{3}{8}$, $\frac{2}{5}$, $\frac{5}{9}$ and $\frac{4}{7}$

Exercise 2.5

1) $\frac{5}{6}$ 2) $\frac{13}{10}$ 3) $\frac{9}{8}$
4) $\frac{11}{20}$ 5) $\frac{17}{8}$ 6) $\frac{167}{120}$
7) $\frac{79}{16}$ 8) $\frac{214}{15}$ 9) $\frac{23}{7}$
10) $\frac{47}{6}$ 11) $\frac{181}{16}$ 12) $\frac{311}{30}$

Exercise 2.6

1) $\frac{1}{6}$ 2) $\frac{2}{15}$ 3) $\frac{1}{6}$
4) $\frac{1}{2}$ 5) $\frac{1}{24}$ 6) $-\frac{1}{8}$
7) $\frac{16}{7}$ 8) $-\frac{21}{8}$ 9) $-\frac{99}{40}$

Exercise 2.7

1) $\frac{19}{8}$ 2) $-\frac{43}{20}$ 3) $-\frac{57}{8}$
4) $\frac{21}{10}$ 5) $\frac{23}{60}$ 6) $\frac{44}{7}$
7) $-\frac{14}{3}$ 8) $\frac{3}{14}$

Exercise 2.8

1) $\frac{8}{15}$ 2) $\frac{15}{28}$ 3) $\frac{14}{27}$
4) $\frac{55}{36}$ 5) $\frac{36}{35}$ 6) $\frac{27}{110}$

Exercise 2.9

1) $\frac{4}{3}$ 2) 4 3) $\frac{7}{16}$
4) $\frac{3}{2}$ 5) $\frac{1}{24}$ 6) 4
7) $\frac{27}{4}$ 8) $\frac{33}{4}$ 9) 12
10) 100 11) 3 12) 2

Exercise 2.10

1) $\frac{3}{5}$ 2) 8 3) $\frac{4}{3}$
4) $\frac{3}{2}$ 5) $\frac{2}{3}$ 6) $\frac{25}{26}$
7) $\frac{6}{5}$ 8) $\frac{23}{6}$

Exercise 2.11

1) $\frac{55}{14}$ 2) 5 3) $\frac{5}{2}$
4) $\frac{5}{6}$ 5) $\frac{2}{3}$ 6) $\frac{5}{2}$
7) $\frac{7}{5}$ 8) $\frac{2}{3}$ 9) $\frac{1}{6}$
10) $\frac{3}{25}$

Self-Test 2

1) b 2) a and e 3) c
4) a 5) d 6) c
7) c 8) e 9) b
10) d 11) b 12) c
13) c 14) b, c 15) b
16) true 17) false 18) true
19) true 20) true 21) true
22) true 23) true 24) true
25) false

ANSWERS TO CHAPTER 3

Exercise 3.1

1) 0.7 2) 0.37 3) 0.589
4) 0.009 5) 0.03 6) 0.017
7) 8.06 8) 24.020 9 9) 50.008

Exercise 3.2

1) 3 2) 11.5
3) 24.04 4) 58.616
5) 54.852 6) 4.12
7) 15.616 8) 0.339
9) 0.812 10) 5.4109
11) 13.49 mm
12) A = 55.59 mm B = 25.13 mm
13) 36.51 mm 14) 1.52 mm
15) A = 8.25 mm, B = 121.98 mm, C = 176.48 mm
16) A = 22.00 mm

Exercise 3.3

1) 41 410 4100
2) 24.2 242, 2420
3) 0.46, 4.6, 46 4) 3.5, 35, 350
5) 1.486, 14.86, 148.6
6) 0.017 53, 0.175 3, 1.753
7) 48.53 8) 9
9) 1700.6 10) 5639.5

Exercise 3.4

1) 0.36, 0.036, 0.003 6
2) 6.419 8, 0.641 98, 0.064 198
3) 0.007, 0.000 7, 0.000 07
4) 51.04, 5.104, 0.510 4
5) 0.035 2, 0.003 52, 0.000 352
6) 0.054 7) 0.002 05
8) 0.004 9) 0.000 008 6
10) 0.062 742 8

Exercise 3.5

1) 743.026 6 2) 0.951 534
3) 0.288 8 4) 7.511 25
5) 0.001 376 6) 280.46 min
7) 12.95 mm 8) 1191 mm

Exercise 3.6

1) 1.33 2) 0.016
3) 189.74 4) 4.106 6
5) 43.2 6) 82.6 ohm
7) 700 8) 19.05 kg
9) 34 000 10) 2916

Exercise 3.7

1) 24.865 8, 24.87, 25
2) 0.008 357, 0.008 36, 0.008 4
3) 4.978 5, 4.98, 5
4) 22 5) 35.60
6) 28 388 000, 28 000 000
7) 4.149 8, 4.150, 4.15
8) 9.20

Exercise 3.8

1) 650.25 mm², 600.25 mm²
2) 79 687 mm², 79 102 mm²
3) 4.4 4) 0.070 1 5) 42
6) 10

Exercise 3.9

1) $200 \times 0.005 = 1$
2) $32 \times 0.25 = 8$
3) $0.7 \times 0.1 \times 2 = 0.14$
4) $80 \div 20 = 4$
5) $0.06 \div 0.003 = 20$
6) $30 \times 30 \times 0.03 = 27$

7) $\frac{0.7 \times 0.006}{0.03} = 0.14$

8) $\frac{30 \times 30}{10 \times 3} = 30$

Exercise 3.10

1) 0.25 2) 0.75
3) 0.375 4) 0.687 5
5) 0.5 6) 0.666 7
7) 0.656 3 8) 0.453 1
9) 1.833 3 10) 2.437 5
11) 0.333 12) 0.778
13) 0.133 14) 0.189
15) 0.356 16) 0.232
17) 0.525 18) 0.384
19) 0.328 20) 0.567

Exercise 3.11

1) $\frac{1}{5}$ 2) $\frac{9}{20}$ 3) $\frac{5}{16}$
4) $2\frac{11}{20}$ 5) $\frac{3}{400}$ 6) $2\frac{1}{8}$
7) 0.000 1 8) 0.001 875

Self-Test 3

1) b 2) b 3) d
4) a 5) c 6) d
7) c 8) d 9) c
10) a 11) false 12) false
13) true 14) false 15) false
16) false 17) true 18) true
19) true 20) true

ANSWERS TO CHAPTER 4

Exercise 4.1

1) (a) $\frac{3}{400}$ (b) $\frac{3}{50}$ (c) 20 (d) $\frac{2}{15}$
2) 1.5 m 3) 148 mm
4) (a) 0.385 ohms (b) 6.18 m
5) 72 mm 6) 50 rev/min
7) 192 rev/min
8) 80 mm and 60 mm
9) 12.9 kg, 2.1 kg
10) 1 m, 3.5 m, 6 m
11) 48 and 32 kg
12) 93.33 kg
13) 42 mm and 56 mm

Exercise 4.2

1) 30% 2) 55% 3) 36%
4) 80% 5) 62% 6) 63%
7) 81.3% 8) 66.7% 9) 72.3%
10) 2.7% 11) 0.32 12) 0.78
13) 0.06 14) 0.24 15) 0.315
16) 0.482 17) 0.025 18) 0.0125
19) 0.0395 20) 0.201

Exercise 4.3

1) (a) 9.6 (b) 21.3 (c) 2.52
2) (a) 16 (b) 16.3 (c) 45.5
3) 1150 mm 4) 73.4, 23.3, 3.3
5) 75 000 kg 6) 70, 20, 10
7) 200, 70, 10 and 1720 kg
8) 44% copper, 8% tin
9) 7% 10) 59%

Self-Test 4

1) true 2) false 3) true
4) true 5) true 6) false
7) true 8) false 9) false
10) true 11) true 12) false
13) true 14) false 15) true
16) false 17) false 18) true
19) true 20) false 21) b
22) d 23) b 24) c
25) a 26) c 27) b
28) c 29) c 30) c

ANSWERS TO CHAPTER 5

Exercise 5.1

1) 15 2) – 12 3) – 32
4) 14 5) – 24 6) 26
7) – 18 8) 23

Exercise 5.2

1) – 5 2) – 9 3) 5
4) 5 5) – 1 6) 0
7) – 4 8) 7

Exercise 5.3

1) 2 2) 3 3) 14
4) 4 5) 1 6) – 5
7) – 5 8) 16

Exercise 5.4

1) – 42 2) – 42 3) 42
4) 42 5) – 48 6) 4
7) 120 8) 9

Exercise 5.5

1) – 3 2) – 3 3) 3
4) 3 5) – 2 6) – 1
7) 2 8) – 1 9) – 2
10) 12 11) – 1 12) – 2
13) – 8 14) – 3

Self-Test 5

1) false 2) true 3) false
4) true 5) true 6) false
7) false 8) true 9) true
10) true

ANSWERS TO CHAPTER 6

Exercise 6.1

1) 2^{11} 2) a^8 3) n^3
4) 3^{11} 5) b^{-3} 6) 10^4
7) z^3 8) 3^{-4} 9) m^4
10) x^{-3} 11) 9^{12} 12) y^{-6}
13) t^8 14) c^{14} 15) a^{-9}
16) 7^{-12} 17) b^{10} 18) s^{-9}
19) 8 20) 1 21) 0.5
22) 8 23) 0.25 24) 100
25) 0.25 26) 0.143 27) 0.04
28) 3 29) 7 30) 25
31) 7 32) 3.375 33) 0.003 91

Exercise 6.2

1) $x^{1/2}$ 2) $x^{4/5}$ 3) $x^{-1/2}$
4) $x^{-1/3}$ 5) $x^{-4/3}$ 6) $x^{-3/2}$
7) $x^{2/3}$ 8) $x^{0.075}$ 9) $x^{2/3}$
10) $x^{1/3}$ 11) $x^{1/3}$ 12) x
13) 5 14) 2 15) 2
16) 2 17) 4 18) 125
19) 8 20) 27 21) 2
22) 4 23) 0.167 24) 0.125
25) 0.031 25 26) 64
27) 3 28) $a^{-13/6}$ 29) $a^{-11/3}$
30) $x^{4.5}$ 31) $b^{1/2}$ 32) $m^{7/4}$
33) $z^{2.3}$ 34) 1 35) $u^{-5/2}$
36) $y^{1/4}$ 37) $n^{1/4}$ 38) $x^{11/14}$
39) $t^{-2/3}$

Exercise 6.3

1) $\log_a n = x$ 2) $\log_2 8 = 3$
3) $\log_5 0.04 = -2$
4) $\log_{10} 0.001 = -3$
5) $\log_x 1 = 0$ 6) $\log_{10} 10 = 1$
7) $\log_a a = 1$ 8) $\log_e 7.39 = 2$
9) $\log_{10} 1 = 0$ 10) 3
11) 3 12) 4 13) 3
14) 9 15) 64 16) 100
17) 1 18) 2 19) 3
20) 0.5 21) 1

Exercise 6.4

1) (a) 1.96×10
(b) 3.85×10^2
(c) $5.987\,6 \times 10^4$
(d) 1.5×10^6
(e) 1.3×10^{-2}
(f) 3.85×10^{-3}
(g) 6.98×10^{-4}
(h) 2.385×10^{-2}

2) (a) 150 (b) 47 000
(c) 3 600 000
(d) 9450 (e) 0.25
(f) 0.004 (g) 0.000 08
(h) 0.04

3) (a) 2.1×10^3
(b) 9.95×10^3
(c) 8.58×10^4

4) (a) 2.1×10^{-2}
(b) 8.72×10^{-3}
(c) 2.11×10^{-4}

Exercise 6.5

1) $1.248\,9 \times 10^4$ 2) 1.858×10^{-1}
3) 1.04×10 4) 4×10^{-5}
5) 2.758×10^3 6) $8.322\,8 \times 10^4$
7) $8.136\,8 \times 10^{-2}$ 8) 8.322×10^{-2}
9) 4.55×10^6 10) 1.5×10^{-2}
11) 4.22×10^{10} 12) 3.4×10^{-2}
13) 5.93 14) 5.8×10

Exercise 6.6

1) 8 km 2) 15 Mg
3) 3.8 Mm 4) 1.891 Gg
5) 7 mm 6) 1.3 μm
7) 28 g 8) 360 mm
9) 64 mg 10) 3.6 mA

ANSWERS TO CHAPTER 7

Exercise 7.1

1) 13.14 2) −11.35
3) 27.4 4) 0.001 49
5) 1.94 6) −4.26
7) 1.277 8) 18.8
9) −2.52 10) 527
11) −22.8 12) −22.8
13) 0.007 58 14) −0.348
15) 0.657 16) 0.549
17) −4.07 18) 3.96
19) 6.61 20) 0.001 29

Exercise 7.2

1) 29.0 2) 12.3
3) 0.0391 4) 0.0160
5) 0.010 3 6) 94.2
7) 42.1 8) 86.3
9) 0.670 10) 2.90
11) 2.54 12) 0.506
13) 17.1 14) 2.84
15) 1.62 16) 0.518
17) 468 18) 3.22
19) 62.6 20) 1.92×10^8
21) 1.18 22) 9.19
23) 0.0171

ANSWERS TO CHAPTER 8

Exercise 8.1

1) (a) 1.706 (b) 0.082 43
(c) 152.3 (d) 4 456 000
2) (a) 0.3544 (b) 0.056 43
(c) 3.400 (d) 47.85
3) (a) 24.4° C (b) 72.0° C
(c) 155.7° F (d) 11.7° F
4) (a) 17.0 lb (b) 730 lb
(c) 8.3 kg (d) 0.19 kg
5) (a) 580 MN/m^2
(b) 2.2 mm
(c) 410 kg
6) (a) 2.5 ohms
(b) 7.9 ohms
(c) 3.1 ohms
7) (a) 110 (b) 60
8) (a) C, B, E; 11 days
(b) G, F, J; 13 days

ANSWERS TO CHAPTER 9

Exercise 9.1

1) $18x$ 2) $2x$ 3) $-3x$
4) $-6x$ 5) $-5x$ 6) $5x$
7) $9a$ 8) $12m$ 9) $5b^2$
10) ab 11) $14xy$ 12) $-3x$
13) $-6x^2$ 14) $7x-3y+6z$
15) $9a^2b-3ab^3+4a^2b^2+11b^4$
16) $1.2x^3+0.3x^2+6.2x-2.8$
17) $9pq-0.1qr$
18) $-0.4a^2b^2-1.2a^3-5.5b^3$
19) $10xy$ 20) $12ab$ 21) $12m$
22) $4pq$ 23) $-xy$ 24) $6ab$
25) $-24mn$ 26) $-12ab$
27) $24pqr$ 28) $60abcd$ 29) $2x$
30) $\frac{-4a}{7b}$ 31) $\frac{-5a}{8b}$ 32) $\frac{a}{b}$
33) $\frac{2a}{b}$ 34) $2b$ 35) $3xy$
36) $-2ab$ 37) $2ab$ 38) $\frac{7ab}{3}$
39) a^2 40) $-b^2$ 41) $-m^2$
42) p^2 43) $6a^2$ 44) $5X^2$
45) $-15q^2$ 46) $-9m^2$ 47) $9pq^2$
48) $-24m^3n^4$ 49) $-21a^3b$
50) $10q^4r^6$ 51) $30mnp$ 52) $-75a^3b^2$
53) $-5m^5n^4$

Exercise 9.2

1) $3x+12$ 2) $2a+2b$
3) $9x+6y$ 4) $\frac{x}{2}-\frac{1}{2}$
5) $10p-15q$ 6) $7a-21m$
7) $-a-b$ 8) $-a+2b$
9) $-3p+3q$ 10) $-7m+6$
11) $-4x-12$ 12) $-4x+10$
13) $-20+15x$ 14) $2k^2-10k$
15) $-9xy-12y$ 16) $ap-aq-ar$
17) $4abxy-4acxy+4dxy$
18) $3x^4-6x^3y+3x^2y^2$
19) $-14P^3+7P^2-7P$
20) $2m-6m^2+4mn$
21) $5x+11$ 22) $14-2a$
23) $x+7$ 24) $16-17x$
25) $7x-11y$ 26) $\frac{7y}{6}-\frac{3}{2}$
27) $-8a-11b+11c$
28) $7x-2x^2$ 29) $3a-9b$
30) $-x^3+18x^2-9x-15$

Exercise 9.3

1) x^2+3x+2 2) x^2+4x+3
3) $x^2+9x+20$ 4) $2x^2+11x+15$
5) $3x^2+25x+42$
6) $5x^2+21x+4$ 7) $6x^2+16x+8$
8) $10x^2+17x+3$ 9) $21x^2+41x+10$
10) x^2-4x+3 11) x^2-6x+8
12) $x^2-9x+18$ 13) $2x^2-9x+4$
14) $3x^2-11x+10$
15) $4x^2-33x+8$ 16) $6x^2-16x+8$
17) $6x^2-17x+5$ 18) $21x^2-29x+10$
19) x^2+2x-3 20) $x^2+5x-14$
21) $x^2-2x-15$ 22) $2x^2+x-10$
23) $3x^2+13x-30$
24) $3x^2+23x+30$
25) $6x^2+x-15$ 26) $12x^2+4x-21$
27) $6x^2-x-15$ 28) $3x^2+5xy+2y^2$
29) $2p^2-7pq+3q^2$
30) $6v^2-5uv-6u^2$
31) $6a^2+ab-b^2$ 32) $5a^2-37a+42$
33) $6x^2-xy-12y^2$
34) x^2+2x+1 35) $4x^2+12x+9$
36) $9x^2+42x+49$
37) x^2-2x+1 38) $9x^2-30x+25$
39) $4x^2-12x+9$ 40) $4a^2+12ab+9b^2$
41) $x^2+2xy+y^2$ 42) $P^2+6PQ+9Q^2$
43) $a^2-2ab+b^2$
44) $9x^2-24xy+16y^2$
45) $4x^2-y^2$
46) a^2-9b^2
47) $4m^2-9n^2$
48) x^4-y^2

Exercise 9.4

1) p^2q 2) ab^2 3) $3mn$
4) b 5) $3xyz$ 6) $2(x+3)$
7) $4(x-y)$ 8) $5(x-1)$
9) $4x(1-2y)$ 10) $m(x-y)$
11) $x(a+b+c)$ 12) $\frac{1}{2}\left(x-\frac{y}{4}\right)$
13) $5(a-2b+3c)$ 14) $ax(x+1)$
15) $\pi r(2r+h)$ 16) $3y(1-3y)$
17) $ab(b^2-a)$ 18) $xy(xy-a+by)$
19) $5x(x^2-2xy+3y^2)$
20) $3xy(3x^2-2xy+y^4)$
21) $I_0(1+\alpha t)$ 22) $\frac{1}{3}\left(x-\frac{y}{2}+\frac{z}{3}\right)$
23) $a(2a-3b)+b^2$
24) $x(x^2-x+7)$
25) $\frac{m^2}{pn}\left(1-\frac{m}{n}+\frac{m^2}{pn}\right)$

Exercise 9.5

1) $(x+y)(a+b)$ 2) $(p-q)(m+n)$
3) $(ac+d)^2$ 4) $(2p+q)(r-2s)$
5) $2(a-b)(2x+3y)$
6) $(x^2+y^2)(ab-cd)$
7) $(mn-pq)(3x-1)$
8) $(k^2l-mn)(l-1)$

Exercise 9.6

1) $6a^2$ 2) $2x^2y$ 3) m^2n^2
4) $2abc^2$ 5) $2(x+1)$ 6) $x^2(a+b)^2$
7) $(a+b)(a-b)$ 8) $x(1-x)(x+1)$
9) $(x+2)(x-2)^2$
10) $(x-2)(x+2)$
11) $6(a+b)(a-b)$
12) $6(a+b)^2(a^2+b^2)$

Exercise 9.7

1) $\frac{1}{ab}$ 2) $\frac{b}{a}$ 3) $\frac{x^2}{y^2}$
4) $\frac{xy}{2}$ 5) $\frac{1}{ab}$ 6) $c(a+b)$
7) $1-x^2$ 8) $\frac{c}{(a-b)^2}$ 9) $\frac{x+y}{xy}$
10) $\frac{a+1}{a}$ 11) $\frac{m-n}{n}$ 12) $\frac{b-c^2}{c}$
13) $\frac{ad-bc}{bd}$ 14) $\frac{ac-1}{bc}$
15) $\frac{a^2-1}{a}$ 16) $\frac{1+y+xy}{xy}$
17) $\frac{12+x^2}{4x}$ 18) $\frac{3de+2ce-5cd}{cde}$
19) $\frac{ad+cb+bd}{bd}$ 20) $\frac{2h-5f-3g}{6fgh}$
21) $\frac{8y-16}{(y+3)(y-5)}$ 22) $\frac{4-2x}{x(x+2)}$
23) $\frac{x^2+x-1}{x^2(x-1)}$ 24) $\frac{x^2+1}{(1-x)(1+x)}$
25) $\frac{2}{2-x}$ 26) $\frac{2x}{x-2}$

Exercise 9.8

1) $1+\frac{b}{a}$ 2) $\frac{1}{b}-\frac{1}{a}$ 3) $\frac{1}{c}+1$
4) $\frac{x}{2}+\frac{y}{2x}$ 5) $\frac{a}{bc}-\frac{1}{c}+\frac{1}{b}$

6) $\frac{(x-1)}{(x+1)}+1$
7) $y+\frac{y^2}{x(1-a)}$
8) $\frac{1}{(x-y)}+\frac{1}{x}$
9) $\frac{1}{(a-b)}-\frac{1}{(a+b)}$
10) $\frac{7d-3c}{4c}$
11) $\frac{x}{1+x}$
12) $\frac{a^2}{a^2-1}$
13) $\frac{1}{2(4a+3)}$
14) $\frac{m+1}{1-m}$
15) $\frac{3(3t+5)}{(1-6t)}$
16) $\frac{R_1R_2}{R_1+R_2}$
17) $\frac{a(ac-b)}{c(a+c)}$
18) $\frac{by+cx}{c(xy+b)}$
19) $\frac{b(b-a)}{ab-1}$
20) $\frac{a+b}{b}$
21) $-\frac{2}{x}$
22) $m = 5$
23) $x = -\frac{29}{5}$
24) $x = 2$
25) $x = \frac{45}{8}$
26) $x = -2$
27) $x = -15$
28) $x = \frac{50}{47}$
29) $m = -1.5$
30) $x = \frac{15}{28}$
31) $m = 1$
32) $x = 2.5$
33) $t = 6$
34) $x = 4.2$
35) $y = -70$
36) $x = \frac{5}{3}$
37) $x = 13$
38) $x = -10$
39) $m = \frac{25}{26}$
40) $y = \frac{9}{7}$
41) $x = \frac{25}{3}$
42) $x = 3.5$
43) $x = 20$
44) $x = 13$
45) -53
46) $x = 4$
47) $p = \frac{13}{4}$
48) $m = 3$
49) $x = \frac{15}{4}$
50) $x = \frac{7}{2}$

Self-Test 6

1) True 2) True 3) False
4) True 5) True 6) False
7) True 8) True 9) True
10) False 11) True 12) True
13) True 14) False 15) False
16) True 17) False 18) True
19) False 20) True 21) True
22) False 23) False 24) True
25) True 26) False 27) False
28) True 29) False 30) True
31) True 32) False 33) True
34) True 35) True 36) True
37) False 38) False 39) True
40) False 41) False 42) True
43) True 44) False 45) True
46) True 47) False 48) True
49) False 50) False 51) True
52) False 53) True 54) True
55) False

ANSWERS TO CHAPTER 10

Exercise 10.1

1) $x = 5$ 2) $t = 7$ 3) $q = 2$
4) $x = 20$ 5) $q = -3$ 6) $x = 3$
7) $y = 6$ 8) $m = 12$ 9) $x = 2$
10) $x = 3$ 11) $p = 4$ 12) $x = -2$
13) $x = -1$ 14) $x = 4$ 15) $x = 2$
16) $x = 6$ 17) $m = 2$ 18) $x = -8$
19) $d = 6$ 20) $x = 5$ 21) $x = 3$

Exercise 10.2

1) $(x-5)$ years 2) $(3a+8b)$ pence
3) $(5x+y+z)$ hours
4) $2(l+b)$ mm 5) £$(a-x)$; £$(b+x)$
6) £$(Mx+Ny+Pz)$
7) £$(49u+3v)$ 8) $\frac{100mN}{100M+mn}$
9) £$\left(u+\frac{xy}{100}\right)$ 10) $a(m-a)$

Exercise 10.3

1) £120 and £132
2) 6, 9 and 10 m
3) 52°, 82° and 46°
4) 10×18 mm
5) £500 and £3500
6) 6, 8 and 12 m 7) 22.5×55 m
8) 90p and 180p
9) 25 and 50 ℓ/min
10) £80 and £100

Self-Test 7

1) True 2) False 3) True
4) True 5) False 6) False
7) False 8) False 9) True
10) True 11) False 12) True
13) False 14) False 15) False
16) True 17) False 18) False
19) True 20) False 21) True
22) False 23) False 24) False
25) False 26) a 27) c
28) b 29) b 30) a, b
31) c, d 32) d 33) b
34) c 35) c 36) c

ANSWERS TO CHAPTER 11

Exercise 11.1

1) 1, 2 2) 4, 5 3) 4, 1
4) 7, 3 5) $\frac{1}{2}, \frac{3}{4}$ 6) 7, 10
7) 3, 2 8) 5, 2

Exercise 11.2

1) 1.5, 2 2) 0.2, 1.3, 3.7
3) £144, £128 4) £0.80, £1.60
5) 3, 7 6) 0.3, 0.2, 4.7
7) £0.50, £1.50 8) 15 and 12
9) 9 and 7 10) 40 and 75

Self-Test 8

1) c 2) a 3) b
4) d 5) c 6) a
7) a 8) c 9) a, b
10) b, c

ANSWERS TO CHAPTER 12

Exercise 12.1

1) 22.0 2) 8 3) 56.5
4) 3750 5) 540 6) 4
7) 30.0144 8) 17 600

Exercise 12.2

1) 7.5 2) 0.5 3) 4
4) 45 5) 5 6) 80
7) 1.71 8) 2.67

Exercise 12.3

1) $\frac{C}{\pi}$ 2) $\frac{s}{\pi n}$ 3) $\frac{c}{P}$
4) $\frac{A}{\pi r}$ 5) $\frac{v^2}{2g}$ 6) $\frac{I}{PT}$
7) $\frac{a}{x}$ 8) $\frac{E}{I}$ 9) ax
10) $\frac{PV}{R}$ 11) $\frac{0.866}{d}$
12) $\frac{ST}{s}$ 13) $\frac{EI}{M}$
14) $\frac{4V}{\pi d^2}$ 15) $\frac{lT}{G\theta}$
16) $\frac{v-u}{a}$ 17) $\frac{n-p}{c}$
18) $\frac{y-b}{a}$ 19) $5(y-17)$
20) $\frac{L-a}{d}+1$ 21) $\frac{b-a}{c}$
22) $\frac{B-D}{1.28}$ 23) $\frac{R(V-2)}{V}$
24) $C(R+r)$ 25) $\frac{A}{\pi r}-r$
26) $2Cp+n$ 27) $D-\frac{TL}{12}$
28) $\frac{3-5a}{4a}$

Self-Test 9

1) d 2) a 3) d
4) a 5) b 6) b, d
7) c, d 8) c, d 9) c
10) a, c

ANSWERS TO CHAPTER 13

Exercise 13.1

1) 110 marks, £36
2) 1.24 ohms 3) 13.4 newtons
4) 6 kg, 16 kg
5) 140 ohms, 70 volts
10) 1, 3 11) −5, −2
12) −3, 4 13) 4, −3
14) 4, 13 15) 2, −2
16) 3, 4 17) $y = 3x-5$
18) $y = -2x+40$ 19) $y = -3x-4$
20) $y = 0.5x+20$ 21) 2, 3

22) 0.44, −0.82 23) 0.5, 4
24) 0.03, 20, 21.8 ohms
25) 6.44, 15, 8.7

Exercise 13.2

1) (a) 2 (b) 10 (c) 8
2) (a) $\frac{1}{6}$ (b) 1.67 (c) 30
3) (a) 12 (b) 2 (c) 2
4) 1.82 ohms 5) 24.1 m^3
6) 62.1 mm 7) 435 K
8) (a) 12 amperes (b) 24 ohms

ANSWERS TO CHAPTER 14

Exercise 14.1

1) 135° 2) 54° 3) 60°
4) 63° 5) 18° 6) 135°
7) 288° 8) 288° 9) 108°
10) 90° 11) 28° 37′ 12) 69° 23′
13) 14° 22′34″ 14) 62°48′11″
15) 179°11′25″ 16) 21°3′
17) 22°48′ 18) 7°43′56″
19) 5°54′50″ 20) 36°58′11″
21) 17.433° 22) 28.127°
23) 83.003° 24) 87.975°
25) 28°23′28″ 26) 47°27′36″
27) 58°56′20″ 28) 5°0′58″

Exercise 14.2

1) 20° 2) 100° 3) 35°
4) 70°, 110°, 110°, 70°
5) 54° 7) 65° 8) 230°; 32°
9) 65° 10) 80° 11) 100°
12) 34° 13) 53°31′ 14) 110°
15) (a) $37\frac{1}{2}°$ (b) 25°

Self-Test 10

1) d 2) a 3) b
4) a 5) b, d 6) b, d
7) b, d 8) c 9) c

ANSWERS TO CHAPTER 15

Exercise 15.1

1) (a) 3.61 m (b) 6.98 m
(c) 22.5 mm
2) 38.4 mm 3) 32.5 mm
4) 20.4 mm 5) 37.9 mm
6) 64.0 mm 7) 8.60 m, 8.49 m
8) 6.63 mm

Exercise 15.3

1) 48 2) $\frac{AB}{RQ} = \frac{AC}{RP} = \frac{BC}{QP}$
3) 3.2 m, 2.4 m 4) 16 mm
5) $\frac{1}{3}$ 6) 15 mm
7) KB = 2.6 m, XY = 1.5 m
8) $\frac{1}{3}$

Exercise 15.4

1) 39° 2) 53° 3) 62 mm
4) 42° 5) 50 mm 6) 35.4 mm
7) 37° 8) 62.5 mm

Self-Test 11

1) b, c 2) c 3) d
4) a 5) b 6) b
7) d 8) d 9) b
10) a 11) c 12) c
13) a 14) a, c 15) b
16) a, c 17) b 18) b, d
19) d 20) a, d

ANSWERS TO CHAPTER 16

Exercise 16.1

1) 132 m 2) 2200 mm
3) 270 m 4) 19.9 m
5) 880 mm 6) 267 mm
7) 26.46 m 8) 4400 mm
9) 88.0 m 10) 10.2 m
11) 35.0 km 12) 152 mm

Exercise 16.2

1) 38° 2) 120°
3) 61° 4) 59.9 mm
5) 24.0 mm 6) 57.0 mm
7) 103 mm 8) 46.5 mm
9) 40.0 mm 10) 9.95 mm

Self-Test 12

1) b 2) c 3) d
4) b 5) c

ANSWERS TO CHAPTER 17

Exercise 17.1

1) 143°　2) 93°
3) $\alpha = 39°$, $\beta = 105°$　4) 85°
5) $\alpha = 65°$, $\beta = 35°$　6) 32°
7) XT = TR = 50 mm　8) 100°
9) 110°
10) ADB = 20°, $A\hat{B}D$ = 70°

Exercise 17.2

1) (a) 56 m^2　(b) 220 mm^2
(c) 630 m^2
2) 1.04 m^2　3) 1.74 m^2
4) 28.4 m^2　5) 62 m^2
6) 174 m^2
7) (a) 53.6 m^2　(b) 45.0 m^2
(c) 8.52 m^2
8) (a) 1200 mm^2　(b) 275 mm^2
(c) 260 mm^2　(d) 774 mm^2
(e) 1050 mm^2　(f) 1090 mm^2

Exercise 17.3

1) 56 m^2　2) 0.045 5 m^2
3) 4 m　4) 7.20 mm
5) 2340 mm^2

Exercise 17.4

1) 40 m^2　2) 94.2 m^2
3) 10 mm　4) 3060 mm^2
5) 198 mm^2

Exercise 17.5

1) 10 800 mm^2　2) 22.1 m^2
3) 13.4 m^2　4) 962 mm^2
5) 93.5 m^2

Exercise 17.6

1) 616 m^2　2) 385 000 mm^2
3) 25.0 m^2　4) 1390 mm^2
5) 47.6 m^2　6) 30 670 mm^2
7) 141 m^2　8) 581 mm^2

Exercise 17.7

1) 21.5 m^2　2) 5.36 m^2
3) 1370 mm^2　4) 773 mm^2
5) 21 200 mm^2

Exercise 17.8

1) 140 m^3　2) 225 000 mm^3
3) 7 700 000 mm^3
4) 31 400 mm^3　5) 73 300 mm^3
6) 0.008 75 m^3　7) 3860 mm^3
8) 300 000 mm^3　9) 23.1 m^3
10) 0.022 6 m^3

Exercise 17.9

1) 72 m^2　2) 36 m^2
3) 267 mm^2　4) 3080 mm^2
5) 12 m　6) 2000 m^3
7) (a) 29 700 mm^3
(b) 237 000 mm^3
(c) 24 mm, 52.5 mm, 79.5 mm
8) (a) 3.14 m^3　(b) 3.00
9) 30 mm　10) 102 litres

Self-Test 13

1) c　2) c　3) b
4) a　5) d　6) b
7) c, d　8) a, d　9) c
10) d　11) b, c　12) c, d

ANSWERS TO CHAPTER 18

Exercise 18.1

1) (a) 0.500 0　(b) 0.707 1
(c) 0.927 2
2) (a) 19°28′　(b) 48°36′
(c) 46°3′
3) (a) 0.207 9　(b) 0.312 3
(c) 0.964 6　(d) 0.128 5
(e) 0.999 1　(f) 0.003 2
4) (a) 9°　(b) 66°
(c) 81°6′　(d) 4°36′
(e) 78°55′　(f) 47°41′
(g) 2°52′　(h) 15°40′
5) (a) 3.38 m　(b) 10.1 m
(c) 25.9 mm
6) (a) 41°49′　(b) 40°47′
(c) 22°23′
7) 283 mm　8) 0.794 m
9) 216 mm　10) 7.47 m
11) 44°44′, 44°44′, 90°32′

Exercise 18.2

1) (a) 0.965 9 (b) 0.911 4
(c) 0.201 1 (d) 1.000 0
(e) 0.286 3 (f) 0.766 3
2) (a) 24° (b) 70°
(c) 14°42′ (d) 64°36′
(e) 16°32′ (f) 89°31′
(g) 74°52′ (h) 61°58′
3) (a) 93.3 m (b) 2.64 m
(c) 5.29 m
4) (a) 60°42′ (b) 69°20′
(c) 53°19′
5) 66°7′, 66°7′, 47°46′, 38.4 mm
6) 28.8 mm 7) 1.97 m
8) 92°1′, 87.5 mm
9) 4.53 m, 2.11 m, 2.40 m, 5.65 m

Exercise 18.3

1) (a) 0.324 9 (b) 0.634 6
(c) 1.361 3 (d) 0.8229
(e) 0.200 4 (f) 2.658 3
2) (a) 24° (b) 73°
(c) 4°24′ (d) 21°41′
(e) 19°38′ (f) 39°34′
(g) 62°33′ (h) 0°56′
3) (a) 43.5 mm (b) 9.29 m
(c) 4.43 m
4) (a) 59°2′ (b) 15°57′
(c) 22°42′
5) 77.0 cm 6) 27.8 mm
7) 33.3 m 8) 2.86 m
9) 20.9 mm

Exercise 18.4

1) 0.948 4; 0.334 4
2) $\frac{3}{5}$; $\frac{4}{5}$ 3) $\frac{5}{13}$, $\frac{5}{12}$
7) 0.743 1 8) 0.454 0

Exercise 18.5

1) (a) 22.94 m (b) 32.77 m
2) (a) 5.35 m (b) 32°28′
3) 6.16 m; 12.3 m², 16.9 m
4) (a) 0.649 8 (b) 16°23′
5) 17°14′ 6) 72°58′
7) 54.3 mm 8) 17.2 mm
9) 3.76 mm 10) 56.7, 39.5 mm
11) (a) 34.3, 26.2, 14.5, 37.2
(b) 22°55′
12) 90.0 mm

Self-Test 14

1) a 2) c 3) b
4) b 5) c 6) d
7) c 8) c 9) a
10) b 11) a 12) c
13) a 14) b 15) c
16) c

INDEX

Accuracy
- degrees of 51
- using calculator 96

Algebra 122
Angle property of triangles 214
Angles 200
- in circles 240
- and straight lines 200
- types of 204

Areas 251
- unit of 255

Arithmetical signs 9

Bar charts 307
Binomial expressions 128
Brackets 137

Calculator 94
Cancelling 32
Charts 115
Circle 237
- area of 265

Class width 310
Complementary angles 204, 214
Congruent triangles 220
Construction
- of right angles 219
- of triangles 229

Conversion
- fractions to decimals 54
- decimals to fractions 56

Coordinates 179
Cosine of an angle 287
Cylinder 270

Decimal to fraction conversion 56
Decimal system 39
Degrees of accuracy 51
Directed numbers 70
Divisibility
- tests for 14

Electronic calculator 94
Equations
- linear 147
- simultaneous linear 162

Experimental data on graphs 189

Factorising 132
Factors 132
Formulae, transposition 169
Fractions
- algebraic 136
- numeric 23
- partial 139

Fraction to decimal conversion 54
Frequency distributions 309
Frequency polygon 316

Gradient of a straight line 186
Graphs 178
Grouped data 311

Highest common factor
- algebraic terms 131
- numbers 17

Histogram 310

Identities 146
Indices 78
Isoceles triangle 216

Length
- unit of 255

Linear equations 147
Linear simultaneous equations 162
Logarithms 87
Lowest common denominator 27
Lowest common multiple
- algebraic terms 135

Making expressions 154
Multiples 15

Network diagrams 119
Nomographs 116
Numbers in standard form 89

Operations in arithmetic 1

Parallelogram 251
Partial fractions 139
Percentages 64
Pictograms 321
Pie chart 306
Population 308
Powers 78
Proportion 61
Proportional parts 62
Pythagoras' theorem 215

Quadrilaterals 215

Ratio 60
Recording information 306
Recurring decimals 55
Rectangle 256
Relative percentage frequency 320
Rhombus 252
Rough checks for calculations 52, 96
Rounding 51

Sampling 308
Scales of graphs 179
Sequence of arithmetical
operations 18, 126
Similar shapes 273
Similar solids 274
Similar triangles 224
Simultaneous linear equations 162
Sine of an angle 282
Square 258
Standard form 89
Standard notation for a triangle 214
Statistics 306
Straight line law 186

Tables 109
Tangency of circles 244
Tangent of an angle 291
Tangent to a circle 242
Transposition of formulae 172
Trapezium 252
Triangle
areas of 263
Triangles 213
Triginometrical ratios 281

Units, SI 92

Variables 314
Volume 268